Markov Chains

Markov Chains

Analytic and Monte Carlo Computations

Carl Graham

CNRS and Ecole Polytechnique, Palaiseau, France

WILEY

This work is in the Wiley-Dunod Series co-published between Dunod and John Wiley Sons, Ltd.

This work is in the Wiley-Dunod Series co-published between Dunod and John Wiley & Sons, Ltd.

This edition was first published in 2014
© 2014 John Wiley & Sons, Ltd

Registered office
John Wiley & Sons, Ltd, The Atrium, Southern Gate, Chichester, West Sussex PO19 8SQ, United Kingdom

For details of our global editorial offices, customer services, and information about how to apply for permission to reuse the copyright material in this book, please see our web site at www.wiley.com.

Library of Congress Cataloging-in-Publication Data

Graham, C. (Carl)   Markov chains : analytic and Monte Carlo computations / Carl Graham.
      pages cm
  Includes bibliographical references and index.
  ISBN 978-1-118-51707-9 (cloth)
  1. Markov processes. 2. Monte Carlo method. 3. Numerical calculations. I. Title.
  QA274.7.G73 2014
  519.2'33–dc23

                                                                    2013049515

A catalog record for this book is available from the British Library.

ISBN: 978-1-11851707-9

Set in 10/12pt TimesLTStd by Laserwords Private Limited, Chennai, India

1   2014

Contents

Preface

This book was born from my teaching experience in French engineering schools. In these, mathematical tools should be introduced by showing that they provide models allowing for exact or approximate computations for relevant quantities pertaining to realistic phenomena.

I have taught in particular a course on the Markov chains, the theory of which is already old. I had learnt a lot of it by osmosis while studying or doing research on more recent topics in stochastics. Teaching it forced me do delve deeper into it. This allowed me to rediscover the power and finesse of the probabilistic tools based on the work by Kai Lai Chung, Wolfang (or Vincent) Doeblin, Joseph Leo Doob, William Feller, and Andrei Kolmogorov, which laid the ground for stochastic calculus.

I realized that Markov chain theory is actually a very active research field, both for theory and for applications. The derivation of efficient Monte Carlo algorithms and their variants, for instance adaptive ones, is a hot subject, and often these are the only methods allowing tractable computations for treating the enormous quantities of data that scientists and engineers can now acquire. This need notably feeds theoretical studies on long-time rates of convergence for Markov chains.

This book is aimed at a public of engineering school and master students and of applied scientists and engineers, wishing to acquire the pertinent mathematical bases. It is structured and animated by a few classic examples, which are each investigated at different stages of the book and eventually studied exhaustively. These illustrate in real time the newly introduced notions and their qualitative and quantitative uses. It also elaborates on the general matter of Monte Carlo approximation.

This book owes a lot to the forementioned mathematicians, as well as the authors of the too small bibliography. More personal contributions to this book were provided by discussions with other teachers and researchers and by the interaction with the students, notably those in the engineering schools with their pragmatic perspective on mathematics.

Feller's book [3] has always impressed me, notably by the wealth of examples of a practical nature it contains. It represents a compendium of probability theory at his time and can readily be adapted to a modern public. Reading it has lead me to try to push some explicit computations quite far, notably using generating functions, but my efforts are pale with respect to his achievements in this perspective.

List of Figures

Nomenclature

Acronyms

a.s.	almost surely
i.i.d.	independent identically distributed
l.h.s.	left-hand side
r.h.s.	right-hand side
r.v.	random variable
s.t.	such that
w.r.t.	with respect to

Symbols

$\mathbb{C}$	set of complex numbers, of the form $x + iy$ for x and y in $\mathbb{R}$
$\mathbb{N}$	set $\{0, 1, \cdots \}$ of nonnegative integers, 0 included
$\mathbb{R}$	set of real numbers
$\mathbb{Z}$	set of signed integers
$\mathcal{F}$	see σ-field, filtration
$\mathcal{M}$	Banach space of signed measures (for the total variation norm)
$\mathcal{M}_+^1$	closed subset of probability measures
$\mathcal{V}$	basic state space, discrete
θ	shift operator
Ω	basic probability space, with generic element ω
$\|\cdot\|_{\mathrm{var}}$	total variation norm
$\lfloor x \rfloor$	$\sup\{n \in \mathbb{N} : n \leq x\}$ for $x \in \mathbb{R}$
$\lceil x \rceil$	$\inf\{n \in \mathbb{N} : n \geq x\}$ for $x \in \mathbb{R}$
$\otimes$	product, for measures or transition matrices

Introduction

This book is written with a broad spectrum, that allows for different readings at various levels. It tries nevertheless to plunge quickly into the heart of the matter. The basic analytical tool is the maximum principle, which is natural in this setting. It is superficially compared to martingale methods in some instances. The basic probabilistic tool is the Markov property, strong or not.

After the first definitions in Chapter 1, matrix notation is described and random iterative sequences are introduced. An elementary algebraic study is performed, which could be used for a study of finite-state Markov chains. The Doeblin condition and its consequences are very present. The fundamental examples are then introduced.

Probabilistic techniques start in earnest in Chapter 2, which study filtrations, the Markov property, stopping times, and the strong Markov property. The technique of conditioning on the first step of the chain, called "the one step forward method," is then developed. It is applied to Dirichlet problems and, more generally, to the study of first hitting times and locations.

Chapter 3 delves into the analysis of the probabilistic behaviors of the sample paths. This results in the fundamental theorems which link the algebraic notions of invariant laws and measures with sample-path notions such as transience and recurrence (null or positive). Its Complements subsections extend this perspective to the links between nonnegative superharmonic functions and transience-recurrence properties and to Lyapunov function techniques that pursue that perspective, then proceed to time reversal, and finish with the investigation of birth-and-death processes on $\mathbb{N}$.

These sample-path studies lead naturally to the long-time limit theorems in Chapter 4. The sample-path results, such as the Markov chain ergodic theorem and central limit theorem, are classically related to the strong Markov property and regeneration.

This book continues with a short study of the periodicity phenomenon specific to discrete time. The Kolmogorov ergodic theorem is then proved by a coupling technique. Some basics on functional analysis and Dirichlet forms are presented as an introduction to the study of rates of convergence for the instantaneous laws.

Chapter 5 is an opening toward the field of Monte Carlo Markov chain methods. First, the approximate solution of Dirichlet problems is described, starting with the heat equation at equilibrium and ending with general parabolic equations. Then, the use of Monte Carlo methods for the approximation of invariant laws is developed, notably for Gibbs measures for which the direct computation of the normalizing factor (partition function) is impossible in practice. Stochastic optimization methods are introduced in this perspective. This chapter ends with the study of exact simulation methods for the invariant law.

The Appendix introduces the necessary tools for reading the book. It pursues with some notions on convergence in law, notably by introducing the total variation norm as the strong dual norm of L^∞. It ends with some rigorous elements of measure theory, to provide a mathematical framework for Markov chain theory.

1

First steps

1.1 Preliminaries

This book focuses on a class of random evolutions, in discrete time (by successive steps) on a discrete state space (finite or countable, with isolated elements), which satisfy a fundamental assumption, called the *Markov property*. This property can be described informally as follows: the evolution "forgets" its past and is "regenerated" at each step, retaining as sole past information for its future evolution its present state. The probabilistic description of such an evolution requires

- a law (probability measure) for drawing its initial state and

- a family of laws for drawing iteratively its state at the "next future instant" given its "present state," indexed by the state space.

Such a random evolution will be called a *Markov chain*.

Precise definitions can be found in the Appendix, Section A.3, but we give now the probabilistic framework. A probability space $(\Omega, \mathcal{F}, \mathbb{P})$ will be considered throughout. When $\mathcal{V}$ is discrete, usually its measurable structure is given by the collection of all subsets, and all functions with values in $\mathcal{V}$ are assumed to be measurable.

A random variable (r.v.) with values in a measurable state space $\mathcal{V}$ is a measurable function

$$X : \omega \in \Omega \mapsto X(\omega) \in \mathcal{V}.$$

Intuitively, the output $X(\omega)$ varies randomly with the input ω, which is drawn in Ω according to $\mathbb{P}$, and the measurability assumptions allow to assign a probability to events defined through X.

For the random evolutions under investigation, the natural random elements are sequences $(X_n)_{n \in \mathbb{N}}$ taking values in the same discrete state space $\mathcal{V}$, which are called

Markov Chains: Analytic and Monte Carlo Computations, First Edition. Carl Graham.
© 2014 John Wiley & Sons, Ltd. Published 2014 by John Wiley & Sons, Ltd.

(random) *chains* or (discrete time) *processes*. Each X_n should be an r.v., and its law π_n on the discrete space $\mathcal{V}$ is then given by

$$\pi_n(x) = \pi_n(\{x\}) = \mathbb{P}(X_n = x) , \qquad x \in \mathcal{V} ,$$

$$\pi_n(A) = \mathbb{P}(X_n \in A) = \sum_{x \in A} \pi_n(x) , \qquad A \subset \mathcal{V} ,$$

and hence can be identified in a natural way with $(\pi_n(x))_{x \in \mathcal{V}}$.

Finite-dimensional marginals More generally, for any $k \geq 1$ and $n_1, \dots, n_k$ in $\mathbb{N}$, the random vector

$$(X_{n_1}, \dots, X_{n_k})$$

takes values in the discrete space $\mathcal{V}^k$, and its law $\pi_{n_1, \dots, n_k}$ can be identified with the collection of the

$$\pi_{n_1, \dots, n_k}(x_1, \dots, x_k) = \mathbb{P}(X_{n_1} = x_1, \dots, X_{n_k} = x_k) , \qquad x_1, \dots, x_k \in \mathcal{V}.$$

All these laws for $k \geq 1$ and $0 \leq n_1 < \cdots < n_k$ constitute the family of the *finite-dimensional marginals* of the chain $(X_n)_{n \geq 0}$ or of its law.

Law of the chain The r.v.

$$(X_n)_{n \geq 0} : \omega \mapsto (X_n(\omega))_{n \geq 0}$$

takes values in $\mathcal{V}^{\mathbb{N}}$, which is *uncountable* as soon as $\mathcal{V}$ contains at least two elements. Hence, its law *cannot*, in general, be defined by the values it takes on the elements of $\mathcal{V}^{\mathbb{N}}$. In the Appendix, Section A.3 contains some mathematical results defining the law of $(X_n)_{n \geq 0}$ from its finite-dimensional marginals.

Section A.1 contains some more elementary mathematical results used throughout the book, and Section A.2 a discussion on the total variation norm and on weak convergence of laws.

1.2   First properties of Markov chains

1.2.1   Markov chains, finite-dimensional marginals, and laws

1.2.1.1   First definitions

We now provide rigorous definitions.

Definition 1.2.1 *Let $\mathcal{V}$ be a discrete state space. A matrix $P = (P(x, y))_{x,y \in \mathcal{V}}$ is a transition matrix on $\mathcal{V}$, or also a Markovian or stochastic matrix, if*

$$P(x, y) \geq 0 , \qquad \sum_{y \in \mathcal{V}} P(x, y) = 1.$$

A sequence $(X_n)_{n\geq0}$ of $\mathcal{V}$-valued random variables is a Markov chain on $\mathcal{V}$ with matrix P and initial law π_0 if, for every n in $\mathbb{N}$ and $x_0, \ldots, x_n$ in $\mathcal{V}$,

$$\mathbb{P}(X_0 = x_0, \ldots, X_n = x_n) = \pi_0(x_0)P(x_0, x_1)P(x_1, x_2) \cdots P(x_{n-1}, x_n).$$

Note that, by iteration, $(X_n)_{n\geq0}$ is a Markov chain on $\mathcal{V}$ with matrix P if and only if for every n in $\mathbb{N}$ and $x_0, \ldots, x_n, y$ in $\mathcal{V}$,

$$\mathbb{P}(X_0 = x_0, \ldots, X_n = x_n, X_{n+1} = y) = \mathbb{P}(X_0 = x_0, \ldots, X_n = x_n)P(x_n, y),$$

and that this is trivially true if $\mathbb{P}(X_0 = x_0, \ldots, X_n = x_n) = 0$.

Markov chain evolution   A family $P(x, \cdot)$ of laws on $\mathcal{V}$ indexed by $x \in \mathcal{V}$ is defined by

$$P(x, A) = \sum_{x \in A} P(x, y), \qquad A \subset \mathcal{V}.$$

The evolution of $(X_n)_{n\geq0}$ can be obtained by independent draws, first of X_0 according to π_0, and then iteratively of X_{n+1} according to $P(X_n, \cdot)$ for $n \geq 0$ without taking any further notice of the evolution before the present time n or of its actual value.

Inhomogeneous Markov chains   A more general and complex evolution can be obtained by letting the law of the steps depend on the present instant of time, that is, using the analogous formulae with $P(n; x_n, y)$ instead of $P(x_n, y)$; this corresponds to a *time-inhomogeneous* Markov chain, but we will seldom consider this generalization.

Markov chain graph   The graph of the transition matrix P, or of a Markov chain with matrix P, is the oriented marked graph with nodes given by the elements of $\mathcal{V}$ and directed links given by the ordered pairs (x, y) of elements of $\mathcal{V}$ such that $P(x, y) > 0$ marked by the value of $P(x, y)$. The restriction to the graph to $x \neq y$ in $\mathcal{V}$ is of the form [if $P(x, x)P(x, y)P(y, x)P(y, y) \neq 0$]

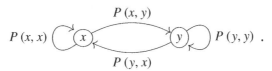

The graph and the matrix are equivalent descriptors for the random evolution. The links from x to x in the graph are redundant as they are marked by $P(x, x) = 1 - \sum_{y \neq x} P(x, y) > 0$, but illustrate graphically the possible transitions from x.

1.2.1.2   Conditional formulations

The last formula in Definition 1.2.1 can be written as

$$\mathbb{P}(X_{n+1} = y \mid X_0 = x_0, \ldots, X_n = x_n) = P(x_n, y) \qquad (1.2.1)$$

which is often used as the definition. Moreover, if f is nonnegative or bounded then

$$\mathbb{E}(f(X_{n+1}) \mid X_0 = x_0, \ldots, X_n = x_n) = \mathbb{E}_{x_n}(f(X_1)) = \sum_{y \in \mathcal{V}} P(x_n, y) f(y).$$

For the sake of mathematical efficiency and simplicity, nonconditional expressions will be stressed, before possibly being translated into equivalent conditional formulations. As an example, Definition 1.2.1 immediately yields by summing over $x_0, \ldots, x_{n-1}$ that

$$\mathbb{P}(X_{n+1} = y \mid X_n = x_n) = \frac{\mathbb{P}(X_n = x_n, X_{n+1} = y)}{\mathbb{P}(X_n = x_n)} = P(x_n, y),$$

which is not quite so obvious starting from (1.2.1).

1.2.1.3   Initial law, instantaneous laws

For a Markov chain $(X_n)_{n \geq 0}$, the law of π_n of X_n is called the instantaneous law at time n and π_0 the initial law. The notations $\mathbb{P}$ and $\mathbb{E}$ implicitly imply that π_0 is given and arbitrary, $\mathbb{P}_\mu$ and $\mathbb{E}_\mu$ for some law μ on $\mathcal{V}$ indicate that $\pi_0 = \mu$, and $\mathbb{P}_x$ and $\mathbb{E}_x$ indicate that $X_0 = x$. By linearity,

$$\mathbb{P}_\mu = \sum_{x \in \mathcal{V}} \mu(x) \mathbb{P}_x, \qquad \mathbb{E}_\mu = \sum_{x \in \mathcal{V}} \mu(x) \mathbb{E}_x.$$

A frequent abuse of notation is to write $\mathbb{P}_x(\cdot) = \mathbb{P}(\cdot \mid X_0 = x)$, and so on.

Lemma 1.2.2 *Let $(X_n)_{n \geq 0}$ be a Markov chain with matrix P and initial law π_0. Then, $\mathbb{P}(X_{n+1} = y \mid X_n = x) = P(x, y)$ for n in $\mathbb{N}$ and x, y in $\mathcal{V}$, and the instantaneous laws $\pi_n = (\pi_n(x))_{x \in \mathcal{V}}$ are given by*

$$\pi_n(x) = \mathbb{P}(X_n = x) = \sum_{x_0, \ldots, x_{n-1} \in \mathcal{V}} \pi_0(x_0) P(x_0, x_1) P(x_1, x_2) \cdots P(x_{n-1}, x)$$

or in matrix notation

$$\pi_n = \pi_{n-1} P = \cdots = \pi_0 P^n.$$

Moreover, $(X_{nk})_{k \geq 0}$ is a Markov chain with matrix the nth matrix power of P

$$P^n = (P^n(x, y))_{x, y \in \mathcal{V}}.$$

Proof: This follows readily from Definition 1.2.1.   ∎

1.2.1.4   Law on the canonical space of the chain

The notions in the Appendix, Section A.3.4, will now be used.

Definition 1.2.1 is actually a statement on the law of the Markov chain $(X_n)_{n \geq 0}$, which it characterizes by giving an explicit expression for its finite-dimensional marginals in terms of its initial law π_0 and transition matrix P.

Indeed, some rather simple results in measure theory show that there is uniqueness of a law on the canonical probability space $\mathcal{V}^{\mathbb{N}}$ with product σ-field having a given finite-dimensional marginal collection.

It is immediate to check that this collection is consistent [with respect to (w.r.t.) projections] and then the Kolmogorov extension theorem (Theorem A.3.10) implies that there is existence of a law $\mathbb{P}_{\pi_0}$ on the canonical probability space $\mathcal{V}^{\mathbb{N}}$ with the product σ-field such that the canonical (projection) process $(X_n)_{n\geq 0}$ has the given finite-dimensional marginal collection, which hence is a Markov chain with initial law π_0 and transition matrix P (see Corollary A.3.11).

The Kolmogorov extension theorem follows from a deep and general result in measure theory, the Caratheodory extension theorem.

1.2.2 Transition matrix action and matrix notation

1.2.2.1 Nonnegative and signed measures, total variation measure, andnorm

A (nonnegative) measure μ on $\mathcal{V}$ is defined by (and can be identified with) a collection $(\mu(x))_{x\in\mathcal{V}}$ of nonnegative real numbers and, in the sense of nonnegative series,

$$\mu : A \subset \mathcal{V} \mapsto \mu(A) := \sum_{x\in A} \mu(x) \in [0, \infty] := \mathbb{R}_+ \cup \{\infty\}.$$

A measure μ is *finite* if its total mass $\mu(\mathcal{V})$ is finite and then $\mu(A) < \infty$ for all $A \subset \mathcal{V}$. A measure is a *probability measure*, or a *law*, if $\mu(\mathcal{V}) = 1$.

For r in $\mathbb{R}$, let $r^+ = \max(r, 0)$ and $r^- = \max(-r, 0)$ denote the nonnegative and nonpositive parts of r, which satisfy $r = r^+ - r^-$ and $|r| = r^+ + r^-$.

For $\mu = (\mu(x))_{x\in\mathcal{V}}$ with $\mu(x) \in \mathbb{R}$, the measures μ^+, μ^-, and $|\mu|$ can be defined term wise. Then,

$$\mu = \mu^+ - \mu^-$$

is the minimal decomposition of μ as a difference of (nonnegative) measures, which have disjoint supports, and

$$|\mu| = \mu^+ + \mu^-$$

is called the *total variation measure* of μ.

If μ is such that $|\mu|$ is finite (equivalently, if both μ^+ and μ^- are finite), then we can extend it to a *signed measure* μ acting on subsets of $\mathcal{V}$ by setting, in the sense of absolutely converging series,

$$\mu : A \subset \mathcal{V} \mapsto \mu(A) := \mu^+(A) - \mu^-(A) = \sum_{x\in A} \mu(x) \in \mathbb{R} ,$$

and we can define its *total variation norm* by

$$\|\mu\|_{var} = |\mu|(\mathcal{V}) = \sum_{x\in\mathcal{V}} |\mu(x)| < \infty.$$

Note that $\mu(A) \leq |\mu|(A) \leq \|\mu\|_{var}$ for all $A \subset \mathcal{V}$.

The space $\mathcal{M} := \mathcal{M}(\mathcal{V})$ of all signed measures, furnished with the total variation norm, is a Banach space, which is isomorphic to the Banach space $\ell^1 := \ell^1(\mathcal{V})$ of summable sequences with its natural norm.

Probability measures or laws   The space of probability measures

$$\mathcal{M}_+^1 := \mathcal{M}_+^1(\mathcal{V}) = \{\mu \in \mathcal{M} : \mu \geq 0, \, \|\mu\|_{var} = 1\}$$

is the intersection of the cone of nonnegative measures with the unit sphere. It is a closed subset of $\mathcal{M}$ and hence is complete for the induced metric.

Some properties of $\mathcal{M}$ and $\mathcal{M}_+^1$ are developed in the Appendix, Section A.2. Note that, according to the definition taken here, nonnegative measures with infinite mass are *not* signed measures.

Complex measures   Spectral theory naturally involves complex extensions. For its purposes, complex measures can be readily defined, and the corresponding space $\mathcal{M}(\mathcal{V}, \mathbb{C})$, where the modulus in $\mathbb{C}$ is again denoted by $|\cdot|$, allows to define a total variation measure $\|\mu\|$ and total variation norm $\|\mu\|_{var} = \|\mu\|(\mathcal{V})$ for μ in $\mathcal{M}(\mathcal{V}, \mathbb{C})$. The Banach space is isomorphic to $\ell^1(\mathcal{V}, \mathbb{C})$. The real and imaginary parts of a complex measure are signed measures.

1.2.2.2   Line and column vectors, measure-function duality

In matrix notation, the functions f from $\mathcal{V}$ to $\mathbb{R}$ are considered as *column* vectors $(f(x))_{x \in \mathcal{V}}$, and nonnegative or signed measures μ on $\mathcal{V}$ as *line* vectors $(\mu(x))_{x \in \mathcal{V}}$, of infinite lengths if $\mathcal{V}$ is infinite. The integral of a function f by a measure μ is denoted by μf, in accordance with the matrix product

$$\mu f = \begin{pmatrix} \cdots & \mu(x) & \cdots \end{pmatrix} \begin{pmatrix} \vdots \\ f(x) \\ \vdots \end{pmatrix} = \sum_{x \in \mathcal{V}} \mu(x) f(x) \, ,$$

defined in $[0, \infty]$ in the sense of nonnegative series if $\mu \geq 0$ and $f \geq 0$ and in $\mathbb{R}$ in the sense of absolutely converging series if $\mu \in \mathcal{M}$ and $f \in L^\infty = L^\infty(\mathcal{V}) = L^\infty(\mathcal{V}, \mathbb{R})$, the Banach space of bounded functions on $\mathcal{V}$ with the uniform norm.

For A subset of $\mathcal{V}$, the indicator function $\mathbb{1}_A$ is defined by

$$\mathbb{1}_A(x) = 1 \, , \quad x \in A \, , \qquad \mathbb{1}_A(x) = 0 \, , \quad x \in \mathcal{V} - A.$$

For x in $\mathcal{V}$, the Dirac mass at x is the probability measure δ_x such that $\delta_x(A) = \mathbb{1}_A(x)$, that is,

$$\delta_x(A) = 1 \, , \quad x \in A \, , \qquad \delta_x(A) = 0 \, , \quad x \in \mathcal{V} - A.$$

For x and y in $\mathcal{V}$, it holds that

$$\delta_x(y) = \mathbb{1}_{\{x\}}(y) = 1 \, , \quad x = y \, , \qquad \delta_x(y) = \mathbb{1}_{\{x\}}(y) = 0 \, , \quad x \neq y \, ,$$

but δ_x will be represented by a line vector and $\mathbb{1}_{\{x\}}$ by a column vector.

If μ is a nonnegative or signed measure, then

$$\mu(A) = \mu \mathbb{1}_A .$$

Duality and total variation norm   A natural duality bracket between the Banach spaces $\mathcal{M}$ and L^∞ is given by

$$(\mu, f) \in \mathcal{M} \times L^\infty \mapsto \mu f \in \mathbb{R} ,$$

and for μ in $\mathcal{M}$, it holds that

$$\mu f \le \|\mu\|_{var} \|f\|_\infty , \quad \|\mu\|_{var} = \mu^+(\mathcal{V}) + \mu^-(\mathcal{V}) = \mu(\mathbb{1}_{\{x\,:\,\mu(x)>0\}} - \mathbb{1}_{\{x\,:\,\mu(x)<0\}}) ,$$

and hence that

$$\|\mu\|_{var} = \max_{f \in L^\infty,\ \|f\|_\infty \le 1} \mu f . \tag{1.2.2}$$

Thus, $\mathcal{M}$ can be identified with a closed subspace of the dual of L^∞ with the strong dual norm, which is the norm as an operator on L^∞.

The space L^∞ can be identified with the space $\ell^\infty = \ell^\infty(\mathcal{V}) = \ell^\infty(\mathcal{V}, \mathbb{R})$ of bounded sequences, and this duality between $\mathcal{M}$ and L^∞ to the natural duality between ℓ^1 and ℓ^∞.

The operations between complex measures and complex functions are performed by separating the real and imaginary parts.

1.2.2.3   Transition matrices, actions on measures, and functions

A matrix P is a transition matrix if and only if each of its line vectors $P(x, \cdot)$ corresponds to a probability measure. Then, its column vector $P(\cdot, y)$ defines a nonnegative function, which is bounded by 1.

A transition matrix can be multiplied on its right by nonnegative functions and on its left by nonnegative measures, or on its right by bounded functions and on its left by signed measures. The order of these operations does not matter. The function Pf, the nonnegative or signed measure μP, and $\mu P f \in \mathbb{R} \cup \{\infty\}$ are given for $x \in \mathcal{V}$ by

$$Pf(x) = P(x, \cdot) f = \sum_{y \in \mathcal{V}} P(x, y) f(y) = \mathbb{E}_x(f(X_1)) ,$$

$$\mu P(x) = \mu P(\cdot, x) = \sum_{z \in \mathcal{V}} \mu(z) P(z, x) = \mathbb{P}_\mu(X_1 = x) ,$$

$$\mu P f = \sum_{x,y \in \mathcal{V}} \mu(x) P(x, y) f(y) = \mathbb{E}_\mu(f(X_1)) ,$$

in which the notations $\mathbb{P}_\mu$ and $\mathbb{E}_\mu$ are used *only* when μ is a *law*.

In matrix notation,

$$Pf = \begin{pmatrix} \ddots & \vdots & \ddots \\ \cdots & P(x,y) & \cdots \\ \ddots & \vdots & \ddots \end{pmatrix} \begin{pmatrix} \vdots \\ f(y) \\ \vdots \end{pmatrix} = \begin{pmatrix} \vdots \\ Pf(x) = \sum_{y \in \mathcal{V}} P(x,y)f(y) \\ \vdots \end{pmatrix},$$

$$\mu P = \begin{pmatrix} \cdots & \mu(z) & \cdots \end{pmatrix} \begin{pmatrix} \ddots & \vdots & \ddots \\ \cdots & P(z,x) & \cdots \\ \ddots & \vdots & \ddots \end{pmatrix}$$

$$= \begin{pmatrix} \cdots & \mu P(x) = \sum_{z \in \mathcal{V}} \mu(z)P(z,x) & \cdots \end{pmatrix},$$

$$\mu Pf = \begin{pmatrix} \cdots & \mu(x) & \cdots \end{pmatrix} \begin{pmatrix} \ddots & \vdots & \ddots \\ \cdots & P(x,y) & \cdots \\ \ddots & \vdots & \ddots \end{pmatrix} \begin{pmatrix} \vdots \\ f(y) \\ \vdots \end{pmatrix}$$

$$= \sum_{x,y \in \mathcal{V}} \mu(x)P(x,y)f(y).$$

Intrinsic notation   The linear mapping

$$P : f \in L^{\infty} \mapsto Pf \in L^{\infty}$$

has matrix P in the canonical basis. Its dual, or adjoint, mapping on $\mathcal{M}$, w.r.t. the duality bracket $(\mu, f) \mapsto \mu f$, is given by

$$P^* : \mu \in \mathcal{M} \mapsto \mu P \in \mathcal{M},$$

and has the adjoint (or transpose) matrix, also denoted by P^*. In order to respect the vector space structure and identify linear mappings and their matrices in the canonical bases, we could write $P^*\mu$ instead of μP and $\langle \mu, f \rangle$ instead of μf and $\langle \mu, Pf \rangle = \langle P^*\mu, f \rangle$ instead of μPf.

1.2.2.4   Transition matrix products, many-step transition

If P and Q are both transition matrices on $\mathcal{V}$, then it is easy to check that the matrix product PQ with generic term

$$PQ(x,y) = \sum_{z \in \mathcal{V}} P(x,z)Q(z,y)$$

is a transition matrix on $\mathcal{V}$. Let

$$P^0 := I \text{ (identity matrix)}, \qquad P^n := P^{n-1}P = PP^{n-1}, \quad n \geq 1.$$

Then,

$$P^n(x,y) = \sum_{x_1,\ldots,x_{n-1} \in \mathcal{V}} P(x,x_1)P(x_1,x_2)\cdots P(x_{n-1},y)$$

is the probability for $(X_{nk})_{k\geq0}$ to go in one step from x to y, and hence for $(X_k)_{k\geq0}$ to do so in n steps. In particular,

$$\pi_0 = \delta_x \Rightarrow \pi_n(y) = P^n(x, y).$$

Chapman–Kolmogorov formula As $P^n = P^k P^{n-k}$ for $0 \leq k \leq n$, this yields the Chapman–Kolmogorov formula

$$P^n(x, y) = \sum_{z\in\mathcal{V}} P^k(x, z)P^{n-k}(z, y).$$

Probabilistic interpretation These algebraic formulae have simple probabilistic interpretations: the probability of going from x to y in n steps can be obtained as the sum of the probabilities of taking every n-step path allowing to do so, as well as the sum over all intermediate positions after k steps.

1.2.3   Random recursion and simulation

Many Markov chains are obtained in a natural way as a random recursion, or random iterative sequence, as follows.

Theorem 1.2.3 *Let $(F_k)_{k\geq1}$ be a sequence of independent identically distributed (i.i.d.) random functions from $\mathcal{V}$ to $\mathcal{V}$. For instance, for some not necessarily discrete space $\mathcal{W}$ and $f : \mathcal{W} \times \mathcal{V} \to \mathcal{V}$ and i.i.d. r.v. $(\xi_k)_{k\geq1}$ with values in $\mathcal{W}$,*

$$F_k : x \in \mathcal{V} \mapsto f(\xi_k, x) \in \mathcal{V}.$$

If X_0 is any $\mathcal{V}$-valued r.v. independent of $(F_k)_{k\geq1}$, then $(X_n)_{n\geq0}$ given for $n \geq 1$ by

$$X_n = F_n(X_{n-1}) = \cdots = F_n \circ \cdots \circ F_1(X_0) ,$$

and more precisely by

$$X_n(\omega) = F_n(\omega)(X_{n-1}(\omega)) = \cdots = F_n(\omega) \circ \cdots \circ F_1(\omega)(X_0(\omega)) , \quad \omega \in \Omega ,$$

is a Markov chain on $\mathcal{V}$ with matrix P given by

$$P(x, y) = \mathbb{P}(F_1(x) = y) , \qquad x, y \in \mathcal{V}.$$

Proof: For $x_0, x_1, \ldots, x_n$ in $\mathcal{V}$, it holds that

$$\mathbb{P}(X_0 = x_0, X_1 = x_1 \ldots, X_n = x_n)$$
$$= \mathbb{P}(X_0 = x_0, F_1(x_0) = x_1, \ldots, F_n(x_{n-1}) = x_n)$$
$$= \mathbb{P}(X_0 = x_0)\mathbb{P}(F_1(x_0) = x_1) \ldots \mathbb{P}(F_1(x_{n-1}) = x_n).$$

Thus, Definition 1.2.1 is satisfied for the above matrix P.   ∎

When a random sequence $(X_n)_{n \in \mathbb{N}}$ is defined in some particular way, there is often a natural interpretation in terms of a random recursion of the previous kind. This allows to prove that a $(X_n)_{n \in \mathbb{N}}$ is a Markov chain, without having to directly check the definition, or even having to explicit its matrix. Moreover, such a pathwise representation for the Markov chain may be used for its study or its simulation.

Any Markov chain, with arbitrary initial law π_0 and transition matrix P, can be thus represented, using an i.i.d. sequence $(\xi_k)_{k \geq 1}$ of uniform r.v. on $\mathcal{W} := [0, 1]$. The state space is first enumerated as $\mathcal{V} = \{x_j : j = 0, 1, \dots\}$.

Let $\omega \in \Omega$. For the initial value, if j is determined by

$$\pi_0(x_0) + \cdots + \pi_0(x_{j-1}) < \xi_0(\omega) \leq \pi_0(x_0) + \cdots + \pi_0(x_{j-1}) + \pi_0(x_j),$$

then $X_0(\omega) = x_j$. For the transitions, for $n \geq 1$, if j is determined by

$$P(X_{n-1}(\omega), x_0) + \cdots + P(X_{n-1}(\omega), x_{j-1})$$
$$< \xi_n(\omega) \leq P(X_{n-1}(\omega), x_0) + \cdots + P(X_{n-1}(\omega), x_{j-1}) + P(X_{n-1}(\omega), x_j),$$

then $X_n(\omega) = x_j$.

In theory, this allows for the simulation of the Markov chain, but in practice this is not necessarily the best way to do so.

Note that this representation yields a construction for an arbitrary Markov chain, starting from a rigorous construction of i.i.d. sequences of uniform r.v. on $[0, 1]$, without having to use the more general Kolmogorov extension theorem (Theorem A.3.10).

Remark 1.2.4 *Very different random recursions can be associated with the same transition matrix, and result in diverse pathwise behavior, notably w.r.t. changes in the initial value.*

1.2.4   Recursion for the instantaneous laws, invariant laws

The instantaneous laws $(\pi_n)_{n \geq 0}$ satisfy the recursion

$$\pi_n = \pi_{n-1} P$$

with solution $\pi_n = \pi_0 P^n$. This is a linear recursion in dimension $\mathrm{Card}(\mathcal{V})$, and the affine constraint $\sum_{x \in \mathcal{V}} \pi_n(x) = 1$ allows to reduce it to an affine recursion in dimension $\mathrm{Card}(\mathcal{V}) - 1$. Note that $P^n(x, y) = \pi_n(y)$ for $\pi_0 = \delta_x$.

An elementary study of this recursion starts by searching for its fixed points. These are the only possible large n limits for the instantaneous laws, for any topology such that $\mu \mapsto \mu P$ is continuous.

By definition, a fixed point for this recursion is a law (probability measure) π such that $\pi = \pi P$, and is called an *invariant law* or a *stationary distribution* for (or of) the matrix P, or the Markov chain $(X_n)_{n \in \mathbb{N}}$.

Stationary chain, equilibrium If π is an invariant law and $\pi_0 = \pi$, then $\pi_n = \pi$ for all n in $\mathbb{N}$, and $(X_{n+k})_{k\in\mathbb{N}}$ is again a Markov chain with initial law π and transition matrix P. Then, $(X_n)_{n\in\mathbb{N}}$ is said to be stationary, or in equilibrium.

Invariant measures and laws In order to find an invariant law, one must:

1. Solve the linear equation $\mu = \mu P$.

2. Find which solutions $\mu \neq 0$ are nonnegative, that is, such that $\mu \geq 0$.

3. Normalize such μ by setting $\pi = \mu / \|\mu\|_{var}$, which is possible only if

$$\|\mu\|_{var} := \sum_{x\in\mathcal{V}} \mu(x) < \infty \qquad \text{(always true for finite } \mathcal{V}).$$

A nonnegative measure $\mu \neq 0$ such that $\mu = \mu P$ is called an *invariant measure*. An invariant measure is said to be unique if it is unique up to a multiplicative factor, that is, if all invariant measures are proportional (to it).

Algebraic interpretation The invariant measures are the *left eigenvectors* for the eigenvalue 1 for the matrix P acting on nonnegative measures, that is, for the adjoint (or transposed) matrix P^* acting on nonnegative vectors. Note that the constant function 1 is a *right eigenvector* for the eigenvalue 1, as

$$\pi 1 = \sum_{x\in\mathcal{V}} \pi(x) = 1.$$

The possible convergence of $(\pi_n)_{n\geq 0}$ to an invariant law π is related to the moduli of the elements of the spectrum of the restriction of the action of P on the signed measures of null total mass.

More generally, the exact or approximate computation of P^n depends in a more or less explicit way on a spectral decomposition of P.

1.3   Natural duality: algebraic approach

An algebraic study based on the natural duality between the space of signed measures $\mathcal{M} \simeq \ell^1$ and the space of bounded functions $L^\infty \simeq \ell^\infty$ will provide some structural results. These results are quite complete for finite state spaces $\mathcal{V}$. The complete study for arbitrary discrete $\mathcal{V}$ will be done later using probabilistic techniques. A reader for which this is the main interest may go directly to Section 1.4.

1.3.1   Complex eigenvalues and spectrum

1.3.1.1   Some reminders

The eigenvalues of the operator

$$P : f \mapsto Pf \quad \text{on } L^\infty$$

are given by all $\lambda \in \mathbb{C}$ such that $\lambda I - P$ is not injective as an operator on $L^\infty(\mathcal{V}, \mathbb{C})$.

The eigenspace of λ is the kernel

$$\ker\ (\lambda I - P) \in L^\infty(\mathcal{V}, \mathbb{C})\ ,$$

and its nonzero elements are called eigenvectors. Hence, f in $L^\infty(\mathcal{V}, \mathbb{C}) - \{0\}$ is an eigenvector of λ if and only if

$$Pf = \lambda f\ ,$$

and then

$$P^n f = \lambda^n f\ ,\qquad n \geq 1.$$

The generalized eigenspace of λ is given by

$$\bigcup_{k \geq 1} \ker\ ((\lambda I - P)^k) \in L^\infty(\mathcal{V}, \mathbb{C}).$$

If it contains strictly the eigenspace, then it contains some f and some eigenvector g such that

$$Pf = \lambda f + g\ ,$$

and then

$$P^n f = \lambda^n f + n\lambda^{n-1} g\ ,\qquad n \geq 1.$$

An eigenvalue is said to be

- *semisimple* if the eigenspace and generalized eigenspace coincide,

- *simple* if these spaces have dimension 1.

Similar definitions involving $\mathcal{M}(\mathcal{V}, \mathbb{C})$ are given for the adjoint operator

$$P^* : \mu \in \mathcal{M} \mapsto \mu P \in \mathcal{M}.$$

Its eigenspaces are often said to be eigenspaces *on the left*, or *left eigenspaces*, of P. Those of

$$P : f \in L^\infty \mapsto Pf \in L^\infty$$

may accordingly be called eigenspaces on the right, or right eigenspaces, of P.
Hence, $\mu \in \mathcal{M}(\mathcal{V}, \mathbb{C})$ is a left eigenvector of λ if and only if

$$\mu P = \lambda \mu\ ,$$

and then

$$\mu P^n = \lambda^n \mu\ ,\qquad n \geq 1.$$

If the generalized left eigenspace of λ contains strictly the eigenspace, then it contains some μ and some left eigenvector v such that

$$\mu P = \lambda \mu + v\ ,$$

and then

$$\mu P^n = \lambda^n \mu + n\lambda^{n-1} v , \qquad n \geq 1.$$

The spectrum $\sigma(P)$ of P on L^∞ is given by

$$\sigma(P) = \{\lambda \in \mathbb{C} : \lambda I - P \text{ not invertible in } L^\infty(\mathcal{V}, \mathbb{C})\}.$$

It is a simple matter to check that, using $\mathcal{M}(\mathcal{V}, \mathbb{C})$ in the definition,

$$\sigma(P^*) = \sigma(P)$$

and mentions of "left" or "right" are useless.

The spectrum $\sigma(P)$ contains both the left and right eigenvectors. In finite dimensions, invertibility of an operator is the same as injectivity, and hence the spectrum, the left eigenspace, and the right eigenspace coincide, but it is not so in general.

If $\lambda \in \mathbb{C}$ is in the spectrum of an operator on a real vector space, such as P or P^*, then $\bar{\lambda}$ is also in the spectrum, and if moreover λ is an eigenvalue, then $\bar{\lambda}$ is an eigenvalue, and the corresponding (generalized) eigenspaces are conjugate.

If $\lambda \in \mathbb{R}$, then $L^\infty = L^\infty(\mathcal{V}, \mathbb{R})$ can be considered instead of $L^\infty(\mathcal{V}, \mathbb{C})$ in all definitions. Moreover, the real and complex (generalized) eigenspaces have same dimension.

1.3.1.2   Algebraic results for transition matrices

Theorem 1.3.1 *Let P be a transition matrix on* $\mathcal{V}$.

1. *The operator* $P : f \mapsto Pf$ *on* $L^\infty(\mathcal{V}, \mathbb{C})$ *and its dual operator* $P^* : \mu \mapsto \mu P$ *on* $\mathcal{M}(\mathcal{V}, \mathbb{C})$ *are bounded and have operator norm* 1.

2. *The spectrum* $\sigma(P)$ *is included in the complex unit disk, every left or right eigenvalue of modulus* 1 *is semisimple, and the constant function* 1 *is a right eigenvector of eigenvalue* 1 *for P.*

3. *If* $\mu \in \mathcal{M}(\mathcal{V}, \mathbb{C})$ *and* $\mu P = \lambda\mu$ *for* $|\lambda| = 1$, *then the total variation measure* $|\mu| = (|\mu|(x))_{x\in\mathcal{V}}$ *satisfies* $|\mu|P = |\mu|$ *and hence is an invariant measure.*

Proof: If $f \in L^\infty(\mathcal{V}, \mathbb{C})$, then clearly

$$\|Pf\|_\infty = \sup_{x\in\mathcal{V}} \left| \sum_{y\in\mathcal{V}} P(x,y)f(y) \right| \leq \|f\|_\infty.$$

If $f \equiv 1$, then $Pf \equiv 1$. If $|\lambda| > 1$, then the series

$$(\lambda I - P)^{-1} = \lambda^{-1}(I - \lambda^{-1}P)^{-1} = \lambda^{-1}(I + \lambda^{-1}P + (\lambda^{-1}P)^2 + \cdots)$$

converges in operator norm on L^∞, which is given by $\|Q\|_{op} = \sup_{\|f\|_\infty \leq 1} \|Qf\|_\infty$.

If an eigenvalue λ is not semisimple, then there exists f in the generalized eigenspace and g in the eigenspace such that

$$P^n f = \lambda^n f + n\lambda^{n-1} g , \qquad n \geq 1 ,$$

and $\|P^n f\|_\infty \leq \|f\|_\infty$ then implies that $|\lambda| < 1$.

The corresponding results for P^* can be obtained in a similar way, or by duality. In particular, P^* has operator norm 1, and hence if $\mu \in M(V, \mathbb{C})$, then

$$\sum_{x \in V} |\mu| P(x) := \| \, |\mu| P \, \|_{var} \leq \| \, |\mu| \, \|_{var} := \sum_{x \in V} |\mu(x)| ,$$

and if moreover $\mu P = \lambda \mu$ for $|\lambda| = 1$, then

$$|\mu| P(x) = \sum_{y \in V} |\mu(y)| |P(y,x)| \geq \left| \sum_{y \in V} \mu(y) P(y,x) \right| = |\mu P(x)| = |\lambda \mu(x)| = |\mu(x)|$$

and necessarily $|\mu| P(x) = |\mu(x)| = |\mu|(x)$ for every x.   ∎

1.3.1.3   Uniqueness for invariant laws and irreducibility

A state x in V is *absorbing* if $P(x,x) = 1$ and then δ_x is an invariant law for P.

If V contains subsets V_i for $i \in I$ such that $P(x, V_i) = 1$ for every $x \in V_i$, these are said to be *absorbing* or *closed*. The restriction of P to each V_i is Markovian; if it has an invariant measure μ_i, then any convex combination $\sum_{i \in I} c_i \mu_i$ is an invariant measure on V, and if the μ_i are laws then $\sum_i c_i \mu_i$ is an invariant law. (By abuse of notation, μ_i denotes the extension of the measure to V vanishing outside of V_i.)

Hence, any uniqueness result for invariant measures or laws requires adequate assumptions excluding the above situation.

The standard hypothesis for this is that of *irreducibility*: a transition matrix P on V is *irreducible* if, for every x and y in V, there exists $i := i(x,y) \geq 1$ such that $P^i(x,y) > 0$. Equivalently, there exists in the oriented graph of the matrix a path covering the whole graph (respecting orientation). This notion will be further developed in due time.

Lemma 1.3.2 *Let $\mathbb{P}$ be an irreducible transition matrix. If a measure μ satisfies $\mu P = \mu$, then either $\mu = 0$ or $\mu > 0$.*

Proof: Assume that there exists a state x such that $\mu(x) > 0$. For any state y, there exists $i \geq 1$ such that $P^i(x,y) > 0$. By iteration $\mu = \mu P = \cdots = \mu P^i$, and hence,

$$\mu(y) = \sum_{z \in V} \mu(z) P^i(z,y) \geq \mu(x) P^i(x,y) > 0.$$

Hence, either $\mu > 0$ or $\mu = 0$.   ∎

Theorem 1.3.3 *Let $\mathbb{P}$ be an irreducible transition matrix. If $P^* : \mu \in \mathcal{M} \mapsto \mu P$ has 1 as an eigenvalue, then it is a simple eigenvalue, and its eigenspace is generated by an invariant law π, which is positive and unique.*

Proof: Let $\mu \neq 0$ be in $\mathcal{M}(\mathcal{V}, \mathbb{R})$ and satisfy $\mu P = \mu$. (This is enough, as $\lambda = 1$ is in $\mathbb{R}$.) Theorem 1.3.1 implies that $|\mu|$ is an invariant measure. As $\mu \neq 0$ and hence $|\mu| \neq 0$, Lemma 1.3.2 yields that $|\mu| > 0$, and an everywhere positive invariant law is given by

$$\pi = \frac{|\mu|}{|\mu|(\mathcal{V})}.$$

Moreover,

$$\mu^+ = \frac{1}{2}(|\mu| + \mu), \qquad \mu^- = \frac{1}{2}(|\mu| - \mu),$$

are invariant measures or are zero and cannot be both zero. Lemma 1.3.2 yields that $\mu^+ > 0$ or $\mu^- > 0$, that is, that $\mu > 0$ or $\mu < 0$.

Hence, if π and π' are two invariant laws, then $\pi - \pi' = (\pi - \pi')P$ and hence either $\pi - \pi' = 0$ or $\pi - \pi' > 0$ or $\pi - \pi' < 0$. As $(\pi - \pi')(\mathcal{V}) = 0$, we conclude that $\pi - \pi' = 0$, hence the invariant law is unique.

The eigenvalue 1 is semisimple (see Theorem 1.3.1), hence it is simple. ∎

This proof heavily uses techniques that are referred under the terminology "the maximum principle," which we will try to explain in Section 1.3.3.

1.3.2   Doeblin condition and strong irreducibility

A transition matrix P is *strongly irreducible* if there exists $i \geq 1$ such that $P^i > 0$.

Theorem 1.3.4 (Doeblin) *Let P be a transition matrix $\mathcal{V}$ satisfying the* Doeblin condition: *there exists $k \geq 1$ and $\varepsilon > 0$ and a law $\hat{\pi}$ on $\mathcal{V}$ such that*

$$P^k(x, y) \geq \varepsilon \hat{\pi}(y), \qquad \forall x, y \in \mathcal{V}.$$

Then, there exists a unique invariant law π, which satisfies $\pi \geq \varepsilon \hat{\pi}$. Moreover, for any $\mu \in \mathcal{M}$ such that $\mu(\mathcal{V}) = 0$, it holds that

$$\|\mu P^n\|_{var} \leq (1 - \varepsilon)^{\lfloor n/k \rfloor} \|\mu\|_{var} \leq 2(1 - \varepsilon)^{\lfloor n/k \rfloor} \|\mu\|_{var}, \qquad n \geq 1,$$

which yields the exponential bounds, uniform on the initial law,

$$\sup_{x \in \mathcal{V}} \sum_{y \in \mathcal{V}} |P^n(x, y) - \pi(y)| \leq \sup_{\pi_0 \in \mathcal{M}^1_+} \|\pi_0 P^n - \pi\|_{var} \leq 2(1 - \varepsilon)^{\lfloor n/k \rfloor}, \qquad n \geq 1.$$

The restriction of P to $\{\pi > 0\} := \{x \in \mathcal{V} : \pi(x) > 0\}$ is an irreducible transition matrix, which is strongly irreducible if $\{\pi > 0\}$ is finite.

Proof: Let us first assume the Doeblin condition to hold for $k = 1$. Let $\mu \in \mathcal{M}$ be such that $\mu(\mathcal{V}) := \sum_{x \in \mathcal{V}} \mu(x) = 0$. Then,

$$\|\mu P\|_{var} = \sum_{y \in \mathcal{V}} \left| \sum_{x \in \mathcal{V}} \mu(x) P(x, y) \right| \qquad \text{(by definition)}$$

$$= \sum_{y \in \mathcal{V}} \left| \sum_{x \in \mathcal{V}} \mu(x)(P(x, y) - \varepsilon \hat{\pi}(y)) \right| \qquad \text{(as } \mu(\mathcal{V}) = 0)$$

$$\leq \sum_{y \in \mathcal{V}} \sum_{x \in \mathcal{V}} |\mu(x)|(P(x, y) - \varepsilon \hat{\pi}(y)) \qquad \text{(as } P(x, y) \geq \varepsilon \hat{\pi}(y))$$

$$\leq \|\mu\|_{var}(1 - \varepsilon) \qquad \text{(changing summation order)}.$$

Moreover,

$$\mu P(\mathcal{V}) = \sum_{y \in \mathcal{V}} \sum_{x \in \mathcal{V}} \mu(x) P(x, y) = \sum_{x \in \mathcal{V}} \mu(x) \sum_{y \in \mathcal{V}} P(x, y) = 0$$

and iteration yields that

$$\|\mu P^n\|_{var} \leq \|\mu\|_{var}(1 - \varepsilon)^n, \qquad n \geq 1.$$

If the Doeblin condition holds for an arbitrary $k \geq 1$, Theorem 1.3.1 and the result for $k = 1$ applied to P^k yield that

$$\|\mu P^n\|_{var} = \|\mu(P^k)^{\lfloor n/k \rfloor} P^{n-k\lfloor n/k \rfloor}\|_{var} \leq \|\mu(P^k)^{\lfloor n/k \rfloor}\|_{var} \leq \|\mu\|_{var}(1 - \varepsilon)^{\lfloor n/k \rfloor}.$$

For any laws π_0 and π_0', it holds that $(\pi_0 - \pi_0')(\mathcal{V}) = 0$, and thus

$$\|\pi_0 P^n - \pi_0' P^n\|_{var} \leq (1 - \varepsilon)^{\lfloor n/k \rfloor} \|\pi_0 - \pi_0'\|_{var} \leq 2(1 - \varepsilon)^{\lfloor n/k \rfloor}$$

and this bound for arbitrary π_0 and $\pi_0' = \pi_0 P$ implies that $(\pi_0 P^n)_{n \geq 0}$ is a Cauchy sequence in the complete metric space $\mathcal{M}_+^1$ (by an exponential series bound).

Hence $(\pi_0 P^n)_{n \geq 0}$ converges to some law π, which by continuity must satisfy $\pi = \pi P$, and hence is an invariant law; this convergence also implies that the invariant law is unique.

Taking $\pi_0' = \pi$ and arbitrary π_0 or $\pi_0 = \delta_x$ for arbitrary x in $\mathcal{V}$ yield the bounds, which are uniform on the initial law. Moreover, for every y,

$$P^n(x, y) = \sum_{z \in \mathcal{V}} P^{n-1}(x, z) P(z, y) \geq \varepsilon \hat{\pi}(y) \sum_{z \in \mathcal{V}} P^{n-1}(x, z) = \varepsilon \hat{\pi}(y)$$

and taking the limit yields that $\pi(y) \geq \hat{\pi}(y)$.

If $\pi(x) > 0$ and $P(x, y) > 0$, then $\pi(y) \geq \pi(x) P(x, y) > 0$, and hence the restriction of P to $\{\pi > 0\}$ is Markovian.

If $\pi(y) > 0$, then, as $\lim_{n \to \infty} P^n(x, y) = \pi(y)$ for every x, there exists some $i(x, y) \geq 1$ such that $P^i(x, y) > 0$ for $i \geq i(x, y)$, hence this restriction is irreducible;

if moreover $\{\pi > 0\}$ is finite, then for $i \geq \max_{x,y \in \{\pi > 0\}} i(x, y)$ it holds that $P^i > 0$ on $\{\pi > 0\}$, and hence the restriction of P is strongly irreducible.   ∎

Note that $\varepsilon \leq 1$ and that $\varepsilon = 1$ only in the trivial case in which $(X_n)_{n \geq 1}$ is a sequence of i.i.d. r.v. of law $\hat{\pi}$.

The Doeblin Condition (or strong irreducibility) is seldom satisfied when the state space $\mathcal{V}$ is infinite. For a finite state space, Section 4.2.1 will give verifiable conditions for strong irreducibility. The following result is interesting in these perspectives.

Corollary 1.3.5 *Let P be a strongly irreducible matrix on a finite state space $\mathcal{V}$. Then, P satisfies the Doeblin Condition (see Theorem 1.3.4) for*

$$k \geq 1 \text{ such that } P^k > 0 \, , \quad \varepsilon = \sum_{y \in \mathcal{V}} \min_{x \in \mathcal{V}} P^k(x, y) > 0 \, , \quad \hat{\pi}(y) = \frac{1}{\varepsilon} \min_{x \in \mathcal{V}} P^k(x, y) \, ,$$

and the conclusions of Theorem 1.3.4 hold with $\pi > 0$ on $\mathcal{V}$.

Proof: The proof is immediate.   ∎

1.3.3   Finite state space Markov chains

1.3.3.1   Perron–Frobenius theorem

If the state space $\mathcal{V}$ is finite, then the vector spaces $\mathcal{M}(\mathcal{V})$ and $L^\infty(\mathcal{V})$ have finite dimension $\mathrm{Card}(\mathcal{V})$.

Then, the eigenvalues and the dimensions of the eigenspaces and generalized eigenspaces of P, which are by definition those of P and P^*, are identical, and the spectrum is constituted of the eigenvalues.

A function f is *harmonic* if $Pf = f$. It is a right eigenvector for the eigenvalue 1.

Theorem 1.3.6 Perron–Frobenius *Let $\mathcal{V}$ be a finite state space and P a transition matrix on $\mathcal{V}$.*

1. *The spectrum $\sigma(P)$ of P is included in the complex unit disk, the eigenvalues with modulus 1 are semisimple, the constant functions are harmonic, and there exists an invariant law π.*

2. *If P is irreducible, then the invariant law π is unique and everywhere positive, the only harmonic functions are constant, and there exists an integer $d \geq 1$, called the period of P, such that the only eigenvalues with modulus 1 the dth complex roots of 1, and these eigenvalues are simple.*

3. *If P is strongly irreducible, then $d = 1$.*

Proof: The beginning of the proof is an application of Theorem 1.3.1. The fact that $P1 = 1$ implies that 1 is an eigenvalue and that the constant functions are harmonic. As the dimension is finite, it further implies that there exists a right eigenvector

$\mu \in \mathcal{M}$ for the eigenvalue 1, that is, satisfying $\mu P = \mu$, and Theorem 1.3.1 implies that the law $\pi = |\mu|/|\mu|(\mathcal{V})$ is invariant.

If P is irreducible, then Theorem 1.3.3 yields that the invariant law π is unique and satisfies $\pi > 0$, hence the eigenvalue 1 is simple, and as the dimension is finite, any harmonic function is a multiple of the constant function 1 and thus is a constant.

If P is strongly irreducible, then Corollary 1.3.5 holds. Let $\lambda \in \mathbb{C}$ satisfy $|\lambda| = 1$, and $\mu \in \mathcal{M}$ be such that $\mu P = \lambda \mu$. Then, for $n \geq 1$,

$$\mu P^n = \lambda^n \mu , \qquad \mu P^n = (\mu - |\mu|\pi)P^n + |\mu|\pi P^n = (\mu - |\mu|\pi)P^n + |\mu|\pi ,$$

and letting n go to infinity in the exponential bounds in Theorem 1.3.4 for $\mu - |\mu|\pi$, which satisfies $(\mu - |\mu|\pi)(\mathcal{V}) = 0$, shows that $\mu = |\mu|\pi$ and that $\lambda = 1$, $i.e.$, that the eigenvalue 1 is simple and that any other eigenvalue has modulus strictly <1.

If P is irreducible, then there exists $d \geq 1$ such that P^d is strongly irreducible on each class of a partition of $\mathcal{V}$, which allows to prove the result on the dth complex roots of 1 (see Section 4.2.1), in particular Definitions 4.2.1 and 4.2.6 and Theorems 4.2.4 and 4.2.7. ∎

Theorem 4.2.7 will provide an extension of these results for infinite $\mathcal{V}$, by wholly different methods. (Saloff-Coste, L. (1997), Section 1.2) provides a detailed commentary on the Doeblin condition and the Perron–Frobenius theorem.

Maximum principle This terminology comes from the following short direct proof of the fact that if the state space $\mathcal{V}$ is finite and P is irreducible, then every harmonic function is constant. If f is harmonic on $\mathcal{V}$, then it attains its maximum in at least a state x. Moreover,

$$\max f = f(x) = P^i f(x) = \sum_{y \in \mathcal{V}} P^i(x, y)f(y) , \qquad i \geq 1 ,$$

and as $P^i(x, \cdot)$ is a probability measure, $f(y) = \max f$ for all y such that $P^i(x, y) > 0$. Thus, irreducibility yields that $f(y) = \max f$ for every $y \in \mathcal{V}$, and thus that f is constant.

1.3.3.2 Computation of the instantaneous and invariant laws

We are now going to solve the recursion for the instantaneous laws $(\pi_n)_{n \geq 0}$, and see how the situation deteriorates in practice very quickly as the size of the state space increases.

The chain with two states Let us denote the states by 1 and 2. There exists $0 \leq a, b \leq 1$ such that the transition matrix P and its graph are given by

$$\begin{pmatrix} 1-a & a \\ b & 1-b \end{pmatrix},$$

The recursion formula $\pi_n = \pi_{n-1}P$ then writes

$$(\pi_n(1), \pi_n(2)) = (\pi_{n-1}(1), \pi_{n-1}(2)) \begin{pmatrix} 1-a & a \\ b & 1-b \end{pmatrix}$$

$$= ((1-a)\pi_{n-1}(1) + b\pi_{n-1}(2), \ a\pi_{n-1}(1) + (1-b)\pi_{n-1}(2))$$

and the affine constraint $\pi_n(1) + \pi_n(2) = 1$ allows to reduce this linear recursion in dimension 2 to the affine recursion, in dimension 1,

$$\pi_n(1) = (1-a-b)\pi_{n-1}(1) + b.$$

If $a = b = 0$, then $P = I$ and every law is invariant, and P is not irreducible as $P^n = I$. Else, the unique fixed point is $\frac{b}{a+b}$, the unique invariant law is $\pi = \left(\frac{b}{a+b}, \frac{a}{a+b}\right)$, and

$$\pi_n(1) = \left(\pi_0(1) - \frac{b}{a+b}\right)(1-a-b)^n + \frac{b}{a+b}$$

and the formula for $\pi_n(2)$ is obtained by symmetry or as $\pi_n(2) = 1 - \pi_n(1)$.

If $a = b = 1$, then $P = \begin{pmatrix} 0 & 1 \\ 1 & 0 \end{pmatrix}$ has eigenvalues 1 and -1, the latter with eigenvector $\begin{pmatrix} 1 \\ -1 \end{pmatrix}$, and the chain alternates between the states 1 and 2 and $\pi_n(1)$ is equal to $\pi_0(1)$ for even n and to $1 - \pi_0(1)$ for odd n.

If $(a, b) \notin \{(0, 0), (1, 1)\}$, then $\lim_{n\to\infty}\pi_n(1) = \frac{b}{a+b}$ and $\lim_{n\to\infty}\pi_n = \pi$ with geometric rate with reason $1 - a - b$.

The chain with three states Let us denote the states by 1, 2, and 3. There exists $a, b, c, d, e,$ and f in $[0, 1]$, satisfying $a + b \leq 1$, $c + d \leq 1$, and $e + f \leq 1$, such that the transition matrix P and its graph are given by

$$\begin{pmatrix} 1-a-b & a & b \\ c & 1-c-d & d \\ e & f & 1-e-f \end{pmatrix},$$

As discussed above, we could reduce the linear recursion in dimension 3 to an affine recursion in dimension 2. Instead, we give the elements of a vectorial computation in dimension 3, which can be generalized to all dimensions. This exploits the fact that 1 is an eigenvalue of P and hence a root of its characteristic polynomial $K(X) = \det(XI - P)$. Hence,

$$K(X) = \begin{vmatrix} X+a+b-1 & -a & -b \\ -c & X+c+d-1 & -d \\ -e & -f & X+e+f-1 \end{vmatrix}$$

$$= (X+a+b-1)(X+c+d-1)(X+e+f-1) - ade - bcf$$

$$- ac(X+e+f-1) - be(X+c+d-1) - df(X+a+b-1)$$

and by developing this polynomial and using the fact that 1 is a root, $K(X)$ factorizes into

$$(X-1)(X^2 + (a+b+c+d+e+f-2)X$$

$$+ ad + ae + af + bc + bd + bf + ce + cf + de - a - b - c - d - e - f + 1).$$

The polynomial of degree 2 is the characteristic polynomial of the affine recursion in dimension 2. It has two possible equal roots λ_1 and λ_2 in $\mathbb{C}$, and if $\lambda_1 \in \mathbb{C} - \mathbb{R}$, then $\lambda_2 = \bar{\lambda}_1$. Their exact theoretical expression is not very simple, as the discriminant of this polynomial does not simplify in general, but they can easily be computed on a case-by-case basis.

In order to compute P^n, we will use the Cayley–Hamilton theorem, according to which $K(P) = 0$ (nul matrix). The Euclidean division of X^n by $K(X)$ yields

$$X^n = Q(X)K(X) + a_n X^2 + b_n X + c_n , \qquad P^n = a_n P^2 + b_n P + c_n I ,$$

and in order to effectively compute a_n, b_n, and c_n, we take the values of the polynomials for the roots of $K(X)$, which yields the linear system

$$\begin{cases} a_n + b_n + c_n &= 1 \\ \lambda_1^2 a_n + \lambda_1 b_n + c_n &= \lambda_1^n \\ \lambda_2^2 a_n + \lambda_2 b_n + c_n &= \lambda_2^n. \end{cases}$$

This system has rank 3 if the three roots are distinct.

If there is a double root λ (be it 1 or $\lambda_1 = \lambda_2$), then two of these equations are identical, but as the double root is also a root of $K'(X)$, a simple derivative yields a third equation $2\lambda a_n + b_n = n\lambda^{n-1}$, which is linearly independent of the two others.

If the three roots of $K(X)$ are equal, then they are equal to 1 and $P^n = P = I$.

If P is irreducible, then there exists a unique invariant law π, given by

$$\pi(1) = \frac{ce + cf + of}{ad + ae + af + bc + bd + bf + ce + cf + of},$$

$$\pi(2) = \frac{ae + af + bf}{ad + ae + af + bc + bd + bf + ce + cf + of},$$

$$\pi(3) = \frac{ad + bc + bd}{ad + ae + af + bc + bd + bf + ce + cf + of}.$$

The chain with a finite number of states Let $d = \text{Card}(\mathcal{V})$. The above-mentioned method can be extended without any theoretical problem.

The Euclidean division $X^n = Q(X)K(X) + a_{n,d-1}X^{d-1} + \cdots + a_{n,1}X + a_{n,0}$ and $K(P) = 0$ yield that

$$P^n = a_{n,d-1}P^{d-1} + \cdots + a_{n,1}P + a_{n,0}I.$$

If $\lambda_1, \ldots, \lambda_r$ are the distinct roots of $K(X)$ and $m_1 \geq 1, \ldots, m_r \geq 1$ are their multiplicities, then

$$K(\lambda_i) = 0, \ldots, K^{(m_i-1)}(\lambda_i) = 0, \qquad 1 \leq i \leq r,$$

is a system of $d = \sum_{i=1}^{r} m_i$ linearly independent equations for the d unknowns $a_{n,d-1}, \ldots, a_{n,0}$, which thus has a unique solution.

The enormous obstacle for the effective implementation of this method for computing P^n is that we must compute the roots of $K(X)$ first. The main information we have is that 1 is a root, and in general, computing the roots becomes a considerable problem as soon as $d \geq 4$. Once the roots are found, solving the linear system and finding the invariant laws is a problem only when d is much larger.

This general method is simpler than finding the reduced Jordan form J for P, which also necessitates to find the roots of the characteristic polynomial $K(X)$, and then solving a linear system to find the change-of-basis matrix M and its inverse M^{-1}. Then, $P^n = (MJM^{-1})^n = MJ^nM^{-1}$, where J^n can be made explicit.

1.4   Detailed examples

We are going to describe in informal manner some problems concerning random evolutions, for which the answers will obviously depend on some data or parameters. We then will model these problems using Markov chains.

These models will be studied in detail all along our study of Markov chains, which they will help to illustrate.

In these descriptions, random variables and draws will be supposed to be independent if not stated otherwise.

1.4.1   Random walk on a network

A particle evolves on a network $\mathcal{R}$, that is, on a discrete additive subgroup such as $\mathbb{Z}^d$. From x in $\mathcal{R}$, it chooses to go to $y = x + (y - x)$ in $\mathcal{R}$ with probability $p(y - x) \geq 0$, which satisfies $\sum_{z \in \mathcal{R}} p(z) = 1$. This can be, for instance, a model for the evolution of an electron in a network of crystals.

Some natural questions are the following:

- Does the particle escape to infinity?

- If yes, at what speed?

- With what probability does it reach a certain subset in finite time?
- What is the mean time for that?

1.4.1.1   Modeling

Let $(\xi_k)_{k \geq 1}$ be a sequence of i.i.d. random variables such that $\mathbb{P}(\xi_1 = z) = p(z)$, and

$$X_n = X_{n-1} + \xi_n = \cdots = X_0 + \xi_1 + \cdots + \xi_n.$$

Theorem 1.2.3 shows that $(X_n)_{n \geq 0}$ is a Markov chain on $\mathcal{R}$, with a transition matrix, which is spatially homogeneous, or invariant by translation, given by

$$P(x, y) = \mathbb{P}(\xi_1 = y - x) = p(y - x), \qquad x, y \in \mathcal{R}.$$

The matrix P restricted to the network generated by all z such that $p(z) > 0$ is irreducible. The constant measures are invariant, as

$$\sum_{x \in \mathcal{R}} P(x, y) = \sum_{x \in \mathcal{R}} p(y - x) = \sum_{z \in \mathcal{R}} p(z) = 1, \qquad \forall y \in \mathcal{R}.$$

If $\mathbb{E}(|\xi_1|) < \infty$, then the strong law of large numbers yields that

$$X_n = n\mathbb{E}(\xi_1) + o(n), \quad \text{a.s.,}$$

and for $\mathbb{E}(\xi_1) \neq 0$ the chain goes to infinity in the direction of $\mathbb{E}(\xi_1)$. The case $\mathbb{E}(\xi_1) = 0$ is problematic, and if $\mathbb{E}(|\xi_1|^2) < \infty$, then the central limit theorem shows that $X_n/\sqrt{n}$ converges in law to $\mathcal{N}(0, \text{Cov}(\xi_1))$, which gives some hints to the long-time behavior of the chain.

Nearest-neighbor random walk For $\mathcal{R} = \mathbb{Z}^d$, this Markov chain is called a nearest-neighbor random walk when $P(x, y) = 0$ for $|x - y| > 1$, and the symmetric nearest-neighbor random walk when $P(x, y) = 1/2d$ for $|x - y| = 1$. These terminologies are used for other regular networks, such as the one in Figure 1.1.

1.4.2   Gambler's ruin

Two gamblers A and B play a game of head or tails. Gambler A starts with a fortune of $a \in \mathbb{N}$ units of money and Gambler B of $b \in \mathbb{N}$ units. At each toss, each gambler makes a bet of 1 unit, Gambler A wins with probability p and loses with probability $q = 1 - p$, and the total of the bets is given to the winner; a gambler thus either wins or loses 1 unit.

The game continues until one of the gamblers is *ruined*: he or she is left with a fortune of 0 units, the global winner with a fortune of $a + b = N$ units, and the game stops. This is illustrated in Figure 1.2.

When $p = q = 1/2$, the game is said to be *fair*, else to be *biased*.

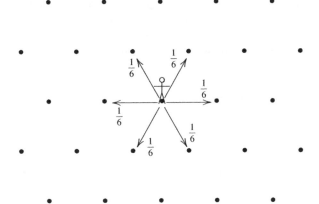

Figure 1.1   Symmetric nearest-neighbor random walk on regular planar triangular network.

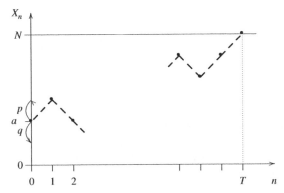

Figure 1.2   Gambler's ruin. Gambler A finishes the game at time T with a gain of b = N − a units starting from a fortune of a units. The successive states of his fortune are represented by the • and joined by dashes. The arrows on the vertical axis give his probabilities of winning or losing at each toss.

As an example, Gambler A goes to a casino (Gambler B). He decides to gamble 1 unit at each draw of red or black at roulette and to stop either after having won a total of b units (what he or she would like to gain) or lost a total of a units (the maximal loss he or she allows himself).

Owing to the 0 and (most usually) the double 0 on the roulette, which are neither red nor black, the game is biased against him, and p is worth either $18/37 \simeq 0.4865$ if there is no double 0 or $18/38 \simeq 0.4737$ if there is one.

From a formal point of view, there is a symmetry in the game obtained by switching a and $b = N - a$ simultaneously with p and $q = 1 - p$. In practice, no casino allows a bias in favor of the gambler, nor even a fair game.

A unilateral case will also be considered, in which $a \in \mathbb{N}$ and $b = N = \infty$. In the casino example, this corresponds to a compulsive gambler, who will stop only when ruined. In all cases, the evolution of the gambler's fortune is given by a nearest-neighbor random walk on $\mathbb{Z}$, stopped when it hits a certain boundary.

Some natural questions are the following:

- What is the probability that Gambler A will be eventually ruined?

- Will the game eventually end ?

- If yes, what is the mean duration of the game?

- What is the law of the duration of the game (possibly infinite) ?

1.4.2.1 Stopped random walk

In all cases, the evolution of the gambler's fortune is given by a nearest-neighbor random walk on $\mathbb{Z}$ stopped when it hits a certain boundary.

1.4.2.2 Modeling

The evolution of the fortune of Gambler A can be represented using a sequence $(\xi_k)_{k \geq 1}$ of i.i.d. r.v. satisfying $\mathbb{P}(\xi_1 = 1) = p$ and $\mathbb{P}(\xi_1 = -1) = q = 1 - p$ by

$$X_n = X_{n-1} + \xi_n \mathbb{1}_{\{0 < X_{n-1} < N\}} , \qquad n \geq 1 ,$$

where X_0 is its initial fortune a, or more generally a r.v. with values in $\{0, 1, \ldots, N\}$ and independent of $(\xi_k)_{k \geq 1}$. Gambler B's fortune at time $n \geq 0$ is $N - X_n$.

Theorem 1.2.3 yields that $(X_n)_{n \geq 0}$ is a Markov chain on $\mathcal{V} = \{0, 1, \ldots, N\}$ with matrix and graph given by

$$P(0, 0) = P(N, N) = 1; \quad P(x, x + 1) = p, \quad P(x, x - 1) = q, \quad 0 < x < N,$$

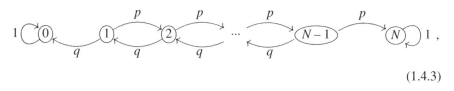

$$(1.4.3)$$

the other terms of P being 0.

The states 0 and $N = a + b$ are absorbing, hence P is not irreducible. The invariant measures μ are of the form $\mu(x) = 0$ if $0 < x < N$ with $\mu(0)$ and $\mu(N)$ arbitrary, and uniqueness does not hold (uniqueness being understood as "up to proportionality").

1.4.3   Branching process: evolution of a population

We study the successive generation of a population, constituted for instance of viruses in an organism, of infected people during an epidemic, or of neutrons during an atomic reaction.

The individuals of one generation disappear in the following, giving birth there each to k descendants with probability $p(k) \geq 0$, with $\sum_{k \in \mathbb{N}} p(k) = 1$. A classic sub-case is that of a binary division: $p(2) = p > 0$ and $p(0) = 1 - p > 0$.

The result of this random evolution mechanism is called a branching process. It is also called a Galton–Watson process; the initial study of Galton and Watson, preceded by a similar study of Bienaymé, bore on family names in Great Britain.

Some natural questions are the following:

- What is the law of the number of individuals in the nth generation?

- Will the population become extinct, almost surely (a.s.), and else with what probability?

- What is the long-time population behavior when it does not become extinct?

1.4.3.1   Modeling

We shall construct a Markov chain $(X_n)_{n \geq 0}$ corresponding to the sizes (numbers of individuals) of the population along the generations.

Let $(\xi_{n,i})_{n,i \geq 1}$ be i.i.d. r.v. such that $\mathbb{P}(\xi_{1,1} = k) = p(k)$ for k in $\mathbb{N}$. We assume that the X_{n-1} individuals of generation $n - 1$ are numbered $i \geq 1$ and that the ith one yields $\xi_{n,i}$ descendants in generation n, so that

$$X_n = \sum_{i=1}^{X_{n-1}} \xi_{n,i} , \qquad n \geq 1.$$

(An empty sum being null by convention.)

Figure 1.3 illustrates this using the genealogical tree of a population, which gives the relationships between individuals in addition to the sizes of the generations, and explains the term "branching."

The state space of $(X_n)_{n \geq 0}$ is $\mathbb{N}$, and Theorem 1.2.3 applied to $\xi_n = (\xi_{n,i})_{i \geq 1}$ yields that it is a Markov chain. The transition matrix is given by

$$P(0,0) = 1 , \qquad P(x,y) = \sum_{k_1 + \cdots + k_x = y} p(k_1) \cdots p(k_x) , \qquad x \geq 1 , y \geq 0 ,$$

and state 0 is absorbing. The matrix is not practical to use under this form.

It is much more practical to use generating functions. If

$$g(s) = \sum_{k \in \mathbb{N}} p(k) s^k = \mathbb{E}(s^{\xi_{1,1}})$$

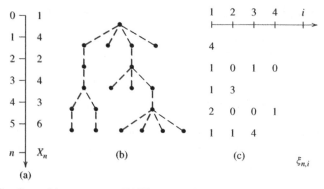

Figure 1.3   Branching process. (b) The genealogical tree for a population during six generations; • represent individuals and dashed lines their parental relations. (a) The vertical axis gives the numbers n of the generations, of which the sizes figure on its right. (c) The table underneath the horizontal axis gives the $\xi_{n,i}$ for $n \geq 1$ and $1 \leq i \leq X_{n-1}$, of which the sum over i yields X_n.

denotes the generating function of the reproduction law, the i.i.d. manner in which individuals reproduce yields that

$$\sum_{y \in \mathcal{V}} P(x, y)s^y := \mathbb{E}_x(s^{X_1}) = \mathbb{E}_1(s^{X_1})^x = g(s)^x.$$

For n in $\mathbb{N}$, let

$$g_n(s) = \sum_{x \in \mathbb{N}} \mathbb{P}(X_n = x)s^x = \mathbb{E}(s^{X_n})$$

denote the generating function of the size of generation n. An elementary probabilistic computation yields that, for $n \geq 1$,

$$g_n(s) = \mathbb{E}\left(s^{\sum_{i=1}^{X_{n-1}} \xi_{n,i}}\right)$$

$$= \sum_{x \in \mathbb{N}} \mathbb{E}\left(s^{\sum_{i=1}^{x} \xi_{n,i}} \mathbb{1}_{\{X_{n-1}=x\}}\right)$$

$$= \sum_{x \in \mathbb{N}} \mathbb{E}\left(s^{\sum_{i=1}^{x} \xi_{n,i}}\right) \mathbb{P}(X_{n-1} = x)$$

$$= \sum_{x \in \mathbb{N}} g(s)^x \mathbb{P}(X_{n-1} = x)$$

and hence that

$$g_n(s) = g_{n-1}(g(s)) = \cdots = g^{on}(g_0(s)) = g(g_{n-1}(s)).$$

We will later see how to obtain this result by Markov chain techniques and then how to exploit it.

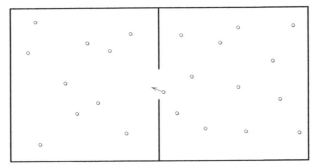

Figure 1.4   The Ehrenfest Urn. Shown at an instant when a particle transits from the right compartment to the left one. The choice of the particle that changes compartment at each step is uniform.

1.4.4   Ehrenfest's Urn

A container (urn, ...) is constituted of two communicating compartments and contains a large number of particles (such as gas molecules). These are initially distributed in the two compartments according to some law, and move around and can switch compartment.

Tatiana and Paul Ehrenfest proposed a statistical mechanics model for this phenomenon. It is a discrete time model, in which at each step a particle is chosen uniformly among all particles and changes compartment. See Figure 1.4.

Some natural questions are the following:

- starting from an unbalanced distribution of particles between compartments, is the distribution of particles going to become more balanced in time?

- In what sense, with what uncertainty, at what rate?

- Is the distribution going to go through astonishing states, such as having all particles in a single compartment, and at what frequency?

1.4.4.1   Microscopic modeling

Let N be the number of molecules, and let the compartments be numbered by 0 and 1.

A microscopic description of the system at time $k \geq 0$ is given by

$$X_k = (X_k^i)_{1 \leq i \leq N} \quad \text{with values in} \quad \{0, 1\}^N ,$$

where the ith coordinate X_k^i is the number of the compartment in which the ith particle is located.

Starting from a sequence $(\xi_k)_{k \geq 1}$ of i.i.d. r.v. which are uniform on $\{1, \ldots, N\}$, and an initial r.v. X_0 independent of this sequence, we define recursively X_k for $k \geq 1$ by changing the coordinate of rank ξ_k of X_{k-1}. This random recursion is a faithful rendering of the particle dynamics.

Theorem 1.2.3 implies that $(X_k)_{k \geq 0}$ is a Markov chain on $\{0,1\}^N$ with matrix given by

$$P(x,y) = \frac{1}{N} \text{ if } \sum_{i=1}^{N} |x^i - y^i| = 1 , \quad P(x,y) = 0 \text{ else.}$$

This is the symmetric nearest-neighbor random walk on the unit hypercube $\{0,1\}^N$. This chain is irreducible.

Invariant law This chain has for unique invariant law the uniform law π with density $\frac{1}{2^N}$.

As the typical magnitude of N is comparable to the Avogadro number 6.02×10^{23}, the number 2^N of configurations is enormously huge. Any computation, even for the invariant law, is of a combinatorial nature and will be most likely untractable.

1.4.4.2 Reduced macroscopic description

According to statistical mechanics, we should take advantage of the symmetries of the system, in order to stop following individual particles and consider collective behaviors instead.

A reduced macroscopic description of the system is the number of particles in compartment 1 at time $k \geq 0$, given in terms of the microscopic description by

$$S_k = \sum_{i=1}^{N} X_k^i , \quad \text{with values in } \{0, 1, \dots, N\} ,$$

The information carried by S_k being less than the information carried by X_k, it is not clear that $(S_k)_{k \geq 0}$ is a Markov chain, but the symmetry of particle dynamics will allow to prove it.

For $x = (x_i)_{1 \leq i \leq N} \in \{0,1\}^N$, let σ^x be the permutation of $\{1, \dots, N\}$ obtained by first placing in increasing order the i such that $x_i = 1$ and then by increasing order the i such that $x_i = 0$. Setting

$$\xi_k' = \sigma^{X_{k-1}}(\xi_k) ,$$

it holds that

$$S_k = S_{k-1} - \mathbb{1}_{\{S_{k-1} \leq \xi_k'\}} + \mathbb{1}_{\{S_{k-1} > \xi_k'\}} , \qquad k \geq 1.$$

For some deterministic f_k and g_k, using the random recursion for $(X_k)_{k \geq 0}$,

$$X_k = f_k(X_0, \xi_1, \dots, \xi_k) , \quad (X_0, \xi_1', \dots, \xi_k') = g_k(X_0, \xi_1, \dots, \xi_k) ,$$

and hence, for all $a \in \{0,1\}^N \times \{1, \dots, N\}^k$ and $z \in \{1, \dots, N\}$,

$$\mathbb{P}((X_0, \xi_1', \dots, \xi_k') = a, \xi_{k+1}' = z)$$

$$= \sum_{(x, z_1, \dots, z_k) \in g_k^{-1}(a)} \mathbb{P}((X_0, \xi_1, \dots, \xi_k) = (x, z_1, \dots, z_k)),$$

$$\sigma^{f_k(x,z_1,\dots,z_k)}(\xi_{k+1}) = z)$$

$$= \sum_{(x,z_1,\dots,z_k)\in g_k^{-1}(a)} \mathbb{P}((X_0,\xi_1,\dots,\xi_k) = (x,z_1,\dots,z_k))\frac{1}{N}$$

$$= \mathbb{P}((X_0,\xi_1',\dots,\xi_k') = a)\frac{1}{N}$$

as ξ_{k+1} is uniform and independent of $X_0,\xi_1,\dots,\xi_k$. Hence, ξ_{k+1}' is uniform on $\{1,\dots,N\}$ and independent of $(X_0,\xi_1',\dots,\xi_k')$. By a simple recursion, we conclude that the $(\xi_k')_{k\geq 1}$ are i.i.d. r.v. which are uniform on $\{1,\dots,N\}$ and independent of X_0 and hence of S_0.

Thus, Theorem 1.2.3 yields that $(S_k)_{k\geq 0}$ is a Markov chain on $\{0,1,\dots,N\}$ with matrix Q and graph given by

$$Q(x,x+1) = \frac{N-x}{N}, \quad Q(x,x-1) = \frac{x}{N}, \quad 0 \leq x \leq N,$$

$$(1.4.4)$$

all other terms of Q being zero. As $(X_k)_{k\geq 0}$ is irreducible on $\{0,1\}^N$, it is clear that $(S_k)_{k\geq 0}$ is irreducible on $\{0,1,\dots,N\}$, and this can be readily checked.

Invariant law As the uniform law on $\{0,1\}^N$, with density $\frac{1}{2^N}$, is invariant for $(X_k)_{k\geq 0}$, a simple combinatorial computation yields that the invariant law for $(S_k)_{k\geq 0}$ is binomial $\mathcal{B}(N,1/2)$, given by

$$\beta = (\beta(x))_{x\in\{0,1,\dots,N\}}, \qquad \beta(x) = \frac{1}{2^N}\binom{N}{x}.$$

This law distributes the particles uniformly in both compartments, and this is preserved by the random evolution.

1.4.4.3   Some computations on particle distribution

At *equilibrium*, that is, under the invariant law, the X_k are uniform on $\{0,1\}^N$ and hence, the X_k^i for $i = 1,\dots,N$ are i.i.d. uniform on $\{0,1\}$. The strong law of large numbers and the central limit theorem then yield that

$$\frac{S_k}{N} \xrightarrow[N\to\infty]{a.s.} \frac{1}{2}, \qquad \sqrt{N}\left(\frac{S_k}{N} - \frac{1}{2}\right) \xrightarrow[N\to\infty]{in\ law} \mathcal{N}(0,(1/2)^2).$$

Hence, as N goes to infinity, the instantaneous proportion of molecules in each compartment converges to $1/2$ with fluctuations of order $1/\sqrt{N}$. For instance,

$$\mathbb{P}_\pi(|S_k - N/2| \geq a\sqrt{N}) \xrightarrow[N\to\infty]{} 2\int_{2a}^\infty e^{-\frac{x^2}{2}}\frac{dx}{\sqrt{2\pi}},$$

and as a numerical illustration, as $\int_{-\infty}^{4,5} e^{-\frac{x^2}{2}}\frac{dx}{\sqrt{2\pi}} \simeq 0,999997$, the choice $2a = 4,5$ and $N = 6 \times 10^{23}$ yields that $a\sqrt{N} \simeq 1,74 \times 10^{12}$ and hence

$$\mathbb{P}_\pi(|S_k - 3\times 10^{23}| \geq 1,74\times 10^{12}) \simeq 6\times 10^{-6}.$$

For an arbitrary initial law, for $1 \leq i \leq N$ and $k \geq 1$,

$$\mathbb{P}(X_k^i = 1) = \mathbb{P}(X_{k-1}^i = 1)\frac{N-1}{N} + \mathbb{P}(X_{k-1}^i = 0)\frac{1}{N}$$

$$= \mathbb{P}(X_{k-1}^i = 1)\frac{N-2}{N} + \frac{1}{N}$$

and the solution of this affine recursion, with fixed point $1/2$, is given by

$$\mathbb{P}(X_k^i = 1) = \frac{1}{2} + \left(\mathbb{P}(X_0^i = 1) - \frac{1}{2}\right)\left(\frac{N-2}{N}\right)^k.$$

Then, at geometric rate,

$$X_k^i \xrightarrow[k\to\infty]{\text{in law}} \frac{1}{2}(\delta_0 + \delta_1), \qquad \mathbb{E}\left(\frac{1}{N}S_k\right) := \frac{1}{N}\sum_{i=1}^N \mathbb{P}(X_k^i = 1) \xrightarrow[k\to\infty]{} \frac{1}{2},$$

The rate $\frac{N-2}{N}$ seems poor, but the time unit should actually be of order $1/N$, and

$$\left(\frac{N-2}{N}\right)^{Nk} \xrightarrow[N\to\infty]{} e^{-2k}.$$

Explicit variance computations can also be done, which show that $\frac{1}{N}S_k$ converges in probability to $1/2$, but in order to go further some tools must be introduced.

1.4.4.4   Random walk, Fourier transform, and spectral decomposition

The Markov chain $(X_k)_{k\geq 0}$ on $\{0,1\}^N$ (microscopic representation) can be obtained by taking a sequence $(U_k)_{k\geq 1}$ of i.i.d. r.v. which are uniform on the vectors of the canonical basis, independent of X_0, and setting

$$X_k = X_{k-1} + U_k \pmod{2}, \qquad k \geq 1.$$

This is a symmetric nearest-neighbor random walk on the additive group $\{0,1\}^N := (\frac{\mathbb{Z}}{2\mathbb{Z}})^N$, and we are going to exploit this structure, according to a technique adaptable to other random walks on groups.

For b and x in $\{0, 1\}^N$ and for vectors $v = (v(x))_{x \in \{0,1\}^N}$ and $w = (w(x))_{x \in \{0,1\}^N}$, the canonical scalar products will be respectively denoted by

$$b \cdot x := \sum_{1 \leq i \leq N} b_i x_i , \qquad \langle v, w \rangle := \sum_{x \in \{0,1\}^N} v(x) w(x).$$

Let us associate to each r.v. X on $\{0, 1\}^N$ its characteristic function, which is the (discrete) Fourier transform of its law π_X given by, with the notation $e_b = ((-1)^{b \cdot x})_{x \in \{0,1\}^N}$,

$$F_X : b \in \{0, 1\}^N \mapsto \mathbb{E}((-1)^{b \cdot X}) = \sum_{x \in \{0,1\}^N} \mathbb{P}(X = x)(-1)^{b \cdot x} = \langle \pi_X, e_b \rangle.$$

For $b \neq c$ in $\{0, 1\}^N$ and any $1 \leq i \leq N$ such that $b_i \neq c_i$,

$$\langle e_b, e_b \rangle = \sum_{x \in \{0,1\}^N} (-1)^{2b \cdot x} = 2^N ,$$

$$\langle e_b, e_c \rangle = \sum_{x \in \{0,1\}^N \,:\, x_i = 0} (-1)^{b \cdot x + c \cdot x}(1 + (-1)) = 0 ,$$

hence $(e_b)_{b \in \{0,1\}^N}$ is an orthogonal basis of vectors, each with square product 2^N. This basis could easily be transformed into an orthonormal basis.

Fourier inversion formula   From this follows the inversion formula

$$\pi_X = \frac{1}{2^N} \sum_{b \in \{0,1\}^N} F_X(b) e_b ,$$

$$\pi_X(x) = \frac{1}{2^N} \sum_{b \in \{0,1\}^N} F_X(b)(-1)^{b \cdot x} = \frac{1}{2^N} \langle F_X, e_x \rangle , \quad x \in \{0, 1\}^N.$$

Fourier transform and eigenvalues   For $b \in \{0, 1\}^N$, setting

$$\lambda(b) = F_{U_1}(b) = \frac{1}{N} \sum_{i=1}^{N} (-1)^{b_i} = \frac{1}{N} \sum_{i=1}^{N} (1 - 2b_i) = \frac{N - 2 \sum_{i=1}^{N} b_i}{N} ,$$

it holds that

$$F_{X_k}(b) = \lambda(b) F_{X_{k-1}}(b) = \cdots = \lambda(b)^k F_{X_0}(b)$$

and thus e_b is an eigenvector for the eigenvalue $\lambda(b)$ for the transition matrix. There are $N + 1$ distinct eigenvalues

$$\lambda_j = \frac{N - 2j}{N} , \qquad 0 \leq j \leq N ,$$

of which the eigenspace of dimension $\binom{N}{j}$ is generated by the e_b such that b has exactly j terms taking the value 1.

Spectral decomposition of the transition matrix This yields the spectral decomposition of P in an orthogonal basis and

$$\pi_{X_k} = \frac{1}{2^N} \sum_{b,y \in \{0,1\}^N} \lambda(b)^k \langle \pi_{X_0}, e_b \rangle e_b \,,$$

$$\pi_{X_k}(x) = \frac{1}{2^N} \sum_{b,y \in \{0,1\}^N} \lambda(b)^k \pi_{X_0}(y)(-1)^{b \cdot (x+y)} \,, \qquad x \in \{0,1\}^N.$$

Long-time behavior This yields that $F_{X_k}(b)$ is $\mathcal{O}((\frac{N-2}{N})^k)$ for $b \notin \{0,1\}$ (constant vectors) and that

$$F_{X_k}(0) = 1 \,, \quad F_{X_k}(1) = (-1)^k F_{X_0}(1) = (-1)^k(\mathbb{P}(S_0 \in 2\mathbb{N}) - \mathbb{P}(S_0 \in 2\mathbb{N} + 1)).$$

For $\alpha = 2p - 1 \in [-1,1] \iff p = \frac{1+\alpha}{2} \in [0,1]$, let F_α be such that $F_\alpha(0) = 1$, $F_\alpha(1) = \alpha$ and $F_\alpha(b) = 0$ for $b \notin \{0,1\}$, and π_α be the corresponding laws. Then, $F_\alpha = pF_1 + (1-p)F_{-1}$ and thus $\pi_\alpha = p\pi_1 + (1-p)\pi_{-1}$ and

$$\pi_1(x) = \frac{1}{2^N}(1 + (-1)^{1 \cdot x}) \,, \quad \pi_{-1}(x) = \frac{1}{2^N}(1 - (-1)^{1 \cdot x}) \,, \qquad x \in \{0,1\}^N \,,$$

that is, π_1 and π_{-1} are the uniform laws respectively on

$$\left\{ x \in \{0,1\}^N : 1 \cdot x = \sum_{i=1}^N x_i \in 2\mathbb{N} \right\} \,, \quad \left\{ x \in \{0,1\}^N : \sum_{i=1}^N x_i \in 2\mathbb{N} + 1 \right\}$$

Let

$$\alpha = \mathbb{P}(S_0 \in 2\mathbb{N}) - \mathbb{P}(S_0 \in 2\mathbb{N} + 1) = 2\mathbb{P}(S_0 \in 2\mathbb{N}) - 1.$$

The law π_α is the mixture of π_1 and π_{-1}, which respects the probability that S_0 be in $2\mathbb{N}$ and in $2\mathbb{N} + 1$, and $\pi_{-\alpha}$ the mixture that interchanges these. If X_0 is of law π_α, then X_k is of law $\pi_{(-1)^k\alpha}$ and thus

$$F_{X_k}(b) - F_{(-1)^k\alpha}(b) = \lambda(b)^k(F_{X_0}(b) - F_\alpha(b))$$

and, as $F_{X_0}(0) = F_\alpha(0)$ and $F_{X_0}(1) = F_\alpha(1)$, for the Hilbert norm associated with $\langle \cdot, \cdot \rangle$ it holds that

$$\|\pi_{X_k} - \pi_{(-1)^k\alpha}\|^2 = \frac{1}{2^N} \sum_{b \in \{0,1\}^N} \lambda(b)^{2k}(F_{X_0}(b) - F_\alpha(b))^2$$

$$\leq \left(\frac{N-2}{N}\right)^{2k} \|\pi_{X_0} - \pi_\alpha\|^2.$$

In particular, the law of X_{2n} converges exponentially fast to π_α and the law of X_{2n+1} to $\pi_{-\alpha}$.

Periodic behavior The behavior we have witnessed is related to the notion of periodicity: S_{2n} is even or odd the same as S_0, and S_{2n+1} has opposite parity than S_0. This is obviously related to the fact that -1 is an eigenvalue.

1.4.5   Renewal process

A component of a system (electronic component, machine, etc.) lasts a random life span before failure. It is visited at regular intervals (at times n in $\mathbb{N}$) and replaced appropriately. The components that are used for this purpose behave in i.i.d. manner.

1.4.5.1   Modeling

At time 0, a first component is installed, and the ith component is assumed to have a random life span before replacement given by D_i, where the $(D_i)_{i \geq 1}$ are i.i.d. on $\{1, 2, \ldots\} \cup \{\infty\}$. Let D denote an r.v. with same law as D_i, representing a generic life span. It is often assumed that $\mathbb{P}(D = \infty) = 0$.

Let X_n denote the age of the component in function at time $n \geq 0$, with $X_n = 0$ if it is replaced at that time. Setting

$$T_0 = 0 , \qquad T_k = T_{k-1} + D_k = D_1 + \cdots + D_k , \quad k \geq 1 ,$$

it holds that $X_n = n - T_{k-1}$ sur $T_{k-1} \leq n < T_k$ (Figure 1.5).

The X_n are defined for all $n \geq 0$ as $\mathbb{P}(D \geq 1) = 1$. If $\mathbb{P}(D = \infty) = 0$, then all T_k are finite, a.s., else if $\mathbb{P}(D = \infty) > 0$, then there exists an a.s. finite r.v. K such that $D_K = \infty$ and $T_K = T_{K+1} = \cdots = \infty$.

This natural representation in terms of the life spans is *not* a random recursion of the kind discussed in Theorem 1.2.3. We will give a direct proof that $(X_n)_{n \geq 0}$ is a Markov chain and give its transition matrix.

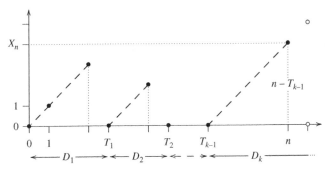

Figure 1.5   Renewal process. The • represent the ages at the discrete instants on the horizontal axis and are linearly interpolated by dashes in their increasing phases. Then, $X_n = n - T_{k-1}$ if $T_{k-1} \leq n < T_k = T_{k-1} + D_k$. The ○ represent the two possible ages at time $n + 1$, which are $X_{n+1} = X_n + 1$ if $D_k > X_n + 1$ and $X_{n+1} = 0$ if $D_k = X_n$.

Note that $\mathbb{P}(X_0 = x_0, \ldots, X_n = x_n) = 0$ except if $x_0 = 0$ and x_k is in $\{0, x_{k-1} + 1\}$ for $1 \leq k \leq n$. These are the only cases to be considered and then

$$\{X_0 = x_0, \ldots, X_n = x_n\} = \{D_1 = d_1, \ldots, D_{k-1} = d_{k-1}, D_k > x_n\},$$

where k is the number of 0 in $(x_0, \ldots, x_n)$ and 0, $t_1 = d_1$, $t_2 = d_1 + d_2$, $\ldots$, $t_{k-1} = d_1 + \cdots + d_{k-1}$ are their ranks (Figure 1.5). This can be written as

$$k = \sum_{i=0}^n \mathbb{1}_{\{x_i = 0\}}, \qquad d_j = \inf\{i \geq 1 : x_{d_1 + \cdots + d_{j-1} + i} = 0\}, \quad 1 \leq j \leq k - 1.$$

The independence of $(D_1, \ldots, D_{k-1})$ and D_k yields that

$$\mathbb{P}(X_0 = x_0, \ldots, X_n = x_n, X_{n+1} = x_n + 1)$$
$$= \mathbb{P}(D_1 = d_1, \ldots, D_{k-1} = d_{k-1}, D_k > x_n, X_{n+1} = x_n + 1)$$
$$= \mathbb{P}(D_1 = d_1, \ldots, D_{k-1} = d_{k-1}, D_k > x_n + 1)$$
$$= \mathbb{P}(D_1 = d_1, \ldots, D_{k-1} = d_{k-1})\mathbb{P}(D_k > x_n + 1)$$
$$= \mathbb{P}(D_1 = d_1, \ldots, D_{k-1} = d_{k-1}, D_k > x_n) \frac{\mathbb{P}(D_k > x_n + 1)}{\mathbb{P}(D_k > x_n)}$$
$$= \mathbb{P}(X_0 = x_0, \ldots, X_n = x_n) \frac{\mathbb{P}(D > x_n + 1)}{\mathbb{P}(D > x_n)}.$$

Moreover,

$$\mathbb{P}(X_0 = x_0, \ldots, X_n = x_n, X_{n+1} = 0)$$
$$= \mathbb{P}(X_0 = x_0, \ldots, X_n = x_n) \frac{\mathbb{P}(D = x_n + 1)}{\mathbb{P}(D > x_n)}$$

is obtained similarly, or by complement to 1.

Hence, the only thing that matters is the age of the component in function, and $(X_n)_{n \geq 0}$ is a Markov chain on $\mathbb{N}$ with matrix P and graph given by

$$P(x, x+1) = \frac{\mathbb{P}(D > x + 1)}{\mathbb{P}(D > x)} = \mathbb{P}(D > x + 1 | D > x) := p_x,$$
$$P(x, 0) = \mathbb{P}(D = x + 1 | D > x) = 1 - p_x, \quad x \in \mathbb{N},$$

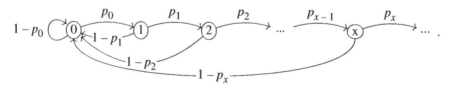

$$(1.4.5)$$

This Markov chain is irreducible if and only if $\mathbb{P}(D > k) > 0$ for every $k \in \mathbb{N}$ and $\mathbb{P}(D = \infty) < 1$.

A mathematically equivalent description Thus, from a mathematical perspective, we can start with a sequence $(p_x)_{x \in \mathbb{N}}$ with values in $[0, 1]$ and assume that a component with age x at an arbitrary time $n \in \mathbb{N}$ has probability p_x to pass the inspection at time $n + 1$, and else a probability $1 - p_x$ to be replaced then. In this setup, the law of D is determined by

$$\mathbb{P}(D > x) = p_0 \cdots p_{x-1} , \qquad x \geq 0.$$

This formulation is not as natural as the preceding one. It corresponds to a random recursion given by a sequence $(\xi_n)_{n \geq 1}$ of i.i.d. uniform r.v. on $[0, 1]$, independent of X_0, and

$$X_n = X_{n-1} + \sum_{x \geq 0} \mathbb{1}_{\{X_{n-1}=x\}} (\mathbb{1}_{\{\xi_n \leq p_x\}} - x \mathbb{1}_{\{\xi_n > p_x\}}) , \qquad n \geq 1.$$

The renewal process is often introduced in this manner, in order to avoid the previous computations. It is an interesting example, as we will discuss later.

Invariant measures and laws An invariant measure $\mu = (\mu(x))_{x \in \mathbb{N}}$ satisfies

$$\sum_{x \geq 0} \mu(x)(1 - p_x) = \mu(0) , \qquad \mu(x-1)p_{x-1} = \mu(x) , \quad x \geq 1 ,$$

thus $\mu(x) = \mu(0)p_0 \cdots p_{x-1} = \mu(0)\mathbb{P}(D > x)$ that yields uniqueness, and existence holds if and only if

$$\mathbb{P}(D = \infty) := \lim_{x \to \infty} p_0 \cdots p_{x-1} = 0.$$

This unique invariant measure can be normalized, in order to yield an invariant law, if and only if it is finite, that is, if

$$\mathbb{E}(D) := \sum_{x \geq 0} \mathbb{P}(D > x) := \sum_{x \geq 0} p_0 \cdots p_{x-1} < \infty ,$$

and then

$$\pi(x) = \frac{\mathbb{P}(D > x)}{\mathbb{E}(D)} = \frac{p_0 \cdots p_{x-1}}{\sum_{y \geq 0} p_0 \cdots p_{y-1}} , \qquad x \geq 0.$$

Renewal process and Doeblin condition A class of renewal processes is one of the rare natural examples of infinite state space Markov chains satisfying the Doeblin condition.

Lemma 1.4.1 *Assume that there exists $m \geq 0$ such that $\inf_{x \geq m}(1 - p_x) > 0$. Then, the Markov chain satisfies the Doeblin condition for $k = m + 1$ and $\varepsilon = \inf_{x \geq m}(1 - p_x)$ and $\hat{\pi} = \delta_0$, and the conclusions of Theorem 1.3.4 hold.*

Proof: This can be checked easily. ∎

1.4.6 Word search in a character chain

A source emits an infinite i.i.d. sequence of "characters" of some "alphabet." We are interested in the successive appearances of a certain "word" in the sequence.

For instance, the characters could be 0 and 1 in a computer system, "red" or "black" in a roulette game, A, C, G, T in a DNA strand, or ASCII characters for a typewriting monkey. Corresponding words could be 01100010, red-red-red-black, GAG, and Abracadabra.

Some natural questions are the following:

- Is any word going to appear in the sequence?

- Is it going to appear infinitely often, and with what frequency?

- What is the law and expectation of the first appearance time?

1.4.6.1 Counting automaton

A general method will be described on a particular instance, the search for the occurrences of the word GAG.

Two different kinds of occurrences can be considered, *without* or *with* overlaps; for instance, GAGAG contains one single occurrence of GAG without overlaps but two with. The case without overlaps is more difficult, and more useful in applications; it will be considered here, but the method can be readily adapted to the other case.

We start by defining a counting automaton with four states Ø, G, GA, and GAG, which will be able to count the occurrences of GAG in any arbitrary finite character chain. The automaton starts in state Ø and then examines the chain sequentially term by term, and:

- In state Ø: if the next state is G, then it takes state G, else it stays in state Ø,

- In state G: if the next state is A, then it takes state GA, if the next state is G, then it stays in state G, else it takes state Ø,

- In state GA: if the next state is G, then it takes state GAG, else it takes state Ø,

- In state GAG: if the next state is G, then it takes state G, else it takes state Ø.

Such an automaton can be represented by a graph that is similar to a Markov chain graph, with nodes given by its possible states and oriented edges between nodes marked by the logical condition for this transition (Figure 1.6).

This automation is now used on a sequence of characters given by an i.i.d. sequence $(\xi_n)_{n\geq 1}$ such that

$$\mathbb{P}(\xi_1 = A) = p_A , \qquad \mathbb{P}(\xi_1 = G) = p_G , \qquad \mathbb{P}(\xi_1 \notin \{A, G\}) = 1 - p_A - p_G ,$$

satisfying $p_A > 0, p_G > 0$, and $p_A + p_G \leq 1$.

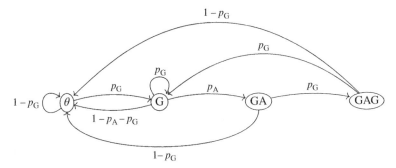

Figure 1.6   Search for the word GAG: Markov chain graph. The graph for the automaton is obtained by replacing p_A by "if the next term is A," p_G by "if the next term is G," $1 - p_G$ by "if the next term is not G," and $1 - p_A - p_G$ by "if the next term is neither A nor G."

Let $X_0 = \emptyset$, and X_n be the state of the automaton after having examined the nth character. Theorem 1.2.3 yields that $(X_n)_{n \geq 0}$ is a Markov chain with graph, given in Figure 1.6, obtained from the automaton graph by replacing the logical conditions by their probabilities of being satisfied.

Markovian description   All relevant information can be written in terms of $(X_n)_{n \geq 0}$. For instance, if $T_0 = 0$ and T_i denotes the time of the ith occurrence (complete, without overlaps) of the word for $i \geq 1$, and N_k the number of such occurrences taking place before $k \geq 1$, then

$$T_i = \inf\{n > T_{i-1} : X_n = \text{GAG}\}, \qquad N_k = \sum_{n=1}^{k} \mathbb{1}_{\{X_n = \text{GAG}\}}.$$

The transition matrix $P = (P(x, y))_{x,y \in \{\emptyset, G, GA, GAG\}}$ is irreducible and has for unique invariant law

$$\pi = \left(\frac{1 - p_G - p_A p_G^2}{1 + p_A p_G}, \frac{p_G}{1 + p_A p_G}, \frac{p_A p_G}{1 + p_A p_G}, \frac{p_A p_G^2}{1 + p_A p_G} \right).$$

Occurrences with overlaps   In order to search for the occurrences *with* overlaps, it would suffice to modify the automaton by considering the overlaps inside the word. For the word GAG, we need only modify the transitions from state GAG: if the next term is G, then the automaton should take state G, and if the next term is A, then it should take state GA, else it should take state $\emptyset$. For more general overlaps, this can become very involved.

1.4.6.2   Snake chain

We describe another method for the search for the occurrences *with* overlaps of a word $c_1 \cdots c_\ell$ of length $\ell \geq 1$ in an i.i.d. sequence $(\xi_n)_{n \geq 1}$ of characters from some alphabet $\mathcal{V}$.

Setting $Z_n = (\xi_n, \ldots, \xi_{n+\ell-1})$, then

$$T_i = \inf\{n > T_{i-1} : Z_{n-\ell+1} = (c_1, \ldots, c_\ell)\}$$

is the time of the ith occurrence of the word, $i \geq 1$ (with $T_0 = 0$), and

$$N_k = \sum_{n=\ell}^{k} \mathbb{1}_{\{Z_{n-\ell+1}=(c_1,\ldots,c_\ell)\}}$$

is the number of such occurrences before time $k \geq \ell$. In general, $(Z_n)_{n\geq 1}$ is not i.i.d., but it will be seen to be a Markov chain.

More generally, let $(Y_n)_{n\geq 0}$ be a Markov chain on $\mathcal{V}$ of arbitrary matrix P and $Z_n = (Y_n, \ldots, Y_{n+\ell-1})$ for $n \geq 0$. Then, $(Z_n)_{n\geq 0}$ is a Markov chain on $\mathcal{V}^\ell$ with matrix P_ℓ with only nonzero terms given by

$$P_\ell((x_1, \ldots, x_\ell), (x_2, \ldots, x_\ell, y)) = P(x_\ell, y) , \qquad x_1, \ldots, x_\ell, y \in \mathcal{V} ,$$

called the snake chain of length ℓ for $(Y_n)_{n\geq 0}$. The proof is straightforward if the conditional formulation is avoided.

Irreducibility If P is irreducible, then P_ℓ is irreducible on its natural state space

$$\mathcal{V}_\ell = \{(x_1, \ldots, x_\ell) \in \mathcal{V}^\ell : P(x_1, x_2) \cdots P(x_{\ell-1}, x_\ell) > 0\}.$$

Invariant Measures and Laws If μ is an invariant measure for $(Y_n)_{n\geq 0}$, then μ_ℓ given by

$$\mu_\ell(y_1, \ldots, y_\ell) = \mu(y_1)P(y_1, y_2) \cdots P(y_{\ell-1}, y_\ell)$$

is immediately seen to be an invariant measure for $(Z_n)_{n\geq 0}$. If further μ is a law, then μ_ℓ is also a law.

In the i.i.d. case where $P(x, y) = p(y)$, the only invariant law for $(Y_n)_{n\geq 0}$ is given by $\pi(y) = p(y)$, and the only invariant law for $(Z_n)_{n\geq 0}$ by the product law $\pi_\ell(y_1, \ldots, y_\ell) = p(y_1) \cdots p(y_\ell)$.

1.4.7   Product chain

Let P_1 and P_2 be two transition matrices on $\mathcal{V}_1$ and $\mathcal{V}_2$, and the matrices $P_1 \otimes P_2$ on $\mathcal{V}_1 \times \mathcal{V}_2$ have generic term

$$P_1 \otimes P_2((x^1, x^2), (y^1, y^2)) = P_1(x^1, y^1)P_2(x^2, y^2).$$

Then, $P_1 \otimes P_2$ is a transition matrix on $\mathcal{V}_1 \times \mathcal{V}_2$, as in the sense of product laws,

$$P_1 \otimes P_2((x^1, x^2), \cdot) = P_1(x^1, \cdot) \otimes P_2(x^2, \cdot).$$

see below for more details. See Figure 1.7.

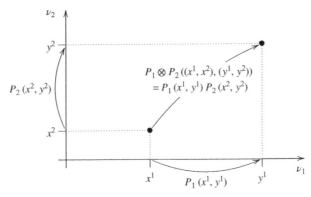

Figure 1.7   Product chain. The first and second coordinates are drawn independently according to P_1 and P_2.

The Markov chain $(X_n^1, X_n^2)_{n \geq 0}$ with matrix $P_1 \otimes P_2$ is called the product chain. Its transitions are obtained by independent transitions of each coordinate according, respectively, to P_1 and P_2. In particular, $(X_n^1)_{n \geq 0}$ and $(X_n^2)_{n \geq 0}$ are two Markov chains of matrices P_1 and P_2, which conditional on (X_0^1, X_0^2) are independent, and

$$(P_1 \otimes P_2)^n = P_1^n \otimes P_2^n , \qquad n \geq 0.$$

1.4.7.1   Invariant measures and laws

Immediate computations yield that if μ_1 is an invariant measure for P_1 and μ_2 for P_2, then the product measure $\mu_1 \otimes \mu_2$ given by

$$(\mu_1 \otimes \mu_2)(x_1, x_2) = \mu_1(x_1)\mu_2(x_2)$$

is invariant for $P_1 \otimes P_2$. Moreover,

$$\|\mu_1 \otimes \mu_2\|_{var} = \|\mu_1\|_{var} \times \|\mu_2\|_{var}$$

and thus if μ_1 and μ_2 are laws then $\mu_1 \otimes \mu_2$ is a law.

1.4.7.2   Irreducibility problem

The matrix $P = \begin{pmatrix} 0 & 1 \\ 1 & 0 \end{pmatrix}$ on $\{1, 2\}$ is irreducible and has unique invariant law the uniform law, whereas a Markov chain with matrix $P \otimes P$ alternates either between $(1, 1)$ and $(2, 2)$ or between $(1, 2)$ and $(2, 1)$, depending on the initial state and is not irreducible on $\{1, 2\}^2$. The laws

$$\frac{1}{2}(\delta_{(1,1)} + \delta_{(2,2)}) , \qquad \frac{1}{2}(\delta_{(1,2)} + \delta_{(2,1)}) ,$$

are invariant for $P \otimes P$ and generate the space of invariant measures.

All this can be readily generalized to an arbitrary number of transition matrices.

Exercises

1.1   The space station, 1   An aimless astronaut wanders within a space station, schematically represented as follows:

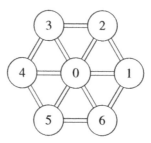

The space station spins around its center in order to create artificial gravity in its periphery. When the astronaut is in one of the peripheral modules, the probability for him to go next in each of the two adjacent peripheral modules is twice the probability for him to go to the central module. When the astronaut is in the central module, the probability for him to go next in each of the six peripheral modules is the same.

Represent this evolution by a Markov chain and give its matrix and graph. Prove that this Markov chain is irreducible and give its invariant law.

1.2   The mouse, 1   A mouse evolves in an apartment, schematically represented as follows:

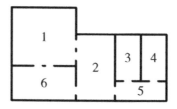

The mouse chooses uniformly an opening of the room where it is to go into a new room. It has a short memory and forgets immediately where it has come from.

Represent this evolution by a Markov chain and give its matrix and graph. Prove that this Markov chain is irreducible and give its invariant law.

1.3   Doubly stochastic matrices   Let $P = (P(x, y))_{x,y \in V}$ be a doubly stochastic matrix on a state space V: by definition,

$$\sum_{x \in V} P(x, y) = \sum_{y \in V} P(x, y) = 1.$$

a) Find a simple invariant measure for P.

b) Prove that P^n is doubly stochastic for all $n \geq 1$.

c) Prove that the transition matrix for a random walk on a network is doubly stochastic.

1.4   The Labouchère system, 1   In a game where the possible gain is equal to the wager, the probability of gain p of the player at each draw typically satisfies $p \leq 1/2$ and even $p < 1/2$, but is usually close to $1/2$, as when betting on red or black at roulette. In this framework, the Labouchère system is a strategy meant to provide a means for earning in a secure way a sum $S \geq 1$ determined in advance.

The sum S is decomposed arbitrarily as a sum of $k \geq 1$ positive terms, which are put in a list. The strategy then transforms recursively this list, until it is empty.

At each draw, if $k \geq 2$ then the sum of the first and last terms of the list are wagered, and if $k \geq 1$ then the single term is wagered. If the gambler wins, he or she removes from the list the terms concerned by the wager. If the gambler loses, he or she retains these terms and adds at the end of the list a term worth the sum just wagered. The game stops when $k = 0$, and hence, the sum S has been won.

(Martingale theory proves that in realistic situations, for instance, if wagers are bounded or credit is limited, then with a probability close to 1 the sum S is indeed won, but with a small probability a huge loss occurs, large enough to prevent the gambler to continue the game and often to ever gamble in the future.)

a) Represent the list evolution by a Markov chain $(L_n)_{n \geq 0}$ on the set

$$\mathcal{V} = \bigcup_{k \geq 0} \mathbb{N}^k$$

of words of the form $n_1 \cdots n_k$. Describe its transition matrix Q and its graph. Prove that if L_n reaches $\emptyset$ (the empty word), then the gambler wins the sum S.

b) Let X_n be the length of the list (or word) L_n for $n \geq 0$. Prove that $(X_n)_{n \geq 0}$ is a Markov chain on $\mathbb{N}$ and give its matrix P and its graph.

1.5   Three-card Monte   Three playing cards are lined face down on a cardboard box at time $n = 0$. At times $n \geq 1$, the middle card is exchanged with probability $p > 0$ with the card on the right and with probability $q = 1 - p > 0$ with the one on the left.

a) Represent the evolution of the three cards by a Markov chain $(Y_n)_{n \geq 0}$. Give its transition matrix Q and its graph. Prove that $(Y_n)_{n \geq 0}$ is irreducible. Find its invariant law ρ.

b) The cards are the ace of spades and two reds. Represent the evolution of the ace of spades by a Markov chain $(X_n)_{n\geq 0}$. Give its transition matrix P and its graph. Prove that it is irreducible. Find its invariant law π.

c) Compute P^n in terms of the initial law π_0 and p and $n \geq 1$. Prove that the law π_n of X_n converges to π as n goes to infinity, give an exponential convergence rate for this convergence, and find for which value of p the convergence is fastest.

1.6 Andy, 1 If Andy is drunk one evening, then he has one odd in ten to end up in jail, in which case will remain sober the following evening. If Andy is drunk one evening and does not end up in jail, then he has one odd in two to be drunk the following evening. If Andy stays sober one evening, then he has three odds out of four to remain sober the following evening.

It is assumed that $(X_n)_{n\geq 0}$ constitutes a Markov chain, where $X_n = 1$ if Andy on the n-th evening is drunk and ends up in jail, $X_n = 2$ if Andy then is drunk and does not end up in jail, and $X_n = 3$ if then he remains sober.

Give the transition matrix P and the graph for $(X_n)_{n\geq 0}$. Prove that P is irreducible and compute its invariant law. Compute P^n in terms of $n \geq 0$. What is the behavior of X_n when n goes to infinity?

1.7 Squash Let us recall the original scoring system for squash, known as English scoring. If the server wins a rally, then he or she scores a point and retains service. If the returner wins a rally, then he or she becomes the next server but no point is scored. In a game, the first player to score 9 points wins, except if the score reaches 8-8, in which case the returner must choose to continue in either 9 or 10 points, and the first player to reach that total wins.

A statistical study of the games between two players indicates that the rallies are won by Player A at service with probability $a > 0$ and by Player B at service with probability $b > 0$, each in i.i.d. manner.

The situation in which Player A has i points, Player B has j points, and Player L is at service is denoted by (i, j, L) in $\mathcal{V} = \{0, 1, \ldots, 10\}^2 \times \{A, B\}$.

a) Describe the game by a Markov chain on $\mathcal{V}$, assuming that if the score reaches 8-8 then they play on to 10 points (the play up to 9 can easily be deduced from this), in the two following cases: (i) all rallies are considered and (ii) only point scoring is considered.

b) Trace the graphs from arriving at 8-8 on the service of Player A to end of game.

c) A game gets to 8-8 on the service of Player A. Compute in terms of a and b the probability that Player B wins according to whether he or she elects to go to 9 or 10 points. Counsel Player B on this difficult choice.

1.8 Genetic models, 1 Among the individuals of a species, a certain gene can appear under $K \geq 2$ different forms called alleles.

In a **microscopic** (individual centered) model for a population of $N \geq 1$ individuals, these are arbitrarily numbered, and the fact that individual i carries allele a_i is coded by the state

$$(a_i)_{1 \leq i \leq N} \in \{1, \dots, K\}^N.$$

A **macroscopic** representation only retains the numbers of individuals carrying each allele, and the state space is

$$\{(n_1, \dots, n_K) \in \mathbb{N}^K : n_1 + \dots + n_K = N\}.$$

We study two simplified models for the reproduction of the species, in which the population size is fixed, and where the selective advantage of every allele a w.r.t. the others is quantified by a real number $c(a) > 0$.

1. **Synchronous: Fisher–Wright model**: at each step, the whole population is replaced by its descendants, and in i.i.d. manner, each new individual carries allele a with a probability proportional both to $c(a)$ and to the number of old individuals carrying allele a.

2. **Asynchronous: Moran model**: at each step, an uniformly chosen individual is replaced by a new individual, which carries allele a with a probability proportional both to $c(a)$ and to the number of old individuals carrying allele a.

   a) Explain how to obtain the macroscopic representation from the microscopic representation

   b) Prove that each pair representation-model corresponds to a Markov chain. Give the transition matrices and the absorbing states.

1.9 Records   Let $(X_i)_{i \geq 1}$ be i.i.d. r.v. such that $\mathbb{P}(X_i = 1) = p > 0$ and $\mathbb{P}(X_i = 0) = 1 - p > 0$, and R_n be the greatest number of consecutive 1 observed in $(X_1, \dots, X_n)$.

a) Show that $(R_n)_{n \geq 0}$ is not a Markov chain.

b) Let $X_0 := 0$, and

$$D_n = \inf\{k \geq 0 : X_{n-k} = 0\}, \qquad n \geq 0.$$

Prove that $(D_n)_{n \geq 0}$ is a Markov chain and give its transition matrix P. Prove that there exists a unique invariant law π and compute it.

c) Let $k \geq 0$,

$$S_k = \inf\{n \geq 0 : D_n = k\}, \qquad Z_n = D_n \text{ if } n \leq S_k, \text{ else } Z_n = D_n.$$

Prove that $(Z_n)_{n \geq 0}$ is a Markov chain on $\{0, 1, \dots, k\}$ and give its transition matrix P_k.

d) Express $\mathbb{P}(R_n \geq k)$ in terms of Z_n, then of P_k. Deduce from this the law of R_n.

e) What is the probability of having at least 5 consecutive heads among 100 fair tosses of head-or-tails? One can use the fact that for $p = 1/2$,

$$
P_5^{100} = \begin{pmatrix}
0,09659 & 0,04913 & 0,02499 & 0,01271 & 0,00647 & 0,81011 \\
0,09330 & 0,04746 & 0,02414 & 0,01228 & 0,00625 & 0,81658 \\
0,08683 & 0,04417 & 0,02247 & 0,01143 & 0,00581 & 0,82929 \\
0,07412 & 0,03770 & 0,01918 & 0,00976 & 0,00496 & 0,85428 \\
0,04913 & 0,02499 & 0,01271 & 0,00647 & 0,00329 & 0,90341 \\
0 & 0 & 0 & 0 & 0 & 1
\end{pmatrix}.
$$

1.10 Incompressible mixture, 1 There are two urns, N white balls, and N black balls. Initially N balls are set in each urn. In i.i.d. manner, a ball is chosen uniformly in each urn and the two are interchanged. The white balls are numbered from 1 to N and the black balls from $N + 1$ to $2N$. We denote by A_n the r.v. with values in

$$
\mathcal{V} = \{E \subset \{1, \ldots, 2N\} : \mathrm{Card}(E) = N\}
$$

given by the set of the numbers in the first urn just after time $n \geq 0$ and by

$$
S_n = \sum_{i=1}^{N} \mathbb{1}_{\{i \in A_n\}}
$$

the corresponding number of white balls.

a) Prove that $(A_n)_{n \in \mathbb{N}}$ is an irreducible Markov chain on $\mathcal{V}$ and give its matrix P. Prove that the invariant law π is unique and compute it.

b) Do the same for $(S_n)_{n \geq 0}$ on $\{0, 1, \ldots, N\}$, with matrix Q and invariant law σ.

c) For $1 \leq i \leq 2N$ find a recursion for $\mathbb{P}(i \in A_n)$, and solve it in terms of n and $\mathbb{P}(i \in A_0)$. Do likewise for $\mathbb{E}(S_n)$. What happens for large n?

1.11 Branching with immigration The individuals of a generation disappear at the following, leaving there k descendants each with probability $p(k) \geq 0$, and in addition, $i \in \mathbb{N}$ immigrants appear with probability $q(i) \geq 0$, where $\sum_{k \geq 0} p(k) = \sum_{k \geq 0} q(k) = 1$. Let

$$
g(s) = \sum_{k \in \mathbb{N}} p(k)s^k, \quad h(s) = \sum_{k \in \mathbb{N}} q(k)s^k, \quad 0 \leq s \leq 1,
$$

be the generating functions for the reproduction and the immigration laws.

Similarly to Section 1.4.3, using X_0 with values in $\mathbb{N}$ and $\xi_{n,i}$ and ζ_n for $n \geq 1$ and $i \geq 1$ such that $\mathbb{P}(\xi_{n,i} = k) = p(k)$ and $\mathbb{P}(\zeta_n = k) = q(k)$ for k in $\mathbb{N}$, all these r.v. being independent, let us represent the number of individuals in generation $n \in \mathbb{N}$ by

$$X_n = \zeta_n + \sum_{i=1}^{X_{n-1}} \xi_{n,i}.$$

Let G_n be the generating function of X_n.

a) Prove that $(X_n)_{n \in \mathbb{N}}$ is a Markov chain, without giving its transition matrix.

b) Compute G_n in terms of g, h, and G_{n-1}, then of h, g, n, and G_0.

c) If $x = \mathbb{E}(\xi_{n,i}) < \infty$ and $z = \mathbb{E}(\zeta_n) < \infty$, compute $\mathbb{E}(X_n)$ in terms of x, z, n, and $\mathbb{E}(X_0)$.

1.12 Single Server Queue Let $(A_n)_{n \geq 1}$ be i.i.d. r.v. with values in $\mathbb{N}$, with generating function $a(s) = \mathbb{E}(s^{A_1})$ and expectation $m = \mathbb{E}(A_1) < \infty$, and let X_0 be an independent r.v. with values in $\mathbb{N}$. Let

$$X_n = (X_{n-1} - 1)^+ + A_n , \qquad g_n(s) = \mathbb{E}(s^{X_n}).$$

a) Prove that $(X_n)_{n \geq 0}$ is a Markov chain with values in $\mathbb{N}$, which is irreducible if and only if $\mathbb{P}(A_1 = 0)\mathbb{P}(A_1 \geq 2) > 0$.

b) Compute g_n in terms of a and g_{n-1}.

c) It is now assumed that there exists an invariant law π for $(X_n)_{n \geq 0}$, with generating function denoted by g. Prove that

$$g(s)(s - a(s)) = \pi(0)(s - 1)a(s)$$

and that $\pi(0) = 1 - m$.

d) Prove that necessarily $m \leq 1$ and that $m = 1$ only in the trivial case where $\mathbb{P}(A_n = 1) = 1$.

e) Let $\mu = \sum_{x \in \mathbb{N}} \pi(x)x$. Prove that $\mu < \infty$ if and only if $\mathbb{E}(A_1^2) < \infty$, and then that for $\sigma^2 = Var(A_1)$, it holds that $\mu = \frac{1}{2}\left(m + \frac{\sigma^2}{1-m}\right)$.

1.13 Dobrushin mixing coefficient Let P be a transition matrix on $\mathcal{V}$, and

$$\rho_n = \frac{1}{2} \sup_{x,y \in \mathcal{V}} \|P^n(x, \cdot) - P^n(y, \cdot)\|_{var} , \qquad n \in \mathbb{N}.$$

a) Prove that $\rho_n \leq 1$ and that, for all laws μ and $\tilde{\mu}$,

$$\|\mu P^n - \tilde{\mu}P^n\|_{var} \leq 2\rho_n.$$

b) Prove that

$$\rho_{n+m} \leq \rho_n \rho_m , \quad m, n \geq 0 , \qquad \rho_n \leq \rho_k^{\lfloor n/k \rfloor} , \quad k \geq 1.$$

One may use that $\inf_{c \in \mathbb{R}} \sup_{x \in M} |g(x) - c| \leq \frac{1}{2}\sup_{x,y \in M} |g(x) - g(y)|$.

c) Prove that if $k \geq 1$ is such that $\rho_k < 1$, then $(\mu P^n)_{n\in\mathbb{N}}$ is a Cauchy sequence, its limit is an invariant law π, and

$$\|\mu P^n - \pi\|_{var} \leq 2\rho_n \leq 2\rho_k^{\lfloor n/k \rfloor}.$$

d) Assume that P satisfies the Doeblin condition: there exists $k \geq 1$ and $\varepsilon > 0$ and a law $\hat{\pi}$ such that $P^k(x, \cdot) \geq \varepsilon\hat{\pi}$. Prove that $\rho_k \leq 1 - \varepsilon$. Compare with the result in Theorem 1.3.4.

e) Let $(F_i)_{i\geq 1}$ be a sequence of i.i.d. random functions from $\mathcal{V}$ to $\mathcal{V}$, and $X_0^x = x \in \mathcal{V}$ and $X_{n+1}^x = F_{n+1}(X_n^x)$ for $n \geq 0$, so that P is the transition matrix of the Markov chain induced by this random recursion, see Theorem 1.2.3 and what follows. Let

$$T_{x,y} = \inf\{n \geq 0 : \overset{x}{\underset{n}{X}} = X_n^y\}.$$

Prove that

$$\rho_k \leq \sup_{x,y\in\mathcal{V}} \mathbb{P}(T_{x,y} > k).$$

2

Past, present, and future

2.1 Markov property and its extensions

2.1.1 Past σ-field, filtration, and translation operators

Let $(X_n)_{n \in \mathbb{N}}$ be a sequence of random variables. The σ-fields

$$\mathcal{F}_n = \{ \{ (X_0, X_1, \ldots, X_n) \in E \} : E \subset \mathcal{V}^{n+1} \}$$

contain all events that can be observed using exclusively $(X_0, \cdots, X_n)$, that is, the past up to time n included of the sequence. Obviously

$$\mathcal{F}_n \subset \mathcal{F}_{n+1} .$$

A family of nondecreasing σ-fields, such as $(\mathcal{F}_n)_{n \geq 0}$, is called a filtration, and provides a mathematical framework for the accumulation of information obtained by the step-by-step observation of the sequence.

Product σ-field Definition 1.2.1 gives an expression for the probability of any event A in $\bigcup_{n \geq 0} \mathcal{F}_n$ in terms of π_0 and P. The Kolmogorov extension theorem (Theorem A.3.10) in Section A.3.4 of the Appendix then attributes a corresponding probability to any event in the product σ-field of $\mathcal{V}^{\mathbb{N}}$, which is the smallest σ-field containing every $\mathcal{F}_n$, which we denote by $\mathcal{F}_\infty$. The σ-field $\mathcal{F}_\infty$ contains all the information that can be reconstructed from the observation of an arbitrarily large finite number of terms of the sequence $(X_n)_{n \geq 0}$.

As a σ-field is stable under countable intersections, $\mathcal{F}_\infty$ contains events allowing to characterize a.s. convergences, of which we will discuss later.

Markov Chains: Analytic and Monte Carlo Computations, First Edition. Carl Graham.
© 2014 John Wiley & Sons, Ltd. Published 2014 by John Wiley & Sons, Ltd.

Shift operators For $n \geq 0$, the shift operators θ_n act on sequences of r.v. by

$$\theta_n(X_k)_{k\geq 0} = \theta_n(X_0, X_1, \ldots) := (X_{n+k})_{k\geq 0} = (X_n, X_{n+1}, \ldots) .$$

This action extends naturally to any $\mathcal{F}_\infty$-measurable random variable Y and to any event A in $\mathcal{F}_\infty$: if $Y = f(X_0, X_1, \ldots)$ and $A = \{(X_0, X_1, \ldots) \in E\}$ for some measurable (deterministic) function f on $\mathcal{V}^{\mathbb{N}}$ and measurable subset $E \subset \mathcal{V}^{\mathbb{N}}$, then

$$\theta_n : Y = f(X_0, X_1, \ldots) \mapsto \theta_n Y = f(X_n, X_{n+1}, \ldots) ,$$

$$\theta_n : A = \{(X_0, X_1, \ldots) \in E\} \subset \Omega \mapsto \theta_n A = \{(X_n, X_{n+1}, \ldots) \in E\} \subset \Omega .$$

The σ-field $\theta_n \mathcal{F}_\infty$ is included in $\mathcal{F}_\infty$ and contains the events of the future after time n included for the sequence. It corresponds for $(X_n, X_{n+1}, \ldots)$ to what $\mathcal{F}_\infty$ corresponds for $(X_0, X_1, \ldots)$.

2.1.2 Markov property

Definition 1.2.1 is in fact equivalent to the following Markov property, which is apparently stronger as it yields that the past and the future of a Markov chain are independent conditional on its present state.

Theorem 2.1.1 (Markov property) *Let $(X_n)_{n\geq 0}$ be a Markov chain with matrix P. Then, for $n \geq 0$ and $B \in \mathcal{F}_n$ and $x \in \mathcal{V}$ and $A \in \mathcal{F}_\infty$,*

$$\mathbb{P}(B, X_n = x, \theta_n A) = \mathbb{P}(B, X_n = x)\mathbb{P}_x(A) ,$$

and hence in conditional formulation

$$\mathbb{P}(B, \theta_n A \mid X_n = x) = \mathbb{P}(B \mid X_n = x)\mathbb{P}_x(A) , \quad \mathbb{P}(\theta_n A \mid B, X_n = x) = \mathbb{P}_x(A) .$$

Moreover, the shifted chain $\theta_n(X_k)_{k\geq 0} = (X_{n+k})_{k\geq 0}$ is again a Markov chain with matrix P and initial law $\pi_n = \mathcal{L}(X_n)$. Conditional on $X_n = x$, this shifted chain is a Markov chain with matrix P, started from x, and independent of $\mathcal{F}_n$, that is, of $(X_0, \ldots, X_n)$.

Proof: Measure theory yields that the proof follows for general $A \in \mathcal{F}_\infty$ as soon as it is proved for every A in $\bigcup_{k\geq 0} \mathcal{F}_k$. Then, for some appropriate k and E and F,

$$A := \{(X_0, \ldots, X_k) \in E\} , \qquad B := \{(X_0, \ldots, X_n) \in F\} ,$$

and Definition 1.2.1 yields that

$$\mathbb{P}(B, X_n = x) = \sum_{(x_0, \ldots, x_{n-1}, x) \in F} \pi_0(x_0)P(x_0, x_1) \cdots P(x_{n-1}, x) ,$$

$$\mathbb{P}_x(A) = \sum_{(x, x_1 \ldots, x_k) \in E} P(x, x_1)P(x_1, x_2) \ldots P(x_{k-1}, x_k) ,$$

$$\mathbb{P}(B, X_n = x, \theta_n A) = \sum_{\substack{(x_0,\dots,x_{n-1},x)\in F \\ (x,x_{n+1}\cdots,x_{n+k})\in E}} \pi_0(x_0)P(x_0,x_1)\cdots P(x_{n-1},x)$$

$$\times P(x, x_{n+1})\cdots P(x_{n+k-1}, x_{n+k}) ,$$

and an obvious change of indices and summation yields the first formula. This formula expresses that $(X_0, \dots, X_n)$ and $(X_{n+k})_{k\geq 0}$ are independent conditional to $X_n = x$. In particular,

$$\mathbb{P}(X_n = x_0, \dots, X_{n+k} = x_k, X_{n+k+1} = y)$$
$$= \mathbb{P}(X_n = x_0, \dots, X_{n+k} = x_k)P(x_k, y)$$

and hence $(X_{n+k})_{k\geq 0}$ is a Markov chain with matrix P and initial law π_n.   ∎

Thus, conditional to $X_n = x$, the future after time n of the chain is given by a "regeneration" of the chain starting at x and independent of the past. Note that past and future are taken in wide sense and include the present: both B and $\theta_n A$ may contain information on X_n "superseded" by the conditioning on $X_n = x$. All this is illustrated by Figure 2.1, for $T = n$.

These formulae are compact, have rich interpretations, and can readily be extended, for instance, if f and g are nonnegative or bounded then

$$\mathbb{E}(f(X_0, \dots, X_{n-1}, X_n)g(X_n, X_{n+1}, X_{n+2}, \dots) \mid X_n = x)$$
$$= \mathbb{E}(f(X_0, \dots, X_{n-1}, x) \mid X_n = x)\mathbb{E}_x(g(x, X_1, X_2, \dots)) . \qquad (2.1.1)$$

We now proceed to further extend them by replacing deterministic instants $n \in \mathbb{N}$ by an adequate class of random instants.

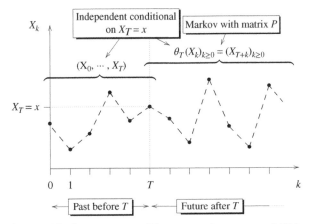

Figure 2.1   Strong Markov property. The successive states of $(X_k)_{k\geq 0}$ are represented by the filled circles and are linearly interpolated by dashed lines and T is a stopping time.

2.1.3   Stopping times and strong Markov property

A stopping time will be defined as a random instant that can be determined "in real time" from the observation of the chain, and hence at which a decision (such as stopping a game) can be taken without further knowledge of the future.

Definition 2.1.2 (Stopping time and its past) *Let T be a r.v. with values in $\mathbb{N} \cup \{\infty\}$. Then T is said to be a stopping time if*

$$\{T = n\} \in \mathcal{F}_n \ , \qquad \forall n \geq 0 \ .$$

Equivalently,

$$\{T \leq n\} \in \mathcal{F}_n \ , \ n \geq 0 \ , \ \text{or} \ \{T > n\} \in \mathcal{F}_n \ , \ n \geq 0 \ .$$

The σ-field of the past before a stopping time T is given by

$$\mathcal{F}_T = \{B \in \mathcal{F}_\infty : B \cap \{T = n\} \in \mathcal{F}_n \ , \ \forall n \geq 0\}$$
$$= \{B \in \mathcal{F}_\infty : B \cap \{T \leq n\} \in \mathcal{F}_n \ , \ \forall n \geq 0\} \ .$$

The equivalences in the definition follow readily from

$$\{T \leq n\} = \bigcup_{0 \leq k \leq n} \{T = k\} = \{T > n\}^c \ , \quad \{T = n\} = \{T \leq n\} - \{T \leq n-1\} \ .$$

We set $X_T = X_n$ and $\theta_T = \theta_n$ on $\{T = n\}$. On $\{T < \infty\}$, it holds that, for $A = \{(X_0, X_1, \ldots) \in E\}$,

$$\theta_T(X_n)_{n \geq 0} = (X_{T+n})_{n \geq 0} \ , \quad \theta_T A = \{(X_T, X_{T+1}, \ldots) \in E\} \ .$$

For $\omega \in \{T < \infty\}$, it holds that $X_T(\omega) = X_{T(\omega)}(\omega)$, and for any $B \in \mathcal{F}_T$, one can determine whether a certain ω is in B by examining only $X_0(\omega), \ldots, X_T(\omega)$.

Trivial example: deterministic times   If T is deterministic, that is, if

$$\exists n \in \mathbb{N} \cup \{\infty\} \ , \qquad \mathbb{P}(T = n) = 1 \ ,$$

then T is a stopping time and $\mathcal{F}_T = \mathcal{F}_n$.

We shall give nontrivial examples of stopping times after the next fundamental result, illustrated in Figure 2.1, which yields that Theorem 2.1.1 and its corollaries, such as (2.1.1), hold conditionally on $\{T < \infty\}$ by replacing n with a stopping time T.

Theorem 2.1.3 (Strong Markov property) *Let $(X_n)_{n\geq 0}$ be a Markov chain with matrix P. Then, for any stopping time T and $B \in \mathcal{F}_T$ and $x \in \mathcal{V}$ and $A \in \mathcal{F}_\infty$,*

$$\mathbb{P}(T < \infty, B, X_T = x, \theta_T A) = \mathbb{P}(T < \infty, B, X_T = x)\mathbb{P}_x(A) .$$

Several corresponding conditional formulations also hold.

Moreover, conditional on $T < \infty$, the shifted chain $\theta_T(X_k)_{k\geq 0} = (X_{T+k})_{k\geq 0}$ is again a Markov chain with matrix P and initial law $\mathcal{L}(X_T \mid T < \infty)$. Conditional on $T < \infty$ and $X_T = x$, this shifted chain is a Markov chain with matrix P started at x and independent of $\mathcal{F}_T$ and, in particular, of $(T, X_0, \ldots, X_T)$.

Proof: Then

$$\mathbb{P}(T < \infty, B, X_T = x, \theta_T A) = \sum_{n\geq 0} \mathbb{P}(T = n, B, X_n = x, \theta_n A) .$$

As $B \in \mathcal{F}_T$, by definition $\{T = n\} \cap B \in \mathcal{F}_n$, and Theorem 2.1.1 yields that

$$\mathbb{P}(T = n, B, X_n = x, \theta_n A) = \mathbb{P}(T = n, B, X_n = x)\mathbb{P}_x(A) .$$

We conclude by summing the series thus obtained.   ∎

A "last hitting time," a "time of maximum," and so on are generally not stopping times, as the future knowledge they imply usually prevents the formula for the strong Markov property to hold.

2.2   Hitting times and distribution

2.2.1   Hitting times, induced chain, and hitting distribution

2.2.1.1   First hitting time and first strict future hitting time

Let E be a subset of $\mathcal{V}$. The r.v., with values in $\mathbb{N} \cup \{\infty\}$, defined by

$$S_E = \inf\{n \geq 0 : X_n \in E\} , \qquad R_E = \inf\{n \geq 1 : X_n \in E\} ,$$

are stopping times, as $\{S_E > n\} = \{X_0 \notin E , \ldots, X_n \notin E\}$ and $\{R_E > n\} = \{X_1 \notin E , \ldots, X_n \notin E\}$. Clearly,

$$R_E = S_E \text{ if } X_0 \notin E , \qquad R_E = 1 + \theta_1 S_E \text{ if } X_0 \in E .$$

When $E = \{x\}$, the notations S_x and R_x are used. The notations S^E and R^E or S^x and R^x, or even S and R if the contest is clear, will also be used. See Figure 2.2 in which $S_0^E = S_E = R_E$.

Classically, S_E is called the (first) hitting time of E by the chain and R_E the (first) strict future hitting time of E by the chain.

2.2.1.2   Successive hitting times and induced chain

The successive hitting times $(S_n^E)_{n\geq0}$ of $E \subset \mathcal{V}$ by the chain $(X_n)_{n\geq0}$ are given by

$$
\begin{cases}
S_0^E = \inf\{k \geq 0 : X_k \in E\} := S_E\,, \\
S_{n+1}^E = \inf\{k > S_n^E : X_k \in E\} = S_n^E + \theta_{S_n^E}R_E\,.
\end{cases}
\tag{2.2.2}
$$

These are stopping times, as $\{S_n^E > m\} = \{\sum_{k=0}^{m} \mathbb{1}_{\{X_k \in E\}} \leq n\}$.

If $S_n^E < \infty$ then we set $X_n^E = X_{S_n^E}$, which is the state occupied at the $(n+1)$th hit of E by the chain. This is illustrated in Figure 2.2.

If $\mathbb{P}_x(R_E < \infty) = 1$ for every $x \in E$, then E is said to be recurrent. Then, the strong Markov property (Theorem 2.1.3) yields that if $\mathbb{P}(S_E < \infty)$, for instance, if $X_0 \in E$ a.s., then $\mathbb{P}(S_n^E < \infty) = 1$ for all $n \geq 0$, and moreover that $(X_n^E)_{n\geq0}$ is a Markov chain on E, which is called the induced chain of $(X_n)_{n\geq0}$ in E.

Cemetery state   In order to define $(X_n^E)_{n\geq0}$ in all generality, we can set $X_n^E = \dagger$ if $S_n^E = \infty$ for a cemetery state $\dagger$ adjoined to $\mathcal{V}$. The strong Markov property implies that $(X_n^E)_{n\geq0}$ thus defined is a Markov chain on the enlarged state space $E \cup \{\dagger\}$, also called the induced chain; one can add that it is killed after its last hit of E.

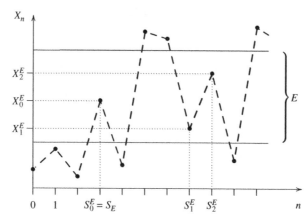

Figure 2.2   Successive hitting times of E and induced chain. The successive states of $(X_n)_{n\geq0}$ are represented by the filled circles and are linearly interpolated, and E corresponds to the points between the two horizontal lines. We see $X_n^E = X_{S_n^E}$ for $n = 0, 1,$ and 2.

2.2.1.3   Hitting distribution and induced chain

For $E \subset \mathcal{V}$, the matrix and function

$$U^E = (U^E(x, y))_{x,y \in \mathcal{V}}, \qquad\qquad U^E(x, y) = \mathbb{P}_x(S_E < \infty, X_{S_E} = y),$$

$$w^E = (w^E(x))_{x \in \mathcal{V}}, \qquad\qquad w^E(x) = \mathbb{P}_x(S_E < \infty) = \sum_{y \in \mathcal{V}} U^E(x, y), \qquad (2.2.3)$$

only depend on P and on E as the starting states are specified. When $E = \{z\}$, the notations U^z and w^z are used, and U and w when the context is clear.

Clearly, $U^E(x, y) = 0$ for $y \notin E$ and U^E is sub-Markovian (nonnegative terms, sum of each line bounded by 1) and is Markovian if and only if $w^E = 1$.

The restriction of U^E to E is Markovian if and only if E is recurrent and then it is the transition matrix of the induced chain $(X_n^E)_{n \geq 0}$. Else, a Markovian extension of $(U^E(x, y))_{x,y \in \mathcal{V}}$ is obtained using a cemetery state $\dagger$ adjoined to $\mathcal{V}$, and setting $U^E(x, \dagger) = 1 - w^E(x)$ for x in $\mathcal{V}$ and $U^E(\dagger, \dagger) = 1$; its restriction to $E \cup \{\dagger\}$ is the transition matrix of the induced chain on the extended state space. The matrix of the chain conditioned to return to E is given by $(U^E(x, y)/w^E(x))_{x,y \in E}$.

The matrix notation conventions in Section 1.2.2 are used for U^E. For x in $\mathcal{V}$, the line vector $U^E(x, \cdot)$ is a subprobability measure, with support E and total mass $w^E(x) \leq 1$, called the (possibly defective) hitting distribution of E starting from x. It associates to $A \subset \mathcal{V}$ the quantity

$$U^E(x, A) = \mathbb{P}_x(S_E < \infty, X_{S_E} \in A) = \mathbb{P}_x(S_{A \cap E} < S_{E-A}) = \sum_{y \in A} U^E(x, y).$$

For $f : E \to \mathbb{R}_+$ or $f : \mathcal{V} \to \mathbb{R}_+$, the function $U^E f$ is given for $x \in \mathcal{V}$ by

$$U^E f(x) = U^E(x, \cdot)f = \mathbb{E}_x \left(f(X_{S_E}) \mathbb{1}_{\{S_E < \infty\}} \right) \in [0, \infty], \qquad (2.2.4)$$

and this can be extended to signed bounded functions, by setting

$$U^E f = U^E f^+ - U^E f^- \in \mathbb{R}.$$

Clearly, $U^E f = U^E(f \mathbb{1}_E)$ and $U^E(x, A) = U^E(x, A \cap E) = U^E \mathbb{1}_A(x)$ and $w^E = U^E 1$. If $x \in E$ then $U^E f(x) = f(x)$ and in particular $U^E(x, A) = \mathbb{1}_{\{x \in A\}}$ and $w^E(x) = 1$.

2.2.2   "One step forward" method, Dirichlet problem

2.2.2.1   General principle

Let $(X_n)_{n \geq 0}$ be a Markov chain with matrix P. The Markov property (Theorem 2.1.1) yields that the shifted chain by one step

$$\theta_1(X_n)_{n \geq 0} = (X_{1+n})_{n \geq 0}$$

is also a Markov chain with matrix P and that conditional on $X_1 = x$ it is started at x and independent of X_0. This is illustrated in Figure 2.1, with $T = 1$.

The "one step forward" method   The "one step forward" method, or the method of conditioning on the first step, exploits this fact. It consists in rewriting quantities of interest for $(X_n)_{n\geq 0}$ in terms of the shifted chain $\theta_1(X_n)_{n\geq 0}$ and then using the above observation to establish equations for these quantities. These equations can then be solved or studied qualitatively and quantitatively.

A natural framework for this method is the study of hitting times and locations of $E \subset V$, as:

- the hitting time $S_E = \inf\{n \geq 0 : X_n \in E\}$ is a function of $(X_n)_{n\geq 0}$, and $\theta_1 S_E = \inf\{n \geq 0 : X_{1+n} \in E\}$ the same function of $\theta_1(X_n)_{n\geq 0} = (X_{1+n})_{n\geq 0}$;

- if $X_0 \in E$ then $S_E = 0$ and $X_{S_E} = X_0$ and if $X_0 \notin E$ then $S_E = 1 + \theta_1 S_E$ and $X_{S_E} = \theta_1 X_{S_E}$ on $\{S_E < \infty\} = \{\theta_1 S_E < \infty\}$.

2.2.2.2   Hitting distribution and Dirichlet problem

This method will provide an equation for U^E, which we are going to study.

Definition 2.2.1 (Dirichlet Problem)   *Let P be a transition matrix on V, and $E \subset V$ be non-empty. A function $u : V \to \mathbb{R}$ is a solution of the Dirichlet problem on $D = V - E$ with boundary data $f : E \to \mathbb{R}$ if u satisfies the linear equation with boundary condition*

$$u = f \text{ on } E , \qquad (I - P)u = 0 \text{ on } V - E . \qquad (2.2.5)$$

The equation $(I - P)u = 0$ should be interpreted in the form $u = Pu$ for extended solutions with values in $[0, \infty]$ with the usual conventions of addition and multiplication given in Section A.3.3.

All that matters about f is its restriction to the boundary of D for P given by

$$\partial_P D = \{y \in E : \exists x \in D, P(x, y) > 0\} ,$$

and the solution is the null extension of $u : D \cup \partial_P D \to \mathbb{R}$.

The r.v. S_E and subprobability measure $U^E(x, \cdot)$ are also called the exit time and distribution of $D = V - E$ by the chain, and one must take care about notation.

Theorem 2.2.2 (Solution of the Dirichlet problem)   *Let P be a transition matrix on V. Consider a nonempty $E \subset V$, a nonnegative or bounded $f : E \mapsto \mathbb{R}$, the corresponding Dirichlet problem (2.2.5), and U^E and w^E in (2.2.4).*

Then, the function $U^E f$ is a solution of the Dirichlet problem, if f is nonnegative in the extended sense with values in $[0, \infty]$, and if f is bounded in the usual sense and bounded by $\|f\|_\infty$. Further, it is the least nonnegative supersolution of the Dirichlet problem, that is,

$$u : V \to [0, \infty] , \quad u \geq f \text{ on } E , \quad u \geq Pu \text{ on } V - E \Rightarrow u \geq U^E f .$$

Lastly, if f is bounded and $\inf w^E > 0$, then $U^E f$ is the unique bounded solution of the Dirichlet problem.

In particular, w^E and $U^E(\cdot, A)$ for $A \subset \mathcal{V}$ are the least nonnegative solutions of the Dirichlet problem for $f = 1$ and $f = \mathbb{1}_A$, and if $\inf w^E > 0$ then they are the unique bounded solutions, and further $w^E = 1$.

Proof: The Dirichlet problem is obtained by applying the "one step forward" method to $U^E f$. Indeed, if $x \in E$ then $S_E = 0$ and hence $U f(x) = f(x)$, and if $x \notin E$ then the Markov property (Theorem 2.1.1) yields that

$$U f(x) = \sum_{y \in \mathcal{V}} \mathbb{E}_x \left(f\left(X_{S_E} \right) \mathbb{1}_{\{S_E < \infty, X_1 = y\}} \right)$$

$$= \sum_{y \in \mathcal{V}} \mathbb{E}_x \left(f\left(\theta_1 X_{S_E} \right) \mathbb{1}_{\{\theta_1 S_E < \infty, X_1 = y\}} \right)$$

$$= \sum_{y \in \mathcal{V}} P(x, y) \mathbb{E}_y \left(f(X_{S_E}) \mathbb{1}_{\{S_E < \infty\}} \right)$$

$$= PU f(x) \,,$$

in $[0, \infty]$ if $f \geq 0$ and in $\mathbb{R}$ with a result bounded by $\|f\|_\infty$ if f is bounded.

If u is a nonnegative supersolution, then $u \geq f = U f$ on E, and if $x \notin E$ then

$$u(x) \geq \sum_{y \in \mathcal{V}} P(x, y) u(y) \geq \sum_{y \in E} P(x, y) f(y) + \sum_{y \notin E} P(x, y) u(y)$$

and by iteration and using that

$$\sum_{x_1, \ldots, x_{k-1} \notin E, x_k \in E} P(x, x_1) \cdots P(x_{k-1}, x_k) f(x_k) = \mathbb{E}_x \left(f\left(X_{S_E} \right) \mathbb{1}_{\{S_E = k\}} \right)$$

it follows that, for $k \geq 1$,

$$u(x) \geq \sum_{n=0}^{k} \mathbb{E}_x(f(X_{S_E}) \mathbb{1}_{\{S_E = n\}}) + \sum_{x_1, \ldots, x_k \notin E} P(x, x_1) \cdots P(x_{k-1}, x_k) u(x_k)$$

and letting k tend to infinity yields that

$$u(x) \geq \sum_{n \in \mathbb{N}} \mathbb{E}_x \left(f\left(X_{S_E} \right) \mathbb{1}_{\{S_E = n\}} \right) = U f(x) \,.$$

For the uniqueness result, let us first assume that u is a bounded nonnegative solution for the Dirichlet problem. Then $u \geq U f$ by the minimality result. Moreover, for every $\varepsilon > 0$, the function

$$w_\varepsilon := w - \varepsilon(u - U f) \leq w$$

is a solution of the Dirichlet problem with boundary data the constant function 1. As $u - U f$ is bounded and $\inf w > 0$, there exists a sufficiently small $\varepsilon > 0$ such that $w_\varepsilon \geq 0$. Then another application of the minimality result yields that $w_\varepsilon \geq U1 = w$

and hence that $u = Uf$. In particular, the constant function 1, which is a solution for $f = 1$, must be equal to $w = U1$.

Now, let u be a signed bounded solution and c a constant such that $u + c \geq 0$. Then $u + c$ is a nonnegative bounded solution of the Dirichlet problem with bounded boundary data $f + c$, and hence

$$u + c = U(f + c) = Uf + cw = Uf + c$$

and thus $u = Uf$. ∎

Remark 2.2.3 *The minimality result is* fundamental *and constitutes a form of the* maximum principle. *Note that the constant function* 1 *is* always *a nonnegative bounded solution of the equation satisfied by* w^E. *The maximum principle is used in many uniqueness proofs, as in the one above.*

Dirichlet problem and recurrence A direct consequence of Theorem 2.2.2 is that an irreducible Markov chain on a finite state space hits every state y starting from any state x, a.s., as irreducibility implies that $w^y(x) > 0$ for all x and hence that $\inf w^y > 0$, and the theorem yields that $w^y \equiv 1$. The strong Markov property implies that if $w^y \equiv 1$ for all y, then the Markov chain visits every state infinitely often, a.s.

We give a more general result. A subset D of $\mathcal{V}$ communicates with another subset E of $\subset \mathcal{V}$ if for every $x \in D$ there exists $y \in E$ and $i = i(x, y) \geq 1$ such that $P^i(x, y) > 0$. This is always the case if P is irreducible and E is nonempty.

Proposition 2.2.4 *Let P be a transition matrix on $\mathcal{V}$ and $E \subset \mathcal{V}$ be nonempty. If $\mathcal{V} - E$ is finite and communicates with E, then there exists $s > 1$ such that $\mathbb{E}(s^{S_E}) < \infty$. In particular, $\mathbb{E}((S_E)^k) < \infty$ for all $k \geq 0$.*

Proof: For $x \in \mathcal{V} - E$, let $y(x) \in E$ and $i(x) = i(x, y(x)) \geq 1$ be such that $P^{i(x)}(x, y) > 0$. Let $I = \max_{x \notin E} i(x) < \infty$ and $p = \min_{x \notin E} P^{i(x)}(x, y) > 0$. The Markov property (Theorem 2.1.1) implies that

$$\mathbb{P}(S_E > (n + 1)I) = \sum_{z \notin E} \mathbb{P}(S_E > nI, X_{nI} = z, S_E > (n + 1)I)$$

$$= \sum_{z \notin E} \mathbb{P}(S_E > nI, X_{nI} = z)\mathbb{P}_z(S_E > I)$$

$$\leq \mathbb{P}(S_E > nI)(1 - p) ,$$

and by iteration $\mathbb{P}(S_E > nI) \leq (1 - p)^n$. Lemma A.1.3 yields that

$$\mathbb{E}(s^{S_E}) = 1 + (s - 1) \sum_{k \in \mathbb{N}} \mathbb{P}(X > k)s^k < \infty$$

for $0 \leq s < (1 - p)^{-1/I}$. ∎

2.2.2.3   Generating functions, joint distribution of hitting time, and location

Let $E \subset \mathcal{V}$ be nonempty. In order to study the (defective) joint distribution of S_E and X_{S_E}, consider for $x, y \in \mathcal{V}$, and $A \subset \mathcal{V}$ the generating functions given for $s \in [0, 1]$ by

$$G^E(x, y, s) = \mathbb{E}_x\left(s^{S_E} \mathbb{1}_{\{S_E < \infty, X_{S_E} = y\}}\right) = \sum_{n \in \mathbb{N}} \mathbb{P}_x\left(S_E = n, X_n = y\right) s^n ,$$

$$G^E(x, A, s) = \sum_{y \in A} G^E(x, y, s) = \sum_{n \in \mathbb{N}} \mathbb{P}_x\left(S_E = n, X_n \in A\right) s^n ,$$

$$h^E(x, s) = \mathbb{E}_x\left(s^{S_E} \mathbb{1}_{\{S_E < \infty\}}\right) = G^E(x, \mathcal{V}, s) , \qquad (2.2.6)$$

which depend only on the matrix P of the chain. Clearly,

$$G^E(x, A, s) = G^E(x, A \cap E, s) .$$

The notations G^z and h^z are used for $E = \{z\}$, and G and h when the context is clear. Then

$$U^E(x, A) = G^E(x, A, 1) , \qquad \mathbb{E}_x\left(s^{S_E} \mid S_E < \infty, X_{S_E} \in A\right) = \frac{G^E(x, A, s)}{U^E(x, A)} ,$$

which allows to study the condition law of S_E given $S_E < \infty$ and $X_{S_E} \in A$.

Theorem 2.2.5 *Let P be a transition matrix on $\mathcal{V}$ and $E \subset \mathcal{V}$ be nonempty. For $A \subset \mathcal{V}$ and $s \in [0, 1]$, the function $G^E(\cdot, A, s)$ is solution of the linear equation with boundary data*

$$g = \mathbb{1}_A \text{ on } E , \qquad (I - sP)g = 0 \text{ on } \mathcal{V} - E ,$$

and is its least nonnegative solution:

$$g \geq \mathbb{1}_{A \cap E} , \quad g \geq sPg \text{ on } \mathcal{V} - E \Rightarrow g \geq G^E(\cdot, A, s) .$$

The equation $(I - sP)g = 0$ can be written $g = sPg$.

Proof: The fact that this provides a solution follows from the "one step forward" method. Indeed, if $x \in E$ then $S_E = 0$ and hence $G(x, A, s) = \mathbb{1}_A$, and if $x \notin E$ then the Markov property (Theorem 2.1.1) yields that

$$G(x, A, s) = \sum_{y \in \mathcal{V}} \mathbb{E}_x\left(s^{S_E} \mathbb{1}_{\{S_E < \infty, X_{S_E} \in A, X_1 = y\}}\right)$$

$$= \sum_{y \in \mathcal{V}} \mathbb{E}_x\left(s^{1 + \theta_1 S_E} \mathbb{1}_{\{\theta_1 S_E < \infty, \theta_1 X_{S_E} \in A, X_1 = y\}}\right)$$

$$= \sum_{y \in \mathcal{V}} P(x, y) \mathbb{E}_y\left(s^{1 + S_E} \mathbb{1}_{\{S_E < \infty, X_{S_E} \in A\}}\right)$$

$$= sPG(\cdot, A, s)(x) .$$

If g in a nonnegative supersolution, then for $x \notin E$ it holds that

$$g(x) \geq s \sum_{y \in \mathcal{V}} P(x,y) g(y) \geq s \sum_{y \in E} P(x,y) \mathbb{1}_A(y) + s \sum_{y \notin E} P(x,y) g(y)$$

and by iteration and using that

$$\sum_{x_1,\dots,x_{k-1} \notin E, x_k \in E} P(x,x_1) \cdots P(x_{k-1},x_k) \mathbb{1}_A(x_k) = \mathbb{P}_x(S_E = k, X_k \in A)$$

it follows that, for $k \geq 1$,

$$g(x) \geq \sum_{n=0}^{k} \mathbb{P}_x(S_E = n, X_n \in A) s^n$$

$$+ s^k \sum_{x_1,\dots,x_k \notin E} P(x,x_1) \cdots P(x_{k-1},x_k) g(x_k)$$

and

$$g(x) \geq \sum_{n \in \mathbb{N}} \mathbb{P}_x(S_E = n, X_n \in A) s^n = G(x,A,s)$$

follows by letting k tend to infinity. ∎

2.2.2.4   Direct computations for expectations

For $x \in \mathcal{V}$ and $A \subset \mathcal{V}$, we have

$$M^E(x,A) := \mathbb{E}_x \left(S_E \mathbb{1}_{\{S_E < \infty, X_{S_E} \in A\}} \right) = \partial_s G^E(x,A,s) \big|_{s=1} \in [0,\infty] \,,$$

$$\mathbb{E}_x \left(S_E \mid S_E < \infty, X_{S_E} \in A \right) = \frac{M^E(x,A)}{U^E(x,A)} \in [0,\infty] \,.$$

Nevertheless, it may be best to work directly on $U^E(x,A)$ and $M^E(x,A)$, and possibly one may be able to solve the corresponding equations and not those for $G^E(x,A,s)$. Let

$$e^E(x) := \mathbb{E}_x(S_E) = \begin{cases} M^E(x,\mathcal{V})\,, & w^E(x) = 1 \,, \\ \infty \,, & w^E(x) < 1 \,. \end{cases}$$

The notations M^y and e^y are used for $E = \{y\}$, and M and e if the context is clear.

Theorem 2.2.6 *Let P be a transition matrix on $\mathcal{V}$ and $E \subset \mathcal{V}$ be nonempty. For $A \subset \mathcal{V}$, the function $M^E(\cdot,A)$ is solution of the affine equation with boundary condition*

$$m = 0 \text{ on } E \,, \qquad m = Pm + U^E(\cdot,A) \text{ on } \mathcal{V} - E \,,$$

and is its least supersolution with values $[0, \infty]$. When m is finite, the affine equation can be written $(I - P)m = U^E(\cdot, A)$. If $\inf w^E > 0$ and m is a bounded solution for this equation, then $m = M^E(\cdot, A)$.

The function e^E satisfies similar properties in which $U^E(\cdot, A)$ is replaced by 1.

Proof: The fact that this provides a solution follows from the "one step forward" method. If $x \in E$ then $S_E = 0$ and hence $M(x, A) = 0$, and if $x \notin E$ then the Markov property (Theorem 2.1.1) and a computation in $[0, \infty]$ yield that

$$M(x, A) = \sum_{y \in \mathcal{V}} \mathbb{E}_x \left(S_E \mathbb{1}_{\{S_E < \infty, X_{S_E} \in A, X_1 = y\}} \right)$$

$$= \sum_{y \in \mathcal{V}} \mathbb{E}_x \left((1 + \theta_1 S_E) \, \mathbb{1}_{\{\theta_1 S_E < \infty, \theta_1 X_{S_E} \in A, X_1 = y\}} \right)$$

$$= U(x, A) + \sum_{y \in \mathcal{V}} P(x, y) \mathbb{E}_y \left(S_E \mathbb{1}_{\{X_{S_E} \in A\}} \right).$$

Let m be a supersolution in $[0, \infty]$. If $x \in E$ then $m \geq 0 = M(x, A)$, and if $x \notin E$ then

$$m(x) \geq U(x, A) + \sum_{y \notin E} P(x, y) m(y)$$

and by iteration and using that

$$\sum_{x_1, \dots, x_k \notin E} P(x, x_1) \cdots P(x_{k-1}, x_k) U(x_k, A) = \mathbb{P}_x(S_E > k, S_E < \infty, X_{S_E} \in A)$$

it follows that, for $k \geq 1$,

$$m(x) \geq \sum_{n=0}^{k-1} \mathbb{P}_x(S_E > n, S_E < \infty, X_{S_E} \in A)$$

$$+ \sum_{x_1, \dots, x_k \notin E} P(x, x_1) \cdots P(x_{k-1}, x_k) m(x_k)$$

and hence, letting k go to infinity, that

$$m(x) \geq \sum_{n \in \mathbb{N}} \mathbb{P}_x(S_E > n, S_E < \infty, X_{S_E} \in A)$$

$$= \mathbb{E}_x(S_E \mathbb{1}_{\{S_E < \infty, X_{S_E} \in A\}}) = M(x, A).$$

Let m be a bounded solution, and $c \geq 0$ a constant such that $m + c \geq 0$. Then $m + c$ is a supersolution and hence $m + c \geq M(\cdot, A)$, and thus $m + c - M(\cdot, A)$ is a bounded solution of the Dirichlet problem (2.2.5) with boundary data given by the constant function 1. Thus, if $\inf w^E > 0$ then Theorem 2.2.2 yields that

$$m + c - M(\cdot, A) = U^E c = c w^E = c$$

and hence that $m = M(\cdot, A)$. Analogous computations yield the results for e^E.   ∎

Remark 2.2.7 *The function with value ∞ on V − E is a solution and must be elimi-nated if possible. Hence, it is often important to check that a solution is nonnegative, or bounded, or that* inf $w^E > 0$, *or that the expectation is finite (using for instance Proposition 2.2.4).*

The "right-hand side" $U^E(\cdot, A)$ or 1 of the affine equation is a solution of the associated linear equation, which corresponds to the Dirichlet problem (2.2.5). This fact can be exploited to find a particular solution of the affine equation.

2.3 Detailed examples

The results in this section will be now applied to some quantities of interest for the detailed examples described in Section 1.4. The main idea will be to solve the equations derived by the "one step forward" method, using the results obtained on the Dirichlet problem and related equations when appropriate.

The description of the Monte Carlo method for the approximate solution of the Dirichlet problem is left for Section 5.1.

2.3.1 Gambler's ruin

See Section 1.4.2. Here $V = \{0, 1, \dots, N\}$ and $E = \{0, N\}$, and the game duration is $S = S_{\{0,N\}} = \inf\{n \geq 0 : X_n \in \{0,N\}\}$. We set $U := U^{\{0,N\}}$. Gambler A is ruined if $S_0 < S_N$ and then Gambler B wins. There is a symmetry between the gamblers by interchanging x with $N - x$ and p with $q = 1 - p$.

2.3.1.1 Finite game and ruin probability

Starting from an initial fortune of $x \in V$, the probability of eventual ruin and even-tual win of Gambler A are given by

$$U(x, 0) = \mathbb{P}_x(S_0 < S_N), \qquad U(x, N) = \mathbb{P}_x(S_N < S_0),$$

and the probability that the game eventually terminates by

$$w(x) = \mathbb{P}_x(S < \infty) = U(x, 0) + U(x, N).$$

Proposition 2.2.4 implies that $w = 1$, but a direct proof will be provided.

Theorem 2.2.2 (or a direct implementation of the "one step forward" method) yields that $U(\cdot, 0)$ and $U(\cdot, N)$ and w are solutions of

$$-qu(x-1) + u(x) - pu(x+1) = 0, \qquad 1 \leq x \leq N - 1, \qquad (2.3.7)$$

with respective boundary conditions $U(0, 0) = 1$ and $U(N, 0) = 0$, $U(0, N) = 0$ and $U(N, N) = 1$, and $w(0) = w(N) = 1$.

This is a linear second-order recursion, with characteristic polynomial

$$pX^2 - X + q = p(X - 1)(X - q/p)$$

with roots 1 and q/p, which can possibly be equal.

Biased game This is the case $p \neq q$, in which the roots are distinct. The general solution is given by $u(x) = \alpha + \beta(q/p)^x$, hence $w = 1$ and

$$U(x, 0) = \frac{(q/p)^N - (q/p)^x}{(q/p)^N - 1} , \qquad U(x, N) = \frac{(q/p)^x - 1}{(q/p)^N - 1} .$$

Fair game This is the case $p = q = 1/2$, in which 1 is a double root. The general solution is given by $u(x) = \alpha + \beta x$, hence $w = 1$ and

$$U(x, 0) = \frac{N - x}{N} , \qquad U(x, N) = \frac{x}{N} .$$

These are the limits for p going to $1/2$ of the values for the biased game.

2.3.1.2 Mean game duration

Theorem 2.2.6 yields that the function $M(\cdot, y) = M^{\{0,N\}}(\cdot, y)$ for $y \in \{0, N\}$ satisfies

$$-qm(x - 1) + m(x) - pm(x + 1) = U(x, y) , \qquad 1 \leq x \leq N - 1 ,$$

with $m(0) = m(N) = 0$ and that $e(x) = \mathbb{E}_x(S) = M(x, 0) + M(x, N)$ satisfies the same equation in which $U(x, y)$ is replaced by 1.

The general solution of such an affine equation is given by the sum of a particular solution and of the general solution to the related linear equation (2.3.7), the latter being already known. The r.h.s. $U(x, y)$ is one of these general solutions, a fact to be exploited in order to find a particular solution.

Biased game A particular solution of the form $m(x) = x(\alpha + \beta(q/p)^x)$ satisfies

$$-qm(x - 1) + m(x) - pm(x + 1) = (q - p)\alpha + (p - q)\beta(q/p)^x ,$$

which yields for the r.h.s. $U(x, 0)$ that $\alpha = \frac{1}{q-p}\frac{(q/p)^N}{(q/p)^N-1}$ and $\beta = \frac{1}{q-p}\frac{1}{(q/p)^N-1}$, for $U(x, N)$ that $\alpha = \beta = \frac{1}{p-q}\frac{1}{(q/p)^N-1}$, and for 1 that $\alpha = \frac{1}{q-p}$ and $\beta = 0$, the sum of the previous quantities. The null boundary conditions allow to conclude that

$$M(x, 0) = \frac{x((q/p)^N + (q/p)^x) - 2N(q/p)^N U(x, N)}{(q - p)((q/p)^N - 1)} ,$$

$$M(x, N) = \frac{x(1 + (q/p)^x) - N(1 + (q/p)^N)U(x, N)}{(p - q)((q/p)^N - 1)} ,$$

$$e(x) = \frac{x - NU(x, N)}{q - p} = U(x, 0)\frac{x}{q - p} + U(x, N)\frac{N - x}{p - q} ,$$

as soon as we have eliminated the possibility of infinite solutions. This is a delicate point in Theorem 2.2.6, see the remark thereafter.

As $\mathcal{V}$ is finite, Proposition 2.2.4 allows to conclude. A more direct proof is that the solution for the r.h.s. 1 (the formula for e) is nondecreasing from the value 0 at state 0 to a certain state, then is nonincreasing down to the value 0 at state N, and hence is nonnegative, and the minimality result in Theorem 2.2.6 allows to eliminate the infinite solution.

It can be noted that $e(x)$ is the mean, weighted by the probabilities of ruin and of win, of two quantities of opposite signs, which essentially correspond to a motion with speed $|p - q|$. This is called a ballistic phenomenon.

Fair game   A particular solution of the form $m(x) = x^2(\alpha + \beta x)$ satisfies

$$-\frac{1}{2}m(x - 1) + m(x) - \frac{1}{2}m(x + 1) = -\alpha - 3\beta x$$

which yields for $U(x, 0)$ that $\alpha = -1$ and $\beta = \frac{1}{3N}$, for $U(x, N)$ that $\alpha = 0$ and $\beta = -\frac{1}{3N}$, and for 1 that $\alpha = -1$ and $\beta = 0$ (the sums). The null boundary conditions allow to conclude that

$$M(x, 0) = \frac{x(N - x)(2N - x)}{3N} , \qquad \frac{M(x, 0)}{U(x, 0)} = \frac{x(2N - x)}{3} ,$$

$$M(x, N) = \frac{x(N - x)(N + x)}{3N} , \qquad \frac{M(x, N)}{U(x, N)} = \frac{N^2 - x^2}{3} ,$$

$$e(x) = x(N - x) = U(x, 0)x^2 + U(x, N)(N - x)^2 ,$$

where the infinite solution can be eliminated by Proposition 2.2.4 or remarking that these solutions are nonnegative.

The mean time taken to reach a distance d is of order d^2, which corresponds to a diffusive phenomenon.

2.3.1.3   Generating function

Theorem 2.2.5 yields that for every $s \in [0, 1]$ the functions $G(\cdot, 0, s)$ and $G(\cdot, N, s)$ and $h(\cdot, s) = G(\cdot, 0, s) + G(\cdot, N, s)$ satisfy

$$-sqg(x - 1) + g(x) - spg(x + 1) = 0 , \qquad 1 \le x \le N - 1 , \qquad (2.3.8)$$

with boundary conditions $G(0, 0, s) = 1$ and $G(N, 0, s) = 0$, $G(0, N, s) = 0$ and $G(N, N, s) = 1$, and $h(0, s) = h(N, s) = 1$, respectively. The characteristic polynomial for this second-order linear recursion is

$$psX^2 - X + qs .$$

We avoid treating separately the cases $p \neq q$ and $p = q = 1/2$ by considering first the case $s < 1$, for which $1 - 4pqs^2 > 0$. Then, there are two distinct roots

$$\lambda_+(s) = \frac{1 + \sqrt{1 - 4pqs^2}}{2ps} , \qquad \lambda_-(s) = \frac{1 - \sqrt{1 - 4pqs^2}}{2ps} , \qquad (2.3.9)$$

the general solution is of the form $x \mapsto \alpha_+(s)\lambda_+(s)^x + \alpha_-(s)\lambda_-(s)^x$, and hence

$$G(x, 0, s) = \frac{\lambda_+(s)^N \lambda_-(s)^x - \lambda_-(s)^N \lambda_+(s)^x}{\lambda_+(s)^N - \lambda_-(s)^N} ,$$

$$G(x, N, s) = \frac{\lambda_+(s)^x - \lambda_-(s)^x}{\lambda_+(s)^N - \lambda_-(s)^N} ,$$

$$h(x, s) = \frac{(1 - \lambda_-(s)^N)\lambda_+(s)^x - (1 - \lambda_+(s)^N)\lambda_-(s)^x}{\lambda_+(s)^N - \lambda_-(s)^N} .$$

Using $p + q = 1$ and

$$\sqrt{1 - 4pq} = \sqrt{1 - 4p + 4p^2} = \sqrt{(1 - 2p)^2} = |1 - 2p| = |q - p| ,$$

the expressions for $U(x, 0) = G(x, 0, 1)$ and $U(x, N) = G(x, N, 1)$ and $w(x) = h(x, 1) = 1$ can be easily recovered, which is left as an exercise.

Explicit joint law A Taylor expansion yields the joint law of S and of the result of the game (ruin $X_S = 0$, win if $X_S = N$), and hence the law of S. For $j \in \mathbb{N}$,

$$\lambda_+(s)^j - \lambda_-(s)^j = \frac{1}{(2ps)^j} \sum_{i=0}^{j} \binom{j}{i} \left(\left(\sqrt{1 - 4pqs^2}\right)^i - \left(-\sqrt{1 - 4pqs^2}\right)^i \right)$$

$$= \frac{2\sqrt{1 - 4pqs^2}}{(2ps)^j} \sum_{k=0}^{\lfloor j/2 \rfloor - 1} \binom{j}{2k + 1} \left(1 - 4pqs^2\right)^k$$

which yields expressions for $G(x, 0, s)$ and $G(x, N, s)$ as rational functions of explicit polynomials, from which (Feller, W. (1968), XIV.5) derives the formula

$$\mathbb{P}_x(S = n, X_S = 0) = \frac{2^{n+1}}{N} p^{\frac{n-x}{2}} q^{\frac{n+x}{2}} \sum_{1 \leq k < N/2} \cos^{n-1} \frac{k\pi}{N} \sin \frac{k\pi}{N} \sin \frac{kx\pi}{N} .$$

Moments Expectations, variances, and higher moments of $S\mathbb{1}_{\{X_S=0\}}$ and $S\mathbb{1}_{\{X_S=N\}}$ and S can be derived by appropriate Taylor expansions at the point 1 for $G(x, 0, s)$ and $G(x, N, s)$ and $h(x, s)$. Computations are tedious by hand but can be performed quickly by appropriate software.

2.3.2   Unilateral hitting time for a random walk

Consider a compulsive Gambler A, which gambles as long as he is not ruined, starting from an initial fortune x. This corresponds to taking $N = \infty$ and $E = \{0\}$ above. Now, the state space for $(X_n)_{n \geq 0}$ is $\mathbb{N}$, and

$$P(0, 0) = 1 , \qquad P(x, x + 1) = p , \quad P(x, x - 1) = q = 1 - p , \quad x \geq 1 .$$

The game duration is S_0, and we set $U(x) = U^0(x, 0) = \mathbb{P}_x(S_0 < \infty)$ and $M(x) = M^0(x, 0) = \mathbb{E}_x(S_0 \mathbb{1}_{\{S_0 < \infty\}})$. Proposition 2.2.4 does no longer apply.

2.3.2.1   Probability and mean duration for ruin

Taking limits   By monotone convergence (Theorem A.3.2),

$$U(x) = \lim_{N \to \infty} U^{\{0,N\}}(x, 0) , \qquad M(x) = \lim_{N \to \infty} M^{\{0,N\}}(x, 0) .$$

- If $p < q$ (game biased against the gambler), then $U(x) = 1$ and $M(x) = \frac{x}{q-p}$, and Gambler A is ruined in finite time at "speed" $q - p = 1 - 2p > 0$.

- If $p = q = 1/2$ (fair game), then $U(x) = 1$ and $M(x) = \infty$, and Gambler A is eventually ruined, but the mean duration for that is infinite.

- If $p > q$ (game biased toward the gambler), then $U(x) = (q/p)^x < 1$ and $M(x) = \frac{x(q/p)^x}{p-q}$ and hence $\frac{M(x)}{U(x)} = \frac{x}{p-q}$, and Gambler A is eventually ruined with probability $(q/p)^x < 1$, and if he does so it happens at "speed" $p - q > 0$. Obviously $\mathbb{E}_x(S_0) = \infty$.

A global expression for this is

$$\mathbb{P}_x(S_0 < \infty) = \min\{1, (q/p)^x\} , \qquad \mathbb{E}(S_0 \mid S_0 < \infty) = \frac{x}{|q - p|} .$$

Direct computation   We seek the least nonnegative solutions for the equations satisfying $U(0) = 1$ and $M(0) = 0$, by considering notably the behavior at infinity.

- If $p < q$ then $U(x) = \alpha + \beta(q/p)^x$ and $M(x) = \frac{x}{q-p} + \alpha + \beta(q/p)^x$ for $x \geq 1$, and hence $U(x) = 1$ and $M(x) = \frac{x}{q-p}$.

- If $p = q = 1/2$ then $U(x) = \alpha + \beta x$ and $M(x) = -x^2 + \alpha + \beta x$ for $x \geq 1$, and hence $U(x) = 1$ and $M(x) = \infty$ for $x \geq 1$.

- If $p > q$ then $U(x) = \alpha + \beta(q/p)^x$ and $M(x) = \frac{x(q/p)^x}{q-p} + \alpha + \beta(q/p)^x$ for $x \geq 1$, and hence $U(x) = (q/p)^x$ and $M(x) = \frac{x(q/p)^x}{q-p}$.

Note that if $p > q$ then the trivial infinite solution of the equations in Theorem 2.2.6 must be accepted, as no other solution is nonnegative. It would have been likewise if we had tried to compute thus $e^0(x) = \mathbb{E}_x(S_0)$ for $p > q$.

Fair game and debt If $p = q = 1/2$ then $X_n = X_{n-1} + \xi_n \mathbb{1}_{\{X_{n-1}>0\}}$ for a sequence $(\xi_k)_{k\geq 1}$ of i.i.d. r.v. such that $\mathbb{P}(\xi_1 = \pm 1) = 1/2$. Assuming $x \neq 0$,

$$\mathbb{E}(\xi_n \mathbb{1}_{\{X_{n-1}>0\}}) = \mathbb{E}(\xi_n \mathbb{1}_{\{\xi_1 > -x, \ldots, \xi_1 + \cdots + \xi_{n-1} > -x\}})$$

$$= \mathbb{E}(\xi_n)\mathbb{P}(\xi_1 > -x, \ldots, \xi_1 + \cdots + \xi_{n-1} > -x) = 0$$

and hence $\mathbb{E}(X_n) = \mathbb{E}(X_{n-1}) = \cdots = \mathbb{E}(X_0) = x \neq 0$, despite the fact that $\mathbb{P}(S_0 < \infty) = 1$ yields that $\lim_{n\to\infty} X_n = X_{S_0} = 0$, a.s. By dominated convergence (Theorem A.3.5), this implies that $\mathbb{E}(\sup_{n\geq 0} X_n) = \infty$.

Point of view of Gambler B The situation seems more reasonable (but less realistic) from the perspective of Gambler B, who ardently desires to gain a sum x and has infinite credit and a compliant adversary. Depending on his probability q of winning at each toss, his eventual win probability and expected time to win are

$$\min\{1, (q/p)^x\}, \qquad \mathbb{E}(S_0 \mid S_0 < \infty) = \frac{x}{|q - p|}.$$

- If $q > p$ then he wins, a.s., after a mean duration of $\frac{x}{q-p}$.

- If $q = p$ then he wins, a.s., but the expected time it takes is infinite, and the expectation of his maximal debt toward Gambler A is infinite.

- If $q < p$ (as in a casino) then his probability of winning is $(q/p)^x < 1$, and if he attains his goal then the expected time for this is $\frac{x}{p-q}$ (else he losses an unbounded quantity of money).

2.3.2.2 Using the generating function

Let $G(x, s) = G^0(x, 0, s)$. Theorem 2.2.5 yields that for $s \in [0, 1]$ the equation satisfied by $G(\cdot, s)$ is the extension of (2.3.8) for $x \geq 1$ with boundary condition $G(0, s) = 1$.

If $s < 1$ then $1 - 4pqs^2 > 0$ and the general solution is given after (2.3.9). The minimality result in Theorem 2.2.5 and the fact that

$$0 \leq 1 - \sqrt{1 - 4pqs^2} \leq 1 + \sqrt{1 - 4pqs^2}$$

yield that

$$G(x, s) = G(1, s)^x = \left(\frac{1 - \sqrt{1 - 4pqs^2}}{2ps}\right)^x.$$

Expectation and variance As $\ln G(x, 1 + \varepsilon) = x \ln G(1, 1 + \varepsilon)$ and $1 - 4pq = (q - p)^2$, it holds that

$$\ln G(1, 1 + \varepsilon) = \ln\left(1 - |q - p|\sqrt{1 - \frac{8pq\varepsilon}{|q - p|^2} - \frac{4pq\varepsilon^2}{|q - p|^2}}\right) - \ln(1 + \varepsilon) - \ln 2p.$$

Classic Taylor expansions yield that, at a precision of order $o(\varepsilon^2)$,

$$\sqrt{1 - \frac{8pq\varepsilon}{|q-p|^2} - \frac{4pq\varepsilon^2}{|q-p|^2}} = 1 - \frac{4pq\varepsilon}{|q-p|^2} - \frac{2pq\varepsilon^2}{|q-p|^2} - \frac{8p^2q^2\varepsilon^2}{|q-p|^4},$$

and, using $1 - |q-p| = p + q - |q-p| = 2\min(p,q) = 2(p \wedge q)$,

$$\ln g\,(1, 1 + \varepsilon) = \ln\left(1 + \frac{2pq\varepsilon}{(p \wedge q)\,|q-p|} + \frac{pq\varepsilon^2}{(p \wedge q)\,|q-p|} + \frac{4p^2q^2\varepsilon^2}{(p \wedge q)\,|q-p|^3}\right)$$

$$+ \ln 2\,(p \wedge q) - \ln\,(1 + \varepsilon) - \ln 2p$$

$$= \ln \frac{p \wedge q}{p} + \left(\frac{2pq}{(p \wedge q)\,|q-p|} - 1\right)\varepsilon$$

$$+ \left(\frac{2pq}{(p \wedge q)\,|q-p|} + \frac{8p^2q^2}{(p \wedge q)\,|q-p|^3} - \frac{4p^2q^2}{(p \wedge q)^2\,|q-p|^2} + 1\right)\frac{\varepsilon^2}{2}.$$

Using $p + q = 1$ and by identification, see (A.1.1), this yields that

- if $p \neq q$ then $\mathbb{P}_x(S_0 < \infty) = \min(1, (q/p)^x)$ and $\mathbb{E}_x(S_0 \mid S_0 < \infty) = \frac{x}{|q-p|}$ and moreover $\mathrm{Var}_x(S_0 \mid S_0 < \infty) = \frac{4pqx}{|q-p|}$,
- if $p = q = 1/2$ then $\mathbb{P}_x(S_0 < \infty) = 1$ and $\mathbb{E}_x(S_0) = \infty$.

Law of game duration The classic Taylor expansion

$$\sqrt{1 - 4pqs^2} = 1 + \sum_{k \geq 1} \binom{1/2}{k} (-4pqs^2)^k = 1 - 2\sum_{k \geq 1} \frac{(2k-2)!}{(k-1)!\,k!}(pqs^2)^k$$

using the generalized binomial coefficient defined by

$$\binom{1/2}{k} = \frac{\frac{1}{2}\left(\frac{1}{2} - 1\right)\left(\frac{1}{2} - 2\right) \cdots \left(\frac{1}{2} - k + 1\right)}{k!} = (-1)^{k-1}\frac{1 \cdot 1 \cdot 3 \cdots (2k-3)}{2^k k!}$$

yields that

$$G(1, s) = \sum_{k \geq 1} \frac{(2k-2)!}{(k-1)!\,k!}p^{k-1}q^k s^{2k-1} = \sum_{k \geq 1} \frac{1}{k}\binom{2k-2}{k-1}p^{k-1}q^k s^{2k-1}.$$

By identification, for $n \geq 0$, it holds that $\mathbb{P}_1(S_0 = 2n) = 0$ and

$$\mathbb{P}_1(S_0 = 2n + 1) = \frac{1}{n+1}\binom{2n}{n}p^n q^{n+1} = \frac{1}{2n+1}\binom{2n+1}{n}p^n q^{n+1}.$$

2.3.2.3   Use of the strong Markov property

The fact that

$$G(x, s) = G(1, s)^x$$

yields that the law of S_0 given that $X_0 = x$ is the law of the sum of x independent r.v. with same law as S_0 given that $X_0 = 1$ and implies that

$$U(x) = U(1)^x , \qquad M(x) = xM(1) .$$

It is actually a consequence of the strong Markov property: in order for Gambler A to loose x units, he must first loose 1 unit and then he starts independently of the past from a fortune of $x - 1$ units; a simple recursion and the spatial homogeneity of a random walk conclude this. A precise formulation is left as an exercise.

2.3.3   Exit time from a box

Extensions of such results to multidimensional random walks are delicate, as the linear equations are much more involved and seldom have explicit solutions. Some guesswork easily allows to find a solution in the following example.

Let $(X_n)_{n \geq 0}$ be the symmetric nearest-neighbor random walk on $\mathbb{Z}^d$ and N_i be in $\mathbb{N}$ for $1 \leq i \leq d$. Consider $E = \mathbb{Z}^d - D$ and $S = S_E = \inf\{n \geq 0 : X_n \notin D\}$, with

$$D = \{(x_1, \dots, x_d) \in \mathbb{Z}^d : 0 < x_i < N_i, 1 \leq i \leq d\} .$$

Theorem 2.2.6 yields that the function $x \in \mathbb{Z}^d \mapsto e(x) = \mathbb{E}_x(S)$ satisfies the affine equation with boundary condition

$$m(x) = 0 \ \text{if} \ x \notin D , \quad m(x) - \frac{1}{2d} \sum_{i=1}^d (m(x + e_i) + m(x - e_i)) = 1 \ \text{if} \ x \in D ,$$

and, considering gambler's ruin, we obtain that

$$\mathbb{E}_x(S) = \sum_{i=1}^d x_i(N_i - x_i) , \qquad x \in D . \tag{2.3.10}$$

Notably, if $x_i = r \in \mathbb{N}$ and $N_i = 2r$ for $1 \leq i \leq d$ then $\mathbb{E}_x(S) = dr^2$ is quadratic in r and linear in the dimension d.

2.3.4   Branching process

See Section 1.4.3. We assume that

$$m = \sum_{x \in \mathbb{N}} xp(x) < \infty ,$$

that is, that the mean number of offspring of an individual is finite, as well as $p(0) > 0$ and $p(0) + p(1) < 1$ to avoid trivial situations. A quantity of interest is the extinction time

$$T = S_0 = \inf\{n \geq 0 : X_n = 0\} .$$

An essential fact is that if $X_0 = x$ then $(X_n)_{n \geq 0}$ can be obtained by the sum of the x chains given by the descendants of each of the initial individuals and that these chains are independent and have same law as $(X_n)_{n \geq 0}$ given that $X_0 = 1$. In particular, we study the case when $X_0 = 1$, the others being deduced easily. This property can be generalized to appropriate subpopulations and is called the branching property.

Notably, if $g_n(s) = \mathbb{E}_1(s^{X_n})$ then $\mathbb{E}_x(s^{X_n}) = g_n(s)^x$ for $n \geq 0$. This, and the "one step forward" method, yields that

$$g_n(s) = \sum_{x \in \mathbb{N}} p(x) \mathbb{E}_x(s^{X_{n-1}}) = \sum_{x \in \mathbb{N}} p(x) g_{n-1}(s)^x = g(g_{n-1}(s)), \qquad (2.3.11)$$

which is the recursion in Section 1.4.3, obtained there by a different method.

2.3.4.1   Probability and rate of extinction

By monotone limit (Lemma A.3.1),

$$\{T < \infty\} = \bigcup_{n \to \infty} \uparrow \{T \leq n\} , \qquad \mathbb{P}(T < \infty) = \lim_{n \to \infty} \uparrow \mathbb{P}(T \leq n) .$$

Moreover, $\mathbb{P}(T \leq n) = \mathbb{P}(X_n = 0) = g_n(0)$, and the recursion (2.3.11) yields that

$$\mathbb{P}(T \leq n) = g(\mathbb{P}(T \leq n - 1)) , \quad n \geq 1 , \qquad \mathbb{P}(T \leq 0) = 0 .$$

This recursion can also be obtained directly by the "one step forward" method: the branching property yields that $\mathbb{P}_x(T \leq n - 1) = \mathbb{P}_1(T \leq n - 1)^x$, and thus

$$\mathbb{P}(T \leq n) = \sum_{x \in \mathbb{N}} p(x) \mathbb{P}_x(T \leq n - 1) = \sum_{x \in \mathbb{N}} p(x) \mathbb{P}(T \leq n - 1)^x = g(\mathbb{P}(T \leq n - 1)) .$$

Thus, $\mathbb{P}(T \leq n)$ solves the recursion $u_n = g(u_{n-1})$ started at $u_0 = 0$, and this nondecreasing sequence converges to $\mathbb{P}(T < \infty)$. As g is continuous on $[0, 1]$, the limit is the least fixed point η of g on $[0, 1]$. Thus, the extinction probability is

$$\mathbb{P}(T < \infty) = \eta = \inf\{s \in [0, 1] : s = g(s)\} .$$

Graphical study   The facts that g and g' are continuous increasing on $[0, 1]$, and $g(0) = p(0) > 0$ and $g(1) = 1$ and $g'(1) = m$, allow to place the graph of g with respect to the diagonal.

- If $m \leq 1$ then the only fixed point for g is $1 = g(1)$, hence the extinction probability is $\mathbb{P}(T < \infty) = \eta = 1$, and the population goes extinct, a.s.

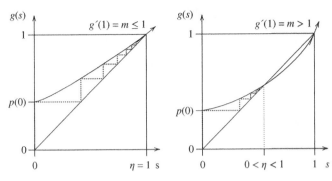

Figure 2.3   Graphical study of $u_n = g(u_{n-1})$ with $u_0 = 0$. (Left) $m \leq 0$, and the sequence converges to 1. (Right) $m > 0$, and the sequence converges to $\eta \in 0, 1$, the unique fixed point of g other than 1.

- If $m > 1$ then there exists a unique fixed point η other than 1, and $0 < \eta < 1$ as $g(0) = p(0) > 0$, hence the extinction probability is $\mathbb{P}(T < \infty) = \eta$.

See Figure 2.3.

Moreover, the strict convexity implies that

$$g'(u_{n-1})(\eta - u_{n-1}) < \eta - u_n = g(\eta) - g(u_{n-1}) < g'(\eta)(\eta - u_{n-1}),$$

which yields some convergence rate results.

- If $m < 1$ then $\mathbb{P}(T > n) < m^n$ for $n \geq 1$ and $\mathbb{E}(T) < \frac{1}{1-m}$.
  Indeed, then $\eta - \mathbb{P}(T \leq n) < m(\eta - \mathbb{P}(T \leq n-1))$ can be written as $\mathbb{P}(T > n) < m\mathbb{P}(T > n-1)$ for $n \geq 1$, and by iteration $\mathbb{P}(T > n) < m^n$ and hence $\mathbb{E}(T) = \sum_{n \geq 0} \mathbb{P}(T > n) < \sum_{n \geq 0} m^n = \frac{1}{1-m}$.

- If $m = 1$ then $\rho^n = o(\mathbb{P}(T > n))$ for every $\rho < 1$.
  Indeed, $\eta = 1$ and thus $\mathbb{P}(T > n) > g'(\mathbb{P}(T \leq n-1))\mathbb{P}(T > n-1)$, which implies the result as $\lim_{n \to \infty} g'(\mathbb{P}(T \leq n-1)) = g'(1) = 1$.

- If $m > 1$ then $0 < \eta - \mathbb{P}(T \leq n) < g'(\eta)^n \eta$ for $n \geq 1$, and $g'(\eta) < 1$.
  Indeed, $\eta - \mathbb{P}(T \leq n) < g'(\eta)(\eta - \mathbb{P}(T \leq n-1)) < \cdots < g'(\eta)^n \eta$ for $n \geq 1$, and it is a simple matter to prove that $g'(\eta) < 1$ (Figure 2.3).

Critical case   The case $m = 1$ is called the critical case and is the most delicate to study. Assume that the number of offspring of a single individual has a variance or equivalently that $g''(1) < \infty$ or that $\sum_{x \geq 0} x^2 p(x) < \infty$. Then $\mathbb{E}(T) = \infty$, and the population goes extinct, a.s., but has an infinite mean life time.

Indeed, $\mathbb{P}(T > n) > g'(\mathbb{P}(T \leq n-1))\mathbb{P}(T > n-1)$ implies for $n \geq 1$ that

$$\mathbb{P}(T > n) > \prod_{k=0}^{n-1} g'(\mathbb{P}(T \leq k)) = \exp\left(\sum_{k=0}^{n-1} \ln g'(1 - \mathbb{P}(T > k))\right)$$

and as $\lim_{n\to\infty} \mathbb{P}(T > n) = 0$ then $\sum_{k=0}^{\infty} \ln g'(1 - \mathbb{P}(T > k)) = -\infty$ with

$$\ln g'(1 - \mathbb{P}(T > k)) = \ln(1 + g'(1 - \mathbb{P}(T > k)) - g'(1))$$
$$= -\mathbb{P}(T > k)g''(1) + o(\mathbb{P}(T > k))$$

and hence $\mathbb{E}(T) = \sum_{n\geq 0} \mathbb{P}(T > n) = \infty$.

2.3.4.2   Mean population size

As $g_n(1 + \varepsilon) = g(g_{n-1}(1 + \varepsilon))$, an order 1 Taylor expansion yields that

$$1 + \mathbb{E}(X_n)\varepsilon + o(\varepsilon) = g(1 + \mathbb{E}(X_{n-1})\varepsilon + o(\varepsilon)) = 1 + m\mathbb{E}(X_{n-1})\varepsilon + o(\varepsilon)$$

and by identification $\mathbb{E}(X_n) = m\mathbb{E}(X_{n-1})$ and hence

$$\mathbb{E}(X_n) = m^n , \qquad n \geq 0 .$$

This can also be directly obtained by the "one step forward" method:

$$\mathbb{E}(X_n) = \sum_{x\in\mathbb{N}} p(x)\mathbb{E}_x(X_{n-1}) = \sum_{x\in\mathbb{N}} p(x)x\mathbb{E}(X_{n-1}) = m\mathbb{E}(X_{n-1}) .$$

A few results follow from this.

- If $m < 1$ then the population mean size decreases exponentially.

- If $m = 1$ (critical case) then $\mathbb{E}(X_n) = 1$ for all n, and $X_n = 0$ for a (random) large enough n, a.s., and by dominated convergence (Theorem A.3.5) $\mathbb{E}(\sup_{n\geq 0} X_n) = \infty$. Hence, the population goes extinct, a.s., but its mean size remains constant, and the expectation of its maximal size is infinite.

- If $m > 1$ then the mean size increases exponentially.

2.3.4.3   Variances and covariances

We assume $\sum_{x\geq 0} x^2 p(x) < \infty$. Let

$$\sigma^2 = \sum_{x\geq 0} x^2 p(x) - m^2 = g''(1) + g'(1) - (g'(1))^2$$

denote the variance of the number of offspring of an individual, and $\sigma_n^2 = \mathrm{Var}(X_n)$ the variance of X_n. As $g_n(1 + \varepsilon) = g(g_{n-1}(1 + \varepsilon))$, using (A.1.1), at a precision of order $o(\varepsilon^2)$,

$$1 + m^n\varepsilon + \left(\sigma_n^2 + m^{2n} - m^n\right)\frac{\varepsilon^2}{2}$$

$$= g\left(1 + m^{n-1}\varepsilon + \left(\sigma_{n-1}^2 + m^{2n-2} - m^{n-1}\right)\frac{\varepsilon^2}{2}\right)$$

$$= 1 + m^n\varepsilon + m\left(\sigma_{n-1}^2 + m^{2n-2} - m^{n-1}\right)\frac{\varepsilon^2}{2} + m^{2n-2}\left(\sigma^2 + m^2 - m\right)\frac{\varepsilon^2}{2}$$

and by identification

$$\sigma_n^2 + m^{2n} - m^n = m(\sigma_{n-1}^2 + m^{2n-2} - m^{n-1}) + m^{2n-2}(\sigma^2 + m^2 - m) \, .$$

Thus, $\sigma_0 = 0$ and

$$\sigma_n^2 = m\sigma_{n-1}^2 + \sigma^2 m^{2n-2} \, , \qquad n \geq 1 \, .$$

Setting $\sigma_n^2 = a_n m^n$, we obtain that $a_0 = 0$ and

$$a_n = a_{n-1} + \sigma^2 m^{n-2} = \cdots = \sigma^2(m^{-1} + \cdots + m^{n-2}) \, ,$$

and hence

$$\operatorname{Var}(X_n) = n\sigma^2 \ \text{ if } \ m = 1 \, , \qquad \operatorname{Var}(X_n) = m^{n-1} \frac{m^n - 1}{m - 1} \sigma^2 \ \text{ if } \ m \neq 1 \, .$$

For $j \geq i \geq 0$, simple arguments yield that

$$\mathbb{E}(X_i X_j) = \sum_{x \in \mathbb{N}} x \mathbb{E}(X_j \mid X_i = x) \mathbb{P}(X_i = x) = \sum_{x \in \mathbb{N}} x \mathbb{E}_x(X_{j-i}) \mathbb{P}(X_i = x)$$

$$= \sum_{x \in \mathbb{N}} x^2 m^{j-i} \mathbb{P}(X_i = x) = m^{j-i} \mathbb{E}(X_i^2) = m^{j-i} \operatorname{Var}(X_i) + m^{i+j}$$

and the covariance and correlation of X_i and X_j are given by

$$\operatorname{Cov}(X_i, X_j) = m^{j-i} \operatorname{Var}(X_i) \, , \qquad \rho(X_i, X_j) = m^{j-i} \sqrt{\frac{\operatorname{Var}(X_i)}{\operatorname{Var}(X_j)}} \, ,$$

and more precisely

$$\rho(X_i, X_j) = \sqrt{\frac{i}{j}} \ \text{ if } \ m = 1 \, , \qquad \rho(X_i, X_j) = \sqrt{\frac{m^{-i} - 1}{m^{-j} - 1}} \ \text{ if } \ m \neq 1 \, .$$

2.3.5   Word search

See Section 1.4.6. The quantity of interest is

$$T = S_{\text{GAG}} = \inf\{n \geq 0 : X_n = \text{GAG}\} \, .$$

The state space is finite and the chain irreducible, hence $\mathbb{P}(T, \infty) < \infty$ by Proposition 2.2.4. The expected value $\mathbb{E}(T)$ and of the generating function $h(s) = \mathbb{E}(s^T)$ can easily be derived by solving the equations yielded by the "one step forward" method, but we leave that as an exercise.

We prefer to describe a more direct method, which is specific to this situation. It explores some possibilities of evolution in the near future, with horizon the word length. The word GAG is constituted of three letters. As $\{T > n\}$ is in $\mathcal{F}_n$ and the sequence $(\xi_k)_{k \geq 1}$ is i.i.d., for $n \geq 0$, it holds that

$$\mathbb{P}(T > n, \xi_{n+1} = \text{G}, \xi_{n+2} = \text{A}, \xi_{n+3} = \text{G}) = \mathbb{P}(T > n) p_{\text{A}} p_{\text{G}}^2$$

and, considering the overlaps within the word GAG and $\{T = n+1\} \subset \mathcal{F}_{n+1}$,

$$\mathbb{P}(T > n, \xi_{n+1} = G, \xi_{n+2} = A, \xi_{n+3} = G)$$
$$= \mathbb{P}(T = n+1, \xi_{n+2} = A, \xi_{n+3} = G) + \mathbb{P}(T = n+3)$$
$$= \mathbb{P}(T = n+1)p_A p_G + \mathbb{P}(T = n+3)$$

and hence

$$\mathbb{P}(T > n) = \frac{\mathbb{P}(T = n+1)}{p_G} + \frac{\mathbb{P}(T = n+3)}{p_A p_G^2}. \tag{2.3.12}$$

Expectation As $\sum_{k\geq 0}\mathbb{P}(T = k) = 1$ and $\mathbb{P}(T < 3) = 0$, it follows that

$$\mathbb{E}(T) = \sum_{n\geq 0}\mathbb{P}(T > n) = \frac{1}{p_G} + \frac{1}{p_A p_G^2}.$$

Generating function Moreover, Lemma A.1.3 yields that $h(s) = \mathbb{E}(s^T) = 1 + (s-1)\sum_{n\geq 0}\mathbb{P}(T > n)s^n$ for $0 \leq s < 1$, and (2.3.12) and $\mathbb{P}(T < 3) = 0$ yield that

$$\sum_{n\geq 0}\mathbb{P}(T > n)s^n = \sum_{n\geq 0}\left(\frac{\mathbb{P}(T = n+1)}{p_G} + \frac{\mathbb{P}(T = n+3)}{p_A p_G^2}\right)s^n$$
$$= \left(\frac{1}{sp_G} + \frac{1}{s^3 p_A p_G^2}\right)h(s),$$

so that eventually

$$h(s) = \left(1 - \frac{s-1}{sp_G} - \frac{s-1}{s^3 p_A p_G^2}\right)^{-1}.$$

As $h(1-) = 1$, this yields $\mathbb{P}(T < \infty) = 1$ again. Moreover, at a precision of order $o(\varepsilon^2)$,

$$\ln h(1 + \varepsilon) = -\ln\left(1 - \frac{\varepsilon}{(1+\varepsilon)p_G} - \frac{\varepsilon}{(1+\varepsilon)^3 p_A p_G^2}\right)$$
$$= -\ln\left(1 - \left(\frac{1}{p_G} + \frac{1}{p_A p_G^2}\right)\varepsilon + \left(\frac{1}{p_G} + \frac{3}{p_A p_G^2}\right)\varepsilon^2\right)$$
$$= \left(\frac{1}{p_G} + \frac{1}{p_A p_G^2}\right)\varepsilon + \left[\left(\frac{1}{p_G} + \frac{1}{p_A p_G^2}\right)^2 - 2\left(\frac{1}{p_G} + \frac{3}{p_A p_G^2}\right)\right]\frac{\varepsilon^2}{2}$$

and (A.1.1) yields that

$$\mathbb{E}(T) = \frac{1}{p_G} + \frac{1}{p_A p_G^2} , \qquad \mathrm{Var}(T) = \frac{1}{p_G^2} + \frac{2}{p_A p_G^3} + \frac{1}{p_A^2 p_G^4} - \frac{1}{p_G} - \frac{5}{p_A p_G^2} .$$

Exercises

2.1   Stopping times   Let $(X_n)_{n \geq 0}$ be a Markov chain on $\mathbb{Z}$, and $k \in \mathbb{N}$. Prove that the following random variables are stopping times.

$$S = \inf\{n \geq 1 : X_n \geq \max\{X_0, \dots, X_{n-1}\} + k\} ,$$

$$T = \inf\left\{ n > k : X_n > \max_{0 \leq i \leq k} X_i \right\} ,$$

$$U = \inf\{n > k : X_n > X_{n-1} > \dots > X_{n-k}\} .$$

2.2   Operations on stopping times   Let S and T be two stopping times.

a) Prove that $S \wedge T$ and $S \vee T$ and $S + \theta_S T$ (with value ∞ if $S = \infty$) are stopping times.

b) Prove that if $S \leq T$ then $\mathcal{F}_S \subset \mathcal{F}_T$, then more generally that $\mathcal{F}_{S \wedge T} = \mathcal{F}_S \cap \mathcal{F}_T$.

c) Prove that if $B \in \mathcal{F}_{S \vee T}$ then $B \cap \{S \leq T\} \in \mathcal{F}_T$ and $B \cap \{T \leq S\} \in \mathcal{F}_S$. Deduce from this that $\mathcal{F}_{S \vee T} = \sigma(\mathcal{F}_S \cup \mathcal{F}_T)$, the least σ-field containing $\mathcal{F}_S \cup \mathcal{F}_T$.

2.3   Induced chain, 1   Let $(X_n)_{n \in \mathbb{N}}$ be a Markov chain on $\mathcal{V}$ with matrix P, having no absorbing state. Let $(S_k, D_k, Y_k)_{k \in \mathbb{N}}$ be defined by $S_0 = 0$ and $D_0 = 0$ and iteratively

$$Y_k = X_{S_k} , \qquad S_{k+1} = \inf\{n > S_k : X_n \neq X_{S_k}\} , \qquad D_{k+1} = S_{k+1} - S_k .$$

Let $(\mathcal{F}_n)_{n \geq 0}$ denotes the filtration generated by $(X_n)_{n \geq 0}$, and $(\mathcal{G}_k)_{k \geq 0}$ the one for $(D_k, Y_k)_{k \geq 0}$: the events in $\mathcal{F}_n$ are of the form $\{(X_0, \dots, X_n) \in E\}$, and those in $\mathcal{G}_k$ of the form $\{(D_0, Y_0, \dots, D_k, Y_k) \in F\}$. Let the matrix $Q = (Q(x, y))_{x, y \in \mathcal{V}}$ and for $x \in \mathcal{V}$, the geometric law $g_x = (g_x(n))_{n \geq 1}$ on $\mathbb{N}^*$ be defined by

$$Q(x, y) = \frac{P(x, y) \mathbb{1}_{\{x \neq y\}}}{1 - P(x, x)} , \qquad g_x(n) = P(x, x)^{n-1}(1 - P(x, x)) .$$

a) Prove that Q is a transition matrix and that P is irreducible if and only if Q is irreducible.

b) Prove that the S_k are stopping times and that $\mathbb{P}(\exists k \geq 1 : S_k = \infty) = 0$.

c) Prove that $(D_k, Y_k)_{k \in \mathbb{N}}$ is a Markov chain with transition matrix given by

$$\mathbb{P}(D_{k+1} = n, Y_{k+1} = z \mid D_k = m, Y_k = y) = g_y(n)Q(y, z) .$$

Prove that $(Y_k)_{k \in \mathbb{N}}$ is a Markov chain with matrix Q. Prove that

$$\mathbb{P}(D_{k+1} = n, Y_{k+1} = z \mid Y_k = y) = g_y(n)Q(y, z) ,$$

$$\mathbb{E}(D_{k+1} \mid Y_k = y) = \mathbb{E}(D_{k+1} \mid D_k = m, Y_k = y) = \frac{1}{1 - P(y, y)} .$$

d) Prove that if U is a stopping time for $(D_k, Y_k)_{k \in \mathbb{N}}$, that is, for $(\mathcal{G}_k)_{k \geq 0}$, then S_U is a stopping time for $(X_n)_{n \in \mathbb{N}}$, that is, for $(\mathcal{F}_n)_{n \geq 0}$.

2.4   **Doeblin coupling**   Let $(X_n^1, X_n^2)_{n \geq 0}$ be a Markov chain on $\mathcal{V} \times \mathcal{V}$ with matrix Q satisfying

$$Q((x^1, x^2), (y^1, y^2)) = Q((x^2, x^1), (y^2, y^1)) ,$$

and

$$T = \inf\{n \geq 0 : X_n^1 = X_n^2\} , \qquad (Z_n^1, Z_n^2) = \begin{cases} (X_n^1, X_n^2) & \text{if } n \leq T , \\ (X_n^2, X_n^1) & \text{if } n > T . \end{cases}$$

Let P be a transition matrix on $\mathcal{V}$ such that there exists $\varepsilon > 0$ and a probability measure $\hat{\pi}$ such that $P(x, y) \geq \varepsilon \hat{\pi}(y)$ for all x; this is the Doeblin condition of Theorem 1.3.4 for $k = 1$.

a) Prove that T is a stopping time.

b) Prove that $(Z_n^1, Z_n^2)_{n \geq 0}$ has same law as $(X_n^1, X_n^2)_{n \geq 0}$. Deduce from this, for instance using (1.2.2), that

$$\|\mathcal{L}(X_n^1) - \mathcal{L}(X_n^2)\|_{var} \leq 2\mathbb{P}(Z_n^1 \neq X_n^2) = 2\mathbb{P}(T > n) .$$

c) Prove that we define transition matrices R on $\mathcal{V}$ and Q on $\mathcal{V} \times \mathcal{V}$ by

$$R(x, y) = \frac{P(x, y) - \varepsilon \hat{\pi}(y)}{1 - \varepsilon} ,$$

$$Q((x^1, x^2), (y^1, y^2)) = \varepsilon \mathbb{1}_{\{y^1 = y^2\}} \hat{\pi}(y^1) + (1 - \varepsilon)R(x^1, y^1)R(x^2, y^2) ,$$

and that $Q((x^1, x^2), (y^1, y^2)) = Q((x^2, x^1), (y^2, y^1))$.

d) Prove that $\mathbb{P}(T > n) \leq \mathbb{P}(X_0^1 \neq X_0^2)(1 - \varepsilon)^n$.

e) Prove that $(X_n^1)_{n \geq 0}$ and $(X_n^2)_{n \geq 0}$ are both Markov chains with matrix P.

f) Conclude that

$$\|\mu P^n - \mu P'^n\|_{var} \le \|\mu - \mu\|_{var}(1 - \varepsilon)^n$$

by using the fact that there exists r.v. X_0^1 and X_0^2 with laws μ and μ such that $2\mathbb{P}(X_0^1 \ne X_0^2) = \|\mu - \mu\|_{var}$ (see Lemma A.2.2).

2.5   The space station, see Exercise 1.1   The astronaut starts from module 1.

a) What is his probability of reaching module 4 before visiting the central module (module 0)?

b) What is his probability of visiting all of the external ring (all peripheral modules and the links between them) before visiting the central module?

c) Compute the generating function, the law, and the expectation of the time that he takes to reach module 4, conditional on the fact that he does so before visiting the central module.

2.6   The mouse, see Exercise 1.2 The mouse starts from room 1. Compute the probability that it reaches room 4 before returning to room 1. Compute the generating function and expectation of the time it takes to reach room 4, conditional on the fact that it does so before returning to room 1.

2.7   Andy, see Exercise 1.6 Andy is just out of jail. Let L be the number of consecutive evenings he spends out of jail, before his first return there. Compute $g(s) = \mathbb{E}(s^L)$ for $0 \le s \le 1$. Compute the law and expectation of L.

2.8   Genetic models, see Exercise 1.8   Let $K = 2$, and Z_n be the number of individuals of allele 1 at time $n \ge 0$. Consider the Dirichlet problem (2.2.5) for $(Z_n)_{n \ge 0}$ on $\mathcal{V} = \{0, \dots, N\}$ for $E = \{0, N\}$.

a) Prove that for asynchronous reproduction the linear equation writes

$$-\frac{c(2)}{c(1) + c(2)}u(x - 1) + u(x) - \frac{c(1)}{c(1) + c(2)}u(x + 1) = 0, \quad 0 < x < N.$$

b) What does this equation remind you of? Compute $\mathbb{P}_x(Z_0 < Z_N)$ and $\mathbb{P}_x(Z_N < Z_0)$ (fixation probability and extinction probability for allele 2).

c) Write the linear equation obtained in the case of synchronous reproduction. Prove that if $c(1) = c(2)$ then $\alpha + \beta x$ solves this equation. Compute $\mathbb{P}_x(Z_0 < Z_N)$ and $\mathbb{P}_x(Z_N < Z_0)$.

2.9   Nearest-neighbor walk, 1-d   Let $(X_n)_{n \ge 0}$ be the random walk on $\mathbb{Z}$ with matrix given by $P(x, x + 1) = p > 0$ and $P(x, x - 1) = 1 - p = q > 0$. Let $R_x = \inf\{n \ge 1 : X_n = x\}$ and $N_x = \sum_{k=0}^{\infty} \mathbb{1}_{\{X_k = x\}}$ for $x \in \mathbb{Z}$, and $D = \sup\{n \ge 0 : X_n = 0\}$ and $M = \inf\{n \ge 0 : X_n = \max_{k \ge 0} X_k\}$.

a) Draw the graph of P. Is this matrix irreducible?

b) Let $x \neq y$ be in $\mathbb{Z}$. Prove, using results in Section 2.3.2, that

$$\mathbb{P}_y(R_x < \infty) = \min\{1, (q/p)^{y-x}\}$$

and that $\mathbb{E}_y(R_x) < \infty$ if and only if $(y - x)(q - p) > 0$.

c) Let x be in $\mathbb{Z}$. Prove that $\mathbb{P}_x(R_x < \infty) = 2\min\{p, q\}$ and $\mathbb{E}_x(R_x) = \infty$.

d) Prove that, for $k \geq 1$,

$$\mathbb{P}_0(N_x \geq k) = \min\{1, (p/q)^x\} \min \{2p, 2q\}^{k-1} .$$

Prove that $\mathbb{P}_0(N_x = \infty) = 1$ if $p = 1/2$ and $\mathbb{P}_0(N_x = \infty) = 0$ if $p \neq 1/2$.

e) Prove that $\lim_{n\to\infty} X_n = \infty$ if $p > 1/2$ and $\lim_{n\to\infty} X_n = -\infty$ if $p < 1/2$, a.s.

f) Prove that $\mathbb{P}_0(D = \infty) = 1$ if $p = 1/2$ and $\mathbb{P}_0(D = \infty) = 0$ if $p \neq 1/2$. In this last case, prove by considering $(X_{D+n})_{n\geq 0}$ that D is not a stopping time.

g) Prove that $\mathbb{P}_0(M = \infty) = 1$ if $p \geq 1/2$ and $\mathbb{P}_0(M = \infty) = 0$ if $p < 1/2$. In this last case, prove that M is not a stopping time.

h) Prove, using results in Section 2.3.2, that

$$\mathbb{E}_x(s^{R_0}) = \begin{cases} \left(\dfrac{1 - \sqrt{1 - 4pqs^2}}{2ps}\right)^x & \text{if } x > 0 , \\[4mm] \left(\dfrac{1 - \sqrt{1 - 4pqs^2}}{2qs}\right)^x & \text{if } x < 0 . \end{cases}$$

i) Prove that $\mathbb{E}_0(s^{R_0}) = 1 - \sqrt{1 - 4pqs^2}$. Deduce from this the law of R_0 when $X_0 = 0$.

2.10 Labouchère system, see Exercise 1.4   Let $(X_n)_{n\geq 0}$ be the random walk on $\mathbb{Z}$ with matrix given by $P(x, x - 2) = p > 0$ and $P(x, x + 1) = 1 - p > 0$, and $T = S_{\{-1,0\}} = \inf\{n \geq 0 : X_n \in \{-1, 0\}\}$.

a) What is the relation of these objects to Exercise 1.4?

b) Draw the graph of P. Is this matrix irreducible?

c) Prove that $u(x) = \mathbb{P}_x(T < \infty)$ satisfies $u(-1) = u(0) = 1$ and

$$-pu(x - 2) + u(x) - (1 - p)u(x + 1) = 0 , \quad x \geq 1 .$$

Prove that this recursion has 1 as a solution, then that its general solution is given for $p \neq 1/3$ by $(\alpha_- \lambda_-^x + \alpha_+ \lambda_+^x + \beta)_{x\geq -1}$ with $\lambda_\pm = \frac{p \pm \sqrt{p(4-3p)}}{2(1-p)}$ and for $p = 1/3$ by $(\alpha_-(-1/2)^x + \alpha_+ x + \beta)_{x\geq -1}$.

d) Prove that if $p > 1/3$ then $-1 < \lambda_- < 0 < 1 < \lambda_+$ and if $p < 1/3$ then $-1 < \lambda_- < 0 < \lambda_+ < 1$.

Deduce from this that, for $x \geq 1$, if $p \geq 1/3$ then $u(x) = 1$ and if $p < 1/3$ then $u(x) = \frac{\lambda_+ + p}{\lambda_+ - \lambda_-}\lambda_+^x - \frac{\lambda_- + p}{\lambda_+ - \lambda_-}\lambda_-^x < 1$.

e) Prove that $m(x) = \mathbb{E}_x(T)$ satisfies $m(-1) = m(0) = 0$ and

$$-pm(x-2) + m(x) - (1-p)m(x+1) = 1 , \quad x \geq 1 .$$

Deduce from this that, for $x \geq 1$, if $p \leq 1/3$ then $m(x) = \infty$ and if $p > 1/3$ then $m(x) = \frac{1}{3p-1}(x + \frac{1-\lambda_-^x}{1-\lambda_-^{-1}})$.

f) What prevents us from computing the generating function of T?

3

Transience and recurrence

3.1 Sample paths and state space

3.1.1 Communication and closed irreducible classes

Let $(X_n)_{n \geq 0}$ be a Markov chain with transition matrix P on $\mathcal{V}$. Consider its first strict future hitting time of $y \in \mathcal{V}$, given by

$$R_y = \inf\{n \geq 1 : X_n = y\} \, .$$

If $\mathbb{P}_x(R_y < \infty) > 0$, then y is said to be *reachable* from x or that x *leads to* y. Equivalently, there exists $i = i(x, y) \geq 1$ such that $P^i(x, y) > 0$, that is, such that (s.t.)

$$\exists x_1, \dots, x_{i-1} \in \mathcal{V} : P(x, x_1)P(x_1, x_2) \cdots P(x_{i-1}, y) > 0 \, ,$$

and these states form an oriented path from x to y through the graph:

This property is denoted by $x \to y$ and its negation by $x \nrightarrow y$.

It is said that x and y *communicate* if $x \to y$ and $y \to x$. This property is denoted by $x \leftrightarrow y$ and its negation by $x \nleftrightarrow y$.

A subset $\mathcal{W}$ of $\mathcal{V}$ is *irreducible* if $x \leftrightarrow y$ for all x and y in $\mathcal{W}$, and the chain or matrix is irreducible if the state space $\mathcal{V}$ is irreducible. This definition coincides with the definition given in Theorem 1.3.3 and is equivalent to the fact that there is an oriented path through all the nodes of the graph.

Markov Chains: Analytic and Monte Carlo Computations, First Edition. Carl Graham.
© 2014 John Wiley & Sons, Ltd. Published 2014 by John Wiley & Sons, Ltd.

A subset $\mathcal{W}$ of $\mathcal{V}$ is *closed* or *absorbing* if

$$x \in \mathcal{W} , \ x \to y \Rightarrow y \in \mathcal{W} .$$

Equivalently, the restriction of the matrix P to $\mathcal{W}$ is again a transition matrix.

If $\mathcal{W}$ is closed and irreducible, then the Markov chain for the restriction of P to $\mathcal{W}$ is irreducible on $\mathcal{W}$; such a $\mathcal{W}$ is called a closed irreducible class. Results made under the hypothesis that P is irreducible can be generalized by being applied to the *restrictions* of P to closed irreducible classes.

3.1.2 Transience and recurrence, recurrent class decomposition

3.1.2.1 Transient and recurrent states, potential matrix

Let $(X_n)_{n \geq 0}$ be a Markov chain on $\mathcal{V}$ with matrix P. For a state x, define the first hitting time, the first strict future hitting time, and the number of hits by

$$S_x = \inf\{n \geq 0 : X_n = x\} , \quad R_x = \inf\{n \geq 1 : X_n = x\} , \quad N_x = \sum_{n \geq 0} \mathbb{1}_{\{X_n = x\}} .$$

The strong Markov property (Theorem 2.1.3) yields that

$$\mathbb{P}(N_x = \infty) = \mathbb{P}(S_x < \infty)\mathbb{P}_x(N_x = \infty) , \quad \mathbb{E}(N_x) = \mathbb{P}(S_x < \infty)\mathbb{E}_x(N_x) , \quad (3.1.1)$$

and that N_x under $\mathbb{P}_x$ is geometric on $\mathbb{N}^*$ as

$$\mathbb{P}_x(N_x > k) = \mathbb{P}_x(R_x < \infty)^k , \qquad k \geq 0 .$$

In this perspective, recall that

$$\mathbb{P}(N_x = \infty) = \lim_{k \to \infty} \downarrow \mathbb{P}(N_x > k) , \qquad \mathbb{E}(N_x) = \sum_{k \geq 0} \mathbb{P}(N_x > k) .$$

The state space $\mathcal{V}$ can then be *partitioned* in a subset $\mathcal{T}$ of *transient* states and a subset $\mathcal{R}$ of *recurrent* states by the alternative

$$x \in \mathcal{T} \iff \mathbb{P}_x(R_x < \infty) < 1 \iff \mathbb{P}_x(N_x = \infty) = 0 \iff \mathbb{E}_x(N_x) < \infty ,$$
$$x \in \mathcal{R} \iff \mathbb{P}_x(R_x < \infty) = 1 \iff \mathbb{P}_x(N_x = \infty) = 1 \iff \mathbb{E}_x(N_x) = \infty .$$

The extreme results

$$x \in \mathcal{T} \Rightarrow \mathbb{E}(N_x) < \infty , \qquad x \in \mathcal{R} \Rightarrow \mathbb{P}(N_x = \infty) = \mathbb{P}(S_x < \infty) , \qquad (3.1.2)$$

and their contradictions follow from (3.1.1). Moreover, $\mathbb{E}(N_x) = \frac{\mathbb{P}(S_x < \infty)}{\mathbb{P}_x(R_x = \infty)}$ with the usual conventions, but this formula will not be very useful.

Potential matrix criterion The formal inverse of $I - P$ is given by the *potential matrix* $\sum_{n \geq 0} P^n$. It holds that

$$\mathbb{E}_x(N_y) = \sum_{n \geq 0} \mathbb{P}_x(X_n = y) = \sum_{n \geq 0} P^n(x, y) .$$

The *potential matrix criterion* is the fact that

$$\sum_{n \geq 0} P^n(x, x) < \infty \iff x \in \mathcal{T} , \qquad \sum_{n \geq 0} P^n(x, x) = \infty \iff x \in \mathcal{R} .$$

Lemma 3.1.1 *Let $(X_n)_{n \geq 0}$ be a Markov chain on $\mathcal{V}$ with matrix P. Then,*

$$y \in \mathcal{T} \Rightarrow \lim_{n \to \infty} P^n(y, y) = 0 \Rightarrow \lim_{n \to \infty} P^n(x, y) = 0 , \ \forall x \in \mathcal{V} .$$

Moreover, if $(X_n)_{n \geq 0}$ has an invariant law π, then $\pi(y) = 0$ for every y s.t. $\lim_{n \to \infty} P^n(y, y) = 0$, and there exists a recurrent state.

Proof: If y is transient, then $\sum_{n \geq 0} P^n(y, y) < \infty$, and thus $\lim_{n \to \infty} P^n(y, y) = 0$. The strong Markov property yields that

$$P^n(x, y) = \sum_{k=0}^n \mathbb{P}_x(S_y = k)P^{n-k}(y, y) \leq \max_{\lfloor n/2 \rfloor \leq i \leq n} P^i(y, y) + \sum_{k=\lfloor n/2 \rfloor}^n \mathbb{P}_x(S_y = k)$$

and both r.h.s. terms go to 0. Moreover, by iteration $\pi = \pi P^n$ and thus

$$\pi(y) = \sum_{x \in \mathcal{V}} \pi(x)P^n(x, y) ,$$

and dominated convergence (Theorem A.3.5) yields that $\pi(y) = 0$. ∎

Lemma 3.1.2 *Let $(X_n)_{n \geq 0}$ be a Markov chain on $\mathcal{V}$ with matrix P. If there exists a finite nonempty subset F of $\mathcal{V}$ s.t. $\mathbb{P}_x(R_F < \infty) = 1$ for every x in F or for every x in $\mathcal{V} - F$, then there exists a recurrent state in F.*

Proof: The strong Markov property yields that the chain hits F infinitely often, so that $\sum_{x \in F} N_x = \infty$. Hence, $\sum_{x \in F} \mathbb{E}(N_x) = \infty$ and, as F is finite, there exists $y \in F$ s.t. $\mathbb{E}(N_y) = \infty$ and then (3.1.2) yields that y is recurrent. ∎

3.1.2.2 Irreducible classes, state space decomposition

Lemma 3.1.3 *Let $(X_n)_{n \geq 0}$ be a Markov chain on $\mathcal{V}$. If $x \to y$ and x is recurrent, then $\mathbb{P}_x(N_y = \infty) = 1$ and y is recurrent. If $x \leftrightarrow y$, then x and y are both simultaneously either transient or recurrent.*

Proof: The successive hitting times $0 = S_0 < S_1 < \cdots$ of x defined in (2.2.2) are finite, $\mathbb{P}_x$-a.s., and

$$N_y = \sum_{k \geq 1} Y_k \,, \qquad Y_k = \sum_{n=S_{k-1}}^{S_k} \mathbb{1}_{\{X_n = y\}} \,.$$

The strong Markov property yields that the Y_k are i.i.d. under $\mathbb{P}_x$, and $x \to y$ yields that $\mathbb{P}_x(Y_1 = 0) < 1$. As

$$\{N_y < \infty\} = \bigcup_{k \geq 0} \uparrow \{Y_{k+1} = Y_{k+2} = \cdots = 0\} \,,$$

by monotone limit (Lemma A.3.1)

$$\mathbb{P}_x(N_y < \infty) = \lim_{k \to \infty} \uparrow \mathbb{P}_x(Y_{k+1} = Y_{k+2} = \cdots = 0) \,,$$

and

$$\mathbb{P}_x(Y_{k+1} = Y_{k+2} = \cdots = 0) = \mathbb{P}_x(Y_1 = Y_2 = \cdots = 0) = 0$$

as $\mathbb{P}_x(Y_1 = Y_2 = \cdots = 0) \leq \mathbb{P}_x(Y_1 = \cdots = Y_n = 0) = \mathbb{P}_x(Y_1 = 0)^n$ for all n and hence $\mathbb{P}_x(N_y = \infty) = 1$. Then, $\mathbb{P}_x(N_y = \infty) = \mathbb{P}_x(S_y < \infty)\mathbb{P}_y(N_y = \infty)$ implies that $\mathbb{P}_y(N_y < \infty) = 1$, so that y is recurrent. It is straightforward to conclude. ∎

The idea behind this proof will eventually yield the Pointwise ergodic theorem. Other proofs of this lemma, which are less straightforward, are in Exercise 3.3.

Irreducible class decomposition Lemma 3.1.3 allows to define that an *irreducible* chain or class is *transient* or *recurrent* if and only if one of its states is *transient* or *recurrent* and then all states will be so. It also implies the following.

- The relation $x \leftrightarrow y$ is an equivalence relation in $\mathcal{R}$, and its equivalence classes are closed and irreducible and are called the *recurrent classes*. Denoting them by $\mathcal{R}_i$ for $i \in I$, $\mathcal{V}$ can be decomposed according to the *partition*

$$\mathcal{V} = \mathcal{T} \cup \mathcal{R} = \mathcal{T} \cup \bigcup_{i \in I} \mathcal{R}_i \,.$$

- If X_0 is in $\mathcal{R}_i$, then $(X_n)_{n \geq 0}$ can be considered to be a recurrent irreducible chain on $\mathcal{R}_i$, which will visit infinitely often every state.

- If X_0 is in $\mathcal{T}$, then either the chain eventually exits $\mathcal{T}$ by entering some $\mathcal{R}_i$ in which it will stay thereafter, or stays in $\mathcal{T}$ and then (3.1.2) yields that the chain goes to *infinity* as it visits each state only a finite number of times (this random number even having finite expectation).

It is usually not a simple matter to detect the transient states for a chain. Usually one starts with exploring the communication properties of the state space, which is a much simpler matter.

- First, the states x s.t. $x \not\rightarrow x$ are found. These are necessarily transient.

- Then, $\{x \in \mathcal{V} : x \leftrightarrow x\}$ is partitioned into equivalence classes for $\leftrightarrow$, as was done for $\mathcal{R}$. These classes are called the *irreducible classes*.

- Among the irreducible classes, those that in addition are closed are identified. The others are subsets of $\mathcal{T}$.

- The restriction of the Markov chain to a closed irreducible class is an irreducible Markov chain, and then the difficult task of determining whether it is transient or recurrent must be handled. The closed irreducible class will be a recurrent class $\mathcal{R}_i$ if and only if this restriction is recurrent, else it will be subset of $\mathcal{T}$.

- Note that (3.1.2) yields that a finite closed irreducible class will be a recurrent class, as the chain visits each transient state only a finite number of times.

3.1.3   Detailed examples

3.1.3.1   Gambler's ruin

See Section 1.4.2. States 0 and N are absorbing, hence $\{0\}$ and $\{N\}$ are recurrent classes. Moreover, $\{1, \dots, N-1\}$ is not closed and is irreducible, hence it must be transient; as this class is finite, (3.1.2) yields that $\mathbb{P}(S_{\{0,N\}} < \infty) = 1$. In the unilateral case in which $N = \infty$, state 0 is absorbing and $\mathbb{N}^*$ is transient, and hence, (3.1.2) yields that

$$\mathbb{P}(S_0 < \infty) + \mathbb{P}(\lim_{n\to\infty} X_n = \infty) = 1 \ .$$

3.1.3.2   Nearest-neighbor walk in one dimension

The nearest-neighbor random walk on $\mathbb{Z}$ goes from x to $x+1$ with probability $p > 0$ and from x to $x-1$ with probability $q = 1 - p > 0$, and its graph is

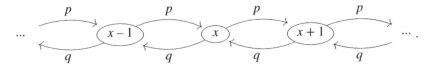

It was studied in Exercise 2.9, and we revisit its main points.

The "one step forward" method yields that

$$\mathbb{P}_0(R_0 < \infty) = p\mathbb{P}_1(S_0 < \infty) + q\mathbb{P}_{-1}(S_0 < \infty)$$

and the results on the unilateral hitting time and symmetry yield that

$$\mathbb{P}_1(S_0 < \infty) = \min\{1, q/p\} \ , \qquad \mathbb{P}_{-1}(S_0 < \infty) = \min\{1, p/q\} \ ,$$

and thus

$$\mathbb{P}_0(R_0 < \infty) = 2\min\{p, q\} \ .$$

Hence, the chain is transient for $p \neq 1/2$ and recurrent for $p = q = 1/2$. In the latter case, the "one step forward" method yields that

$$\mathbb{E}_0(R_0) = 1 + \frac{1}{2}\mathbb{E}_1(S_0) + \frac{1}{2}\mathbb{E}_{-1}(S_0) = \infty .$$

If $p \neq 1/2$, then every state is visited only a finite number of times, and

$$\mathbb{P}(\lim_{n \to \infty} X_n = \infty) + \mathbb{P}(\lim_{n \to \infty} X_n = -\infty) = 1 .$$

Moreover, if $p < 1/2$, then $\mathbb{P}_0(R_{-x} < \infty) = \mathbb{P}_x(R_0 < \infty) = 1$ and hence, $\lim_{n \to \infty} X_n = -\infty$ a.s. By symmetry, if $p > 1/2$ then $\lim_{n \to \infty} X_n = \infty$ a.s.

3.1.3.3   Symmetric random walk in many dimensions

For $d \geq 1$, the symmetric nearest-neighbor random walk on $\mathbb{Z}^d$ has probability $1/2d$ of going from a state x to one of its $2d$ neighbors $x \pm e_i$, where $(e_i)_{1 \leq i \leq d}$ is the canonical basis.

This is one of the rare concrete examples in which the $P^n(x, x)$ can be evaluated in a way that enables to use the Potential matrix criterion. For $n \geq 0$, clearly $P^{2n+1}(0, 0) = 0$, and a classic multinomial law computation yields that

$$P^{2n}(0, 0) = \frac{1}{(2d)^{2n}} \sum_{n_1 + \cdots + n_d = n} \frac{(2n)!}{(n_1!)^2 \cdots (n_d!)^2}$$

$$= \frac{1}{(2d)^{2n}} \binom{2n}{n} \sum_{n_1 + \cdots + n_d = n} \left(\frac{n!}{n_1! \cdots n_d!} \right)^2 .$$

Dimension 1 or 2   We use the Stirling formula $N! \simeq N^N e^{-N} (2\pi N)^{1/2}$. Then,

$$d = 1 \Rightarrow P^{2n}(0, 0) = \frac{1}{2^{2n}} \binom{2n}{n} \simeq \frac{1}{(\pi n)^{1/2}} .$$

As

$$\sum_{n_1 + n_2 = n} \left(\frac{n!}{n_1! \, n_2!} \right)^2 = \sum_{k=0}^{n} \binom{n}{k} \binom{n}{n-k} = \binom{2n}{n} \simeq \frac{2^{2n}}{(\pi n)^{1/2}} ,$$

moreover

$$d = 2 \Rightarrow P^{2n}(0, 0) = \frac{1}{4^{2n}} \binom{2n}{n}^2 \simeq \frac{1}{\pi n} .$$

These equivalents are summable, and the Potential matrix criterion yields that the random walk is recurrent in dimensions $d = 1$ and $d = 2$.

Dimension 3 and more   The multinomial formula

$$\sum_{n_1+\cdots+n_d=n} \frac{n!}{n_1!\cdots n_d!} = \left(\sum_{i=1}^d 1\right)^n = d^n$$

yields the bound

$$P^{2n}(0,0) \le \frac{1}{(2d)^{2n}} \binom{2n}{n} \max_{n_1+\cdots+n_d=n} \frac{n!}{n_1!\cdots n_d!} \sum_{n_1+\cdots+n_d=n} \frac{n!}{n_1!\cdots n_d!}$$

$$\le \frac{1}{(4d)^n} \binom{2n}{n} \max_{n_1+\cdots+n_d=n} \frac{n!}{n_1!\cdots n_d!}$$

and classically if n is a multiple of d, and using the Stirling formula,

$$\frac{1}{(4d)^n} \binom{2n}{n} \max_{n_1+\cdots+n_d=n} \frac{n!}{n_1!\cdots n_d!} = \frac{1}{(4d)^n} \binom{2n}{n} \frac{n!}{(\frac{n}{d}!)^d}$$

$$\simeq \frac{1}{(4d)^n} \frac{2^{2n}}{(\pi n)^{1/2}} \frac{d^{n+d/2}}{(2\pi n)^{\frac{d-1}{2}}} = \sqrt{2}\left(\frac{d}{2\pi n}\right)^{d/2},$$

and this asymptotic bound on $P^{2n}(0,0)$ can be extended to n, which are not multiples of d using the growth of the multinomial terms.

The corresponding series converges for $d \ge 3$, and the Potential matrix criterion yields that the random walk is transient.

3.1.3.4   General random walk in one dimension

Let $(X_n)_{n\ge 0}$ be the random walk on $\mathbb{Z}$ given by $X_n = X_0 + \xi_1 + \cdots + \xi_n$ with X_0 independent of the i.i.d. sequence $(\xi_i)_{i\ge 1}$, with $\mathbb{E}(|\xi_1|) < \infty$ and $\mathbb{E}(\xi_1) = m$.

By translation invariance, all states are simultaneously transient or recurrent. The strong law of large numbers yields that $X_n = nm + o(n)$, a.s., so that if $m > 0$ then $\lim_{n\to\infty} X_n = \infty$ and if $m < 0$ then $\lim_{n\to\infty} X_n = -\infty$. Hence, for $m \ne 0$, all states are transient.

We are now going to show that if $m = 0$ then all states are recurrent. Let us reason by contradiction, assuming that all states are transient, and thus that $\mathbb{E}_0(N_0) < \infty$. For all x,

$$\mathbb{E}_0(N_x) = \mathbb{P}_0(S_x < \infty)\mathbb{E}_x(N_x) \le \mathbb{E}_x(N_x) = \mathbb{E}_0(N_0)$$

and hence,

$$\sum_{|x|<k} \mathbb{E}_0(N_x) \le 2k\mathbb{E}_0(N_0) ,$$

and if $\varepsilon = \frac{1}{6}\mathbb{E}_0(N_0)^{-1} > 0$ then

$$\sum_{|x|<\varepsilon n} \mathbb{E}_0(N_x) \le \frac{n}{3} . \qquad (3.1.3)$$

The law of large numbers yields that there exists n_0 s.t. if $n \geq n_0$ then

$$\sum_{|x|<\varepsilon n} P^n(0,x) := \mathbb{P}(|X_n| < \varepsilon n) > \frac{1}{2} \,,$$

and hence,

$$\sum_{|x|<\varepsilon n} \mathbb{E}_0(N_x) = \sum_{|x|<\varepsilon n} \mathbb{E}_0\left(\sum_{k\geq 0} \mathbb{1}_{\{X_k=x\}}\right) \geq \sum_{k=0}^{n} \sum_{|x|<\varepsilon k} P^k(0,x) > \frac{n-n_0}{2} \,,$$

which contradicts (3.1.3) for large enough n.

3.1.3.5 Branching process

See Section 1.4.3. The state 0 is absorbing and constitutes a recurrent class. If $p(0) > 0$, then $P(x,0) > 0$ and $x \to 0$ for $x \geq 0$, and every state $x \geq 1$ is transient, hence $\mathbb{N}^*$ is the transient class and necessarily

$$\mathbb{P}(S_0 < \infty) + \mathbb{P}(\lim_{n\to\infty} X_n = \infty) = 1$$

so that the population either becomes extinct or explodes.

Any state $x \geq 1$ communicates with infinitely many states if and only if $p(0) + p(1) < 1$. The precise description of the set of states that can be reaches starting from state x depends on $\{k \in \mathbb{N} : p(k) > 0\}$.

If $p(1) = 1$, then every state is absorbing. If $p(0) = 1$, then $P(x,0) = 1$ for all $x \geq 0$. If $p(0) + p(1) = 1$ and $0 < p(0) < 1$, then $P(x, x-k) = p(1)^{x-k}p(0)^k > 0$ for $0 \leq k \leq i$ and $x \to y$ if and only if $x \geq y$.

3.1.3.6 Ehrenfest Urn

See Section 1.4.4. The microscopic and macroscopic chains are irreducible on $\{0,1\}^N$ and $\{0, \ldots, N\}$. Lemma 3.1.2 implies that these chains are recurrent, and hence visit infinitely often every state, even the state in which compartment 1 is empty.

3.1.3.7 Renewal process

See Section 1.4.5. If $y > x$, then $x \to y$ if and only if

$$p_x \cdots p_{y-1} > 0 \,,$$

If $y \leq x$, then in order to visit 0 the chain must visit 0 first, and $x \to y$ if and only if

$$\exists z \geq x : p_z < 1 \,, \qquad p_0 \cdots p_{y-1} > 0 \,.$$

which allows to determine easily the irreducible classes.

Notably, $(X_n)_{n \in \mathbb{N}}$ is irreducible on $\mathbb{N}$ if and only if $p_x > 0$ for every $x \geq 0$ and $p_x < 1$ for infinitely many $x \geq 0$, and then it is recurrent if and only if

$$\mathbb{P}(D = \infty) = \lim_{x \to \infty} p_0 \cdots p_{x-1} = 0 \, .$$

3.1.3.8   Word search

See Section 1.4.6. As in the particular situation that was studied, the matrix is irreducible on a finite state space, and Lemma 3.1.2 yields that there is at least an irreducible state. Hence, the chain is irreducible and recurrent, which notably implies that $\mathbb{P}(T < \infty) = 1$.

3.1.3.9   Snake chain

See Section 1.4.6, where we have seen that if P is irreducible on $\mathcal{V}$, then P_ℓ is irreducible on its natural state space $\mathcal{V}_\ell$. If x_1 is transient for P, then any ℓ-tuple of the form $(x_1, \ldots, x_\ell)$ is clearly transient for the snake chain. If x_1 is recurrent, then any ℓ-tuple in $\mathcal{V}_\ell$ is recurrent for the snake chain: the head of the snake will visit x_1 infinitely often, and at each visit, in i.i.d. manner, there is a probability $P(x_1, x_2) \ldots P(x_{\ell-1}, x_\ell) > 0$ that the original chain visit $x_1, \cdots, x_\ell$ successively, and this event will eventually happen, a.s., by a simple geometric law argument similar to the one in Lemma 3.1.3.

3.1.3.10   Product chain

See Section 1.4.7. We have seen there on a simple example that P_1 and P_2 may well be both irreducible without $P_1 \otimes P_2$ being so. If P_1 of P_2 is transient, then $P_1 \otimes P_2$ is transient. As $(P_1 \otimes P_2)^n = P_1^n \otimes P_2^n$ for $n \geq 0$, the Potential matrix criterion yields that P_1 and P_2 may well be recurrent without $P_1 \otimes P_2$ being so (see Exercise 3.2).

3.2   Invariant measures and recurrence

3.2.1   Invariant laws and measures

3.2.1.1   Invariant laws, stationary chain, and balance equations

Let $(X_n)_{n \geq 0}$ be a Markov chain with matrix P, and π_n be its instantaneous laws. Then, $(\pi_n)_{n \geq 0}$ solves the linear (or affine) recursion $\pi_n = \pi_{n-1} P = \cdots = \pi_0 P^n$ and, under weak continuity assumptions, can converge to some law π only if π is a fixed point for the recursion, and hence only if $\pi = \pi P$.

If a law π is s.t. $\pi = \pi P$, and if $\pi_0 = \pi$, then

$$\pi_n = \pi P^n = \pi P^{n-1} = \cdots = \pi \, , \qquad n \geq 0 \, ,$$

and hence, $(X_{n+k})_{k \geq 0}$ is a Markov chain with matrix P and initial law π and thus has same law as $(X_k)_{k \geq 0}$. The chain is then said to be *in equilibrium* or *stationary*, and π is said to be its *invariant law* or *stationary distribution*.

In order to search for an invariant law, three main steps must be followed in order.

1. Solve the linear equation $\mu = \mu P$.

2. Find which solutions μ are *nonnegative* and nonzero. Such a solution is called an *invariant measure*.

3. For any such invariant measure μ, check whether $\|\mu\|_{\text{var}} < \infty$ (always true if V is finite), and if so normalize μ to obtain an invariant law $\pi = \mu/\|\mu\|_{\text{var}}$.

The linear equation $\mu = \mu P$ for the invariant measure can be written in a number of equivalent ways, among which $\mu(I - P) = 0$. It is practical to use such condensed abstract notations for the invariant measure equation, but it is important to be able to write it as a system if one wants to solve it, for instance as follows.

Global balance (or equilibrium) equations   This is the linear system

$$\mu(x)(1 - P(x,x)) = \sum_{y \neq x} \mu(y)P(y,x) , \qquad x \in V . \qquad (3.2.4)$$

It can be interpreted as a balance equation on the graph between all which "leaves" x and all which "enters" x, in strict sense.

The same balance reasoning taken in wide sense yields $\mu = \mu P$, and we obtain this equivalent version of the invariant measure equation using that $1 - P(x,x) = \sum_{y \neq x} P(y,x)$. If μ is an invariant measure then, for every subset A of V, the balance equation

$$\sum_{x \in A} \mu(x)P(x, V - A) = \sum_{y \in V-A} \mu(y)P(y, A)$$

holds for what leaves and enters it. The simple proof is left as an exercise.

As in all developed expressions for $\mu = \mu P$, the global balance system is in general highly coupled and is very difficult to solve. This is why the following system is of interest.

Local balance (or equilibrium) equations   This is the linear system

$$\mu(x)P(x, y) = \mu(y)P(y, x) , \qquad x, y \in V . \qquad (3.2.5)$$

It can be interpreted as a balance equation on the graph between all which goes from x to y and all which goes from y to x.

By summing over $y \neq x$, we readily check that any solution of (3.2.5) is a solution of (3.2.4), but the converse can easily be seen to be false. This system is much less coupled and simpler to solve that the global balance, and often should be tried first, by it often has only the null solution.

This system is also called the *detailed balance equations*, as well as the *reversibility equations*, and the latter terminology will be explained later.

3.2.1.2   Uniqueness, superinvariant measures, and positivity

The space of invariant measures constitutes a positive cone (without the origin).

An invariant measure μ is said to be *unique* if it is so up to a *positive multiplicative constant*, that is, if this space is reduced to the half-line $\{c\mu : c > 0\}$ generated by μ. Then, if $\|\mu\|_{\text{var}} < \infty$, then $\pi = \mu/\|\mu\|_{\text{var}}$ is the *unique* invariant law or else if $\|\mu\|_{\text{var}} = \infty$ then there is *no* invariant law.

A measure μ is said to be *superinvariant* if $\mu \in [0, \infty]$ and $\mu \geq \mu P$. Note that any invariant measure is superinvariant. This notion will be helpful for the uniqueness results. We use the classic conventions for addition and multiplication in $[0, \infty]$ (see Section A.3.3).

Lemma 3.2.1 *Let* $\mu = (\mu(x))_{x\in V}$ *take values in* $[0, \infty]$ *and satisfy* $\mu \geq \mu P$. *If* $x \to y$, *then* $\mu(x) > 0$ *implies that* $\mu(y) > 0$ *and* $\mu(x) = \infty$ *implies that* $\mu(y) = \infty$. *In particular, if* W *is an irreducible class, then either* $0 < \mu < \infty$ *on* W *or* μ *is constant and equal to zero or to infinity on* W.

Proof: By irreducibility, we can find $i \geq 1$ s.t. $P^i(x, y) > 0$. Then, iteratively $\mu \geq \mu P \geq \cdots \geq \mu P^i$ and hence,

$$\mu(y) \geq \sum_{z\in V} \mu(z)P^i(z, y) \geq \mu(x)P^i(x, y)$$

and the implications follow easily. Moreover, if there exists x in W s.t. $\mu(x) = \infty$, then $\mu(y) = \infty$ for every y in W, if there exists y in W s.t. $\mu(y) = 0$, then $\mu(x) = 0$ for every x in W, or else $0 < \mu < \infty$ on W.   ∎

3.2.2   Canonical invariant measure

The strong Markov property naturally leads to decompose a Markov chain started at a *recurrent* state x into its excursions from x, which are i.i.d. The number of visits to a point y during the first excursion can be written indifferently as

$$\sum_{n=0}^{R_x-1} \mathbb{1}_{\{X_n=y\}} = \sum_{n=1}^{R_x} \mathbb{1}_{\{X_n=y\}} \cdot \qquad (\mathbb{P}(R_x < \infty) = 1).$$

These two sums are in correspondence by a step of the chain, and we will see that an invariant measure is obtained by taking expectations.

Theorem 3.2.2 *Let* $(X_n)_{n\geq 0}$ *be a Markov chain, and* $R_x = \inf\{n \geq 1 : X_n = x\}$ *for* x *in* V. *Then,* $\mu_x = (\mu_x(y))_{y\in V}$ *given (with values in* $[0, \infty]$*) by*

$$\mu_x(y) = \mathbb{E}_x\left(\sum_{n=0}^{R_x-1} \mathbb{1}_{\{X_n=y\}}\right) = \sum_{n=0}^{\infty} \mathbb{P}_x(R_x > n, X_n = y)$$

is a superinvariant measure satisfying $\mu_x(x) = 1$ *and* $\|\mu_x\|_{var} = \mathbb{E}_x(R_x) \in [0, \infty]$, *and* $0 < \mu_x(y) < \infty$ *if* $x \to y$ *or else* $\mu_x(y) = 0$. *Moreover, it is an invariant measure if and only if* x *is recurrent, and then it is called the* canonical invariant measure *generated at* x.

Proof: As $\{R_x > n\} \in \mathcal{F}_n$, Theorem 2.1.1 yields that

$$\mu_x P(y) = \sum_{z \in V} \sum_{n=0}^{\infty} \mathbb{P}_x(R_x > n, X_n = z) P(z, y)$$

$$= \sum_{n=0}^{\infty} \sum_{z \in V} \mathbb{P}_x(R_x > n, X_n = z, X_{n+1} = y)$$

$$= \sum_{n=0}^{\infty} \mathbb{P}_x(R_x > n, X_{n+1} = y)$$

$$= \mathbb{E}_x \left(\sum_{n=1}^{R_x} \mathbb{1}_{\{X_n = y\}} \right) \leq \mu_x(y)$$

with equality if and only if $x \neq y$ or $\mathbb{P}_x(R_x < \infty) = 1$. Moreover, $\mu_x(x) = \mathbb{E}_x(1) = 1$ and Lemma 3.2.1 yields that if $x \to y$, then $0 < \mu(y) < \infty$. Clearly, if $x \not\to y$, then $\mu_x(y) = 0$. Moreover, in $[0, \infty]$,

$$\|\mu_x\|_{var} = \sum_{y \in V} \mathbb{E}_x \left(\sum_{n=0}^{R_x - 1} \mathbb{1}_{\{X_n = y\}} \right) = \mathbb{E}_x \left(\sum_{n=0}^{R_x - 1} \sum_{y \in V} \mathbb{1}_{\{X_n = y\}} \right) = \mathbb{E}_x(R_x) .$$

∎

The canonical invariant measure is above all a theoretical tool and is usually impossible to compute. It has just been used to prove that any Markov chain with a recurrent state has an invariant measure. It will be used again in the following uniqueness theorem, the proof of which is due to C. Derman, and uses a minimality result quite similar to Theorem 2.2.2.

Theorem 3.2.3 *Let* P *be an irreducible recurrent transition matrix. Then, the canonical invariant measure is the unique superinvariant measure of* P *and, in particular, its unique invariant measure.*

Proof: Let μ be a superinvariant measure. Let us first prove that $\mu \geq \mu(x)\mu_x$. Clearly, $\mu(x) = \mu(x)\mu_x(x)$. If $y \neq x$, then

$$\mu(y) \geq \sum_{z \in V} \mu(z) P(z, y) = \mu(x) P(x, y) + \sum_{z \neq x} \mu(z) P(z, y)$$

and iteration and use of

$$\sum_{x_1, \cdots, x_{k-1} \neq x} P(x, x_{k-1}) \cdots P(x_1, y) = \mathbb{P}_x(R_x > k, X_k = y)$$

yields for $k \geq 1$ that

$$\mu(y) \geq \mu(x) \sum_{n=0}^{k} \mathbb{P}_x(R_x > n, X_n = y) + \sum_{x_1, \cdots, x_k \neq x} \mu(x_k)P(x_k, x_{k-1}) \cdots P(x_1, y) .$$

Letting k go to infinity yields that

$$\mu(y) \geq \mu(x) \sum_{n=0}^{\infty} \mathbb{P}_x(R_x > n, X_n = y) = \mu(x)\mu_x(y) .$$

Moreover, $\mu_x = \mu_x P$ and thus the measure $\mu = \mu - \mu(x)\mu_x \geq 0$ satisfies $\mu \geq \mu P$ and $\mu(x) = 0$ and then Lemma 3.2.1 yields that $\mu = 0$. ∎

3.2.3 Positive recurrence, invariant law criterion

Let $(X_n)_{n \geq 0}$ be a Markov chain, and $R_x = \inf\{n \geq 1 : X_n = x\}$ for x in $\mathcal{V}$. A recurrent state x is said to be either *null recurrent* or *positive recurrent* according to the alternative

x null recurrent $\iff$ $\mathbb{E}_x(R_x) = \infty$ and $\mathbb{P}_x(R_x < \infty) = 1$,

x positive recurrent $\iff$ $\mathbb{E}_x(R_x) < \infty$ (implying $\mathbb{P}_x(R_x < \infty) = 1$) .

A transition matrix or Markov chain is said to be positive recurrent if all the states are so. The following fundamental result establishes a strong link between this sample path property and an algebraic property.

Theorem 3.2.4 (Invariant law criterion) *An irreducible Markov chain $(X_n)_{n \geq 0}$ is positive recurrent if and only if there exists an invariant law $\pi = (\pi(x))_{x \in \mathcal{V}}$. Then, π is the unique invariant measure, $\pi > 0$, and $\mathbb{E}_x(R_x) = 1/\pi(x)$.*

Proof: Lemma 3.1.3 yields that all states are simultaneously either transient or recurrent. If there exists an invariant law π, then Lemma 3.2.1 yields that $\pi > 0$, and Lemma 3.1.1 that the chain is recurrent. For $x \in \mathcal{V}$, Theorem 3.2.3 yields that $\pi = \pi(x)\mu_x$ is the unique invariant measure, and thus $1 = \pi(x)\|\mu_x\|_{\text{var}}$ and hence, $\mathbb{E}_x(R_x) = \|\mu_x\|_{\text{var}} = 1/\pi(x) < \infty$ so that x is positive recurrent.

Conversely, if x is positive recurrent, then an invariant law $\pi = \mu_x/\mathbb{E}_x(R_x) = \mu_x/\|\mu_x\|_{\text{var}}$ is obtained by normalizing μ_x (see Theorem 3.2.2). ∎

Corollary 3.2.5 *If $x \to y$ and x is positive recurrent, then y is positive recurrent. An irreducible class or chain is positive recurrent as soon as one of its states is so. A recurrent class which is not positive recurrent will be said to be null recurrent, and then all the states will be so.*

Proof: If x is recurrent and $x \to y$, then Lemma 3.1.3 yields that y is recurrent, and clearly x and y are in the same recurrent class, and the chain can be restricted to it and the invariant criterion applied (Theorem 3.2.4). ∎

Remark 3.2.6 *An irreducible chain with an invariant measure of infinite mass cannot be positive recurrent.*

3.2.3.1   Finite state space

The following simple corollary is very important in practice and can readily be proved directly.

Corollary 3.2.7 *A Markov chain with values in a finite state space $\mathcal{V}$ has at least an invariant law and a positive recurrent state. If moreover the chain is irreducible, it is positive recurrent.*

Proof: Lemma 3.1.2 yields that there is at least one recurrent state x. Let μ_x be the canonical invariant measure generated at x. As $\mathcal{V}$ is finite,

$$\mathbb{E}_x(R_x) = \|\mu_x\|_{\mathrm{var}} = \sum_{y \in \mathcal{V}} \mu_x(y) < \infty$$

and hence, x is positive recurrent and $\pi = \mu_x / \|\mu_x\|_{\mathrm{var}}$ is an invariant law.   ∎

3.2.3.2   Mean return time in a finite set

We give an extension of this result, which is quite useful for certain positive recurrence criteria.

Lemma 3.2.8 *Let $(X_n)_{n \geq 0}$ be a Markov chain on $\mathcal{V}$. If there exists a finite nonempty subset F of $\mathcal{V}$ s.t. $\mathbb{E}_x(R_F) < \infty$ for every x in F, then there exists a positive recurrent state in F.*

Proof: Lemma 3.1.2 yields that there is a recurrent state x in F. We restrict the chain to the corresponding recurrent class. So we now consider an irreducible recurrent chain, and Lemma 3.1.3 yields that the sequence $(X_n^F)_{n \geq 0}$ of the successive visits of $(X_n)_{n \geq 0}$ in F, defined in (2.2.2) and thereafter, is well defined (infinite). Corollary 3.2.7 yields that it has an invariant law on the recurrent class, and hence, an invariant law $\hat{\pi}$ on $\mathcal{V}$ obtained by giving weight 0 to the states outside this class. Similarly to the invariant canonical measure, we set, for y in $\mathcal{V}$,

$$\mu(y) = \mathbb{E}_{\hat{\pi}}\left(\sum_{n=0}^{R_F - 1} \mathbb{1}_{\{X_n = y\}}\right) = \sum_{n=0}^{\infty} \mathbb{P}_{\hat{\pi}}(R_F > n, X_n = y) \in [0, \infty]$$

and a computation quite similar to the one in the proof of Theorem 3.2.2 yields that

$$\mu P(y) = \mathbb{E}_{\hat{\pi}}\left(\sum_{m=1}^{R_F} \mathbb{1}_{\{X_m = y\}}\right) = \mu(y) - \mathbb{P}_{\hat{\pi}}(X_0 = y) + \mathbb{P}_{\hat{\pi}}(X_1^F = y) = \mu(y)$$

since $\mathbb{P}_{\hat{\pi}}(X_1^F = y) = \mathbb{P}_{\hat{\pi}}(X_0 = y) = \hat{\pi}(y)$. Hence, μ satisfies $\mu = \mu P$ in $[0, \infty]$, and $\|\mu\|_{\text{var}} = \mathbb{E}_{\hat{\pi}}(R_F) = \sum_{x \in F} \hat{\pi}(x) \mathbb{E}_x(R_F)$. If $\mathbb{E}_x(R_F) < \infty$ for all x in the finite set F, then $\|\mu\|_{\text{var}} < \infty$, and then $\mu / \|\mu\|_{\text{var}}$ is an invariant law, and the invariant law criterion (Theorem 3.2.4) yields that the chain is recurrent positive. ∎

3.2.4   Detailed examples

3.2.4.1   Nearest-neighbor walk in one dimension

The nearest-neighbor random walk on $\mathbb{Z}$ with probability $p > 0$ of going to the right and $1 - p = q > 0$ of going to the left is an irreducible Markov chain (see Section 3.1.3). The local balance equations are given by

$$\mu(x)q = \mu(x - 1)p , \qquad x \in \mathbb{Z} ,$$

and have solution $(\mu(0)(p/q)^x)_{x \in \mathbb{Z}}$. The global balance equations are given by

$$\mu(x) = \mu(x - 1)p + \mu(x + 1)q , \qquad x \in \mathbb{Z} ,$$

and this second-order linear recursion has characteristic polynomial

$$qX^2 - X + p = q(X - 1)(X - p/q)$$

with roots 1 and p/q, possibly equal.

If $p \neq 1/2$, then the invariant measures are of the form $(\alpha + \beta(p/q)^x)_{x \in \mathbb{Z}}$ for all $\alpha \geq 0$ and $\beta \geq 0$ s.t. $(\alpha, \beta) \neq (0, 0)$. There is nonuniqueness of the invariant measure, and Theorem 3.2.3 yields that the random walk is transient.

If $p = q = 1/2$, then the general solution for the global balance equation is $(\alpha + \beta x)_{x \in \mathbb{Z}}$, and the nonnegative solutions are of constant equal to $\alpha > 0$. Hence, the uniform measure is the unique invariant measure, and as it is of infinite mass there is no invariant law. The invariant law criterion (Theorem 3.2.4) yields that this chain is not positive recurrent.

In Section 3.1.3, we have used the study of the unilateral hitting time to prove that for $p = q = 1/2$ the chain is recurrent, and hence, it is null recurrent.

We have given examples, when $p \neq 1/2$ in which there is nonuniqueness of the invariant measure, and when $p = 1/2$ in which the uniqueness result in Theorem 3.2.3 requires the nonnegativity assumption.

3.2.4.2   Symmetric random walk in many dimensions

For the symmetric random walk on $\mathbb{Z}^d$ for $d \geq 1$, the unique invariant measure is uniform. As it has infinite mass, there is no invariant law. The invariant law criterion (Theorem 3.2.4) yields that this chain is not positive recurrent.

We have used the Potential matrix criterion in Section 3.1.3 to prove that in dimension $d = 1$ and $d = 2$ the random walk is recurrent, and hence null recurrent, whereas in dimension $d \geq 3$ it is transient.

3.2.4.3 Nearest-neighbor walk in one dimension reflected at 0

Let us consider the random walk on $\mathbb{N}$ reflected at 0, for which $P(x, x+1) = p > 0$ and $P(x, x-1) = q = 1 - p > 0$ for $x \geq 1$ and at the boundary $P(0,1) = r > 0$ and $P(0,0) = 1 - r \geq 0$, with graph

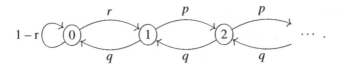

Two cases of particular interest are $r = p$ and $r = 1$.

The global balance equations are given by

$$\mu(0)r = \mu(1)q , \quad \mu(1) = \mu(0)r + \mu(2)q , \quad \mu(x) = \mu(x-1)p + \mu(x+1)q , \quad x \geq 2 .$$

Hence, $\mu(1) = \mu(0)r/q$ and then $\mu(2) = (\mu(1) - \mu(0)r)/q = \mu(0)rp/q^2$, and clearly the recursion then determines $\mu(x)$ for $x \geq 2$, so that there is uniqueness of the invariant measure.

The values $\mu(x)$ for $x \geq 3$ could be determined under the general form $\alpha + \beta(p/q)^x$ if $p \neq q$ or $\alpha + \beta x$ if $p = q = 1/2$, by determining α and β using the values for $x = 1$ and $x = 2$, but it is much simpler to use the local balance equations.

The local balance equations are

$$\mu(0)r = \mu(1)q , \qquad \mu(x-1)p = \mu(x)q , \quad x \geq 2 ,$$

and thus $\mu(1) = \mu(0)r/q$ and $\mu(x) = \mu(1)(p/q)^{x-1}$ for $x \geq 2$, and hence,

$$\mu(x) = \mu(0) \left(\frac{r}{q}\right) \left(\frac{p}{q}\right)^{x-1} = \mu(0) \left(\frac{r}{p}\right) \left(\frac{p}{q}\right)^{x} , \qquad x \geq 1 .$$

For $r = p$, this is a geometric sequence.

If $p < 1/2$, then $\|\mu\|_{\mathrm{var}} < \infty$ and the invariant law criterion (Theorem 3.2.4) yields that the chain is positive recurrent. For $\mu(0) = 1$,

$$\|\mu\|_{\mathrm{var}} = 1 + \frac{r}{q} \sum_{x \geq 1} \left(\frac{p}{q}\right)^{x-1} = 1 + \frac{\frac{r}{q}}{1 - \frac{p}{q}} = \frac{q - p + r}{q - p} ,$$

and the unique invariant law is given by $\pi = \mu/\|\mu\|_{\mathrm{var}} = \pi(0)\mu$. For $r = 1$, it holds that

$$\pi(0) = \frac{1}{2}\left(1 - \frac{p}{q}\right) , \qquad \pi(x) = \frac{1}{2p}\left(1 - \frac{p}{q}\right)\left(\frac{p}{q}\right)^{x} \quad x \geq 1 ,$$

and for $r = p$ we obtain the geometric law $\pi(x) = (1 - \frac{p}{q})(\frac{p}{q})^x$ for $x \in \mathbb{N}$.

If $p \geq 1/2$, then this measure has infinite mass, and the chain cannot be positive recurrent. The results on the unilateral hitting time, or on the nearest-neighbor random walk on $\mathbb{Z}$, yield that the chain is null recurrent if $p = 1/2$ and transient if $p > 1/2$.

3.2.4.4   Ehrenfest Urn

See Section 1.4.4. The chains $(X_n)_{n\in\mathbb{N}}$ on $\{0,1\}^N$ and $(S_n)_{n\in\mathbb{N}}$ on $\{0,\dots,N\}$ are irreducible, and Corollary 3.2.7 yields that they are positive recurrent.

We have seen that the invariant law of $(X_n)_{n\in\mathbb{N}}$ is the uniform law on $\{0,1\}^N$, and deduced from this that the invariant law of $(S_n)_{n\in\mathbb{N}}$ is binomial $B(N,1/2)$, given by

$$\beta(x) = 2^{-N}\binom{N}{x}, \qquad x \in \{0,1,\dots,N\}\ .$$

We also can recover the invariant law of $(S_n)_{n\in\mathbb{N}}$ by solving the local balance equation, given by

$$\frac{x}{N}\mu(x) = \frac{N-x+1}{N}\mu(x-1)\ , \qquad x \in \{1,2,\cdots,N\}\ ,$$

see (1.4.4), and hence,

$$\mu(x) = \frac{N-x+1}{x}\mu(x-1) = \cdots = \frac{N(N-1)\dots(N-x+1)}{x!}\mu(0) = \binom{N}{x}\mu(0)\ .$$

Finding the invariant law β reduces now to computing the normalizing constant

$$\sum_{x=0}^{N}\binom{N}{x} = (1+1)^N = 2^N\ ,$$

and we again find that $\beta(x) = 2^{-N}\binom{N}{x}$.

Such a normalizing problem happens systematically, and may well be untractable, be it the computation of an infinite sum or of a huge combinatorial finite sum.

Mean time to return to vacuum Starting at $S_0 = 0$, the mean waiting time before compartment 1 is empty again is given by

$$\mathbb{E}_0(R_0) = \beta(0)^{-1} = 2^N\ .$$

This is absolutely enormous for N of the order of the Avogadro's number. Compared to it, the duration of the universe is absolutely negligible (in an adequate timescale).

Mean time to return to balanced state Consider state $\lfloor N/2 \rfloor$, which is a well-balanced state. It is the mean value of S_n in equilibrium if N is even, or else the nearest integer below it. According to whether N is even or odd,

$$\beta(\lfloor N/2 \rfloor) = 2^{-N}\binom{N}{\lfloor N/2 \rfloor} = 2^{-N}\frac{N!}{(N/2)!^2} \quad \text{or} \quad 2^{-N}\frac{N!}{(\frac{N-1}{2})!\,(\frac{N+1}{2})!}$$

and the Stirling formula $M! \sim \sqrt{2\pi} \, M^{M+1/2} e^{-M}$ yields that

$$\beta(\lfloor N/2 \rfloor) \sim 2^{-N} \frac{1}{\sqrt{2\pi}} \frac{N^{N+1/2}}{(N/2)^{N+1}} = \sqrt{\frac{2}{\pi N}} . \tag{3.2.6}$$

Thus

$$\mathbb{E}_{\lfloor N/2 \rfloor}(R_{\lfloor N/2 \rfloor}) = \beta(\lfloor N/2 \rfloor)^{-1} \sim \sqrt{N\pi/2} \ll \mathbb{E}_0(R_0) = 2^N ,$$

and $\mathbb{E}_{\lfloor N/2 \rfloor}(R_{\lfloor N/2 \rfloor}$ is even small compared to the number of molecules N, the inverse of which should give the order of magnitude of the time-step.

Refutation of the Refutation of statistical mechanics   This model was given by the Ehrenfest spouses as a refutation of critiques of statistical mechanics based on the fact that such a random evolution would visit all states infinitely often, even the less likely ones.

We see how important it is to obtain a wholly explicit invariant law. In particular it is important to compute the normalizing factor for a known invariant measure, or at least to derive a good approximation for it. This is a classic difficulty encountered in statistical mechanics in order to obtain useful results.

3.2.4.5   Renewal process

See Section 1.4.5. Assume that $(X_n)_{n \geq 0}$ is irreducible. A necessary and sufficient condition for $(X_n)_{n \geq 0}$ to be recurrent is that

$$\mathbb{P}(D = \infty) = \lim_{x \to \infty} p_0 \cdots p_{x-1} = 0 ,$$

and this is also the necessary and sufficient condition for the existence of an invariant measure. We have given an explicit expression for the invariant measure when it exists, and shown that then it is unique.

A necessary and sufficient condition for the existence of an invariant law π is that

$$\mathbb{E}(D) = \sum_{x \to \infty} p_0 \cdots p_{x-1} < \infty ,$$

and we have given an explicit expression for π when it exists. By the invariant law criterion, this is also a necessary and sufficient condition for positive recurrence.

All this is actually obvious, as $D_1 = R_0$ when $X_0 = 0$.

Note that if $\mathbb{P}(D = \infty) > 0$, then the renewal process is an example of an irreducible Markov chain having no invariant measure.

3.2.4.6   Word search

See Section 1.4.6. The chain is irreducible on a finite state space, so that Corollary 3.2.7 of the invariant law criterion yields that it is positive recurrent. Its invariant law was computed at the end of Section 1.2.

3.2.4.7   Snake chain

See Section 1.4.6, where we proved that if P has an invariant law then so does P_ℓ. The invariant law criterion yields that if P is irreducible positive recurrent on $\mathcal{V}$, then so is P_ℓ on its natural state space $\mathcal{V}_\ell$. It is clear that if x_1 is null recurrent for P, then $(x_1, \ldots, x_\ell)$ cannot be positive recurrent for P_ℓ, and hence is null recurrent.

3.2.4.8   Product chain

See Section 1.4.7. One must be well aware that P_1 and P_2 may be irreducible without $P_1 \otimes P_2$ being so.

The decomposition in recurrent classes and the invariant law criterion yield that if for $i \in \{1, 2\}$ every state is positive recurrent for P_i, then there exists invariant laws $\pi_i > 0$ for P_i. Then, $\pi_1 \otimes \pi_2 > 0$ is an invariant law for $P_1 \otimes P_2$, which is hence positive recurrent.

If for $i \in \{1, 2\}$ every state is null recurrent for P_i, then $P_1 \otimes P_2$ may be either transient or null recurrent (see Exercise 3.2).

3.3   Complements

This is a section giving some openings toward theoretical and practical tools for the study of Markov chains in the perspective of this chapter.

3.3.1   Hitting times and superharmonic functions

3.3.1.1   Superharmonic, subharmonic, and harmonic functions

A function f on $\mathcal{V}$ is said to be *harmonic* if $f = Pf$, or equivalently if $(I - P)f = 0$, that is, if it is an eigenvector of P for the eigenvalue 1. A function f is said to be *superharmonic* if $f \geq Pf$ and to be *subharmonic* if $f \leq Pf$.

Note that a function f is *subharmonic* if and only if $-f$ is *superharmonic* and *harmonic* if and only if f is both *superharmonic* and *subharmonic*.

The constant functions are harmonic, and a natural question is whether these are the only harmonic functions. Theorem 2.2.2 will be very useful in this perspective, for instance, it yields that

$$w^E : x \mapsto \mathbb{P}_x(S_E < \infty) \leq 1$$

is the least nonnegative superharmonic function, which is larger than 1 on E.

Theorem 3.3.1 *Let P be an irreducible transition matrix. Then, P is recurrent if and only if the only nonnegative superharmonic functions are the constant functions. In this statement, "nonnegative superharmonic" can be replaced by "lower-bounded superharmonic" or "upper-bounded subharmonic."*

Proof: The "one step forward" method yields that

$$\mathbb{P}_x(R_x < \infty) = \sum_{y \in \mathcal{V}} P(x, y) w^y(x) \, .$$

If x is transient, then $\mathbb{P}_x(R_x < \infty) < 1$ and thus there exists y s.t. $w^y(x) < 1$, whereas $w^y(y) = 1$, so that w^y is a nonnegative superharmonic function which is not constant. Conversely, if f is a nonconstant nonnegative superharmonic function, then there exists x and y s.t. $f(y) > f(x) \geq 0$. By dividing f by $f(y)$, we may assume that $f(y) = 1$. Theorem 2.2.2 yields that $f \geq w^y$ and hence that $1 = f(y) > f(x) \geq w^y(x)$, so that $\mathbb{P}_x(R_x < \infty) < 1$ and x is transient.

A function f is lower-bounded superharmonic if and only if $f - \inf f$ is nonnegative superharmonic and if and only if $-f$ is upper-bounded subharmonic. ∎

This theorem recalls Theorem 3.2.3, and we will develop in Lemma 3.3.9 and thereafter an appropriate duality to deduce one from the other. Care must be taken, as there exists transient transition matrices without any invariant measure, see the renewal process for $\mathbb{P}(D = \infty) > 0$; others with a unique invariant measure, see the nearest-neighbor random walks reflected at 0 on $\mathbb{N}$; and others with nonunique invariant laws, see the nearest-neighbor random walks on $\mathbb{Z}$.

3.3.1.2   Supermartingale techniques

It is instructive to give two proofs of the "only if" part of Theorem 3.3.1 (the most difficult) using supermartingale concepts.

We assume that f is nonnegative superharmonic and that P is irreducible recurrent. A fundamental observation is that if $(X_n)_{n \in \mathbb{N}}$ is a Markov chain and f is superharmonic, then $(f(X_n))_{n \in \mathbb{N}}$ is a supermartingale.

First Proof This uses a deep result of martingale theory, the Doob convergence theorem, which yields that the nonnegative supermartingale $(f(X_n))_{n \in \mathbb{N}}$ converges, a.s. Moreover, as P is recurrent, $(f(X_n))_{n \in \mathbb{N}}$ visits infinitely often $f(x)$ for every x in $\mathcal{V}$. This is only possible if f is constant.

Second Proof (J.L. Doob) This uses more elementary results. Let x and y be in $\mathcal{V}$, and

$$S = S_y = \inf\{n \geq 0 : X_n = y\} .$$

Then, $(f(X_n))_{n \in \mathbb{N}}$ and the stopped process $(f(X_{S \wedge n}))_{n \in \mathbb{N}}$ are supermartingales, and thus

$$\mathbb{E}_x(f(X_{S \wedge n})) \leq f(x) = \mathbb{E}_x(f(X_{S \wedge 0})) ,$$

and as $\mathbb{P}_x(S < \infty) = 1$ by Lemma 3.1.3, the Fatou lemma (Lemma A.3.3) yields that

$$f(y) = \mathbb{E}_x\left(\lim_{n \to \infty} f(X_{S \wedge n})\right) \leq \lim_{n \to \infty} \inf \mathbb{E}_x(f(X_{S \wedge n})) \leq f(x) .$$

Hence, f is constant, as x and y are arbitrary.

In this and the following proofs, we have elected to use results in Markov chain theory such as Lemma 3.1.3 as much as possible but could replace them by the Doob convergence theorem to prove convergences.

3.3.2   Lyapunov functions

The main intuition behind the second proof is that the supermartingale $(f(X_n))_{n\in\mathbb{N}}$ has difficulty going uphill in the mean.

This leads naturally to the notion of Lyapunov function, which is a function of which the behavior on sample paths allows to determine the behavior of the latter: go to infinity and then the chain is transient, come always back to a given finite set and then the chain is recurrent, do so quickly and then the chain is positive recurrent.

We give some examples of such results among a wide variety, in a way greatly inspired by the presentation by Robert, P. (2003).

3.3.2.1   Transience and nonpositive-recurrence criteria

A simple corollary of the "only if" part of Theorem 3.3.1 allows to get rid of a subset E of the state space $\mathcal{V}$ in which the Markov chain behaves "poorly."

Corollary 3.3.2 *Let P be an irreducible transition matrix. If there exists a nonempty subset E of $\mathcal{V}$ and a function ϕ on $\mathcal{V}$ satisfying*

$$\sup_E \phi < \sup \phi < \infty , \quad \mathbb{E}_x(\phi(X_1)) - \phi(x) := P\phi(x) - \phi(x) \geq 0 , \quad x \notin E ,$$

then P is transient.

Proof: Let $\hat{E} := \{x \in \mathcal{V} : \phi(x) \leq \sup_E \phi\}$ and $\hat{\phi}$ with value $\sup_E \phi$ on $\hat{E}$ and equal to ϕ on $\mathcal{V} - E$. Clearly, $\hat{\phi}$ is upper-bounded subharmonic and is not constant, and Theorem 3.3.1 implies that P is transient.   ∎

It is a simple matter to adapt the proof using Theorem 2.2.2 or the second supermartingale proof for Theorem 3.3.1, and this is left as an exercise.

Note that $\hat{\phi} - \sup_E \phi \geq 0$ and that we may assume that $\phi \geq 0$. The functions under consideration are basically upper-bounded subharmonic and nonnegative.

The sequel is an endeavor to replace the upper-bound assumption by integrability assumptions. Let us start with a variant of results due to J. Lamperti.

Theorem 3.3.3 (Lamperti criterion) *Let P be an irreducible transition matrix. If there exists a nonempty subset E of $\mathcal{V}$ and a function ϕ on $\mathcal{V}$ satisfying for some $\varepsilon > 0$ that*

1. $\mathbb{E}_x(\phi(X_1)) - \phi(x) := P\phi(x) - \phi(x) \geq \varepsilon , \quad x \notin E ,$

2. $\displaystyle\sup_{x\notin E} \mathbb{E}_x((\phi(X_1) - \phi(x))^2) := \sup_{x\notin E} \sum_{y\in\mathcal{V}} P(x,y)(\phi(y) - \phi(x))^2 < \infty ,$

then P is transient.

Proof: We may assume that $\phi \geq 0$ by replacing ϕ with $\hat{\phi} - \sup_E \phi$, see the previous proof. For $a > 0$, we define an approximation of ϕ by

$$\phi_a : x \in \mathcal{V} \mapsto a - \frac{a^2}{a + \phi(x)} = \frac{a\phi(x)}{a + \phi(x)}, \qquad \sup_E \phi_a < \sup \phi_a < a .$$

For $x \notin E$, using twice that $\frac{1}{a+\phi(x)} - \frac{1}{a+\phi(y)} = \frac{\phi(y)-\phi(x)}{(a+\phi(x))(a+\phi(y))}$,

$$P\phi_a(x) - \phi_a(x) = \frac{a^2}{a + \phi(x)} \sum_{y \in \mathcal{V}} P(x,y) \frac{\phi(y) - \phi(x)}{a + \phi(y)}$$

$$= \frac{a^2}{(a + \phi(x))^2} \left[(P\phi - \phi)(x) - \sum_{y \in \mathcal{V}} P(x,y) \frac{(\phi(y) - \phi(x))^2}{a + \phi(y)} \right]$$

$$\geq \frac{a^2}{(a + \phi(x))^2} \left(\varepsilon - \frac{1}{a} \sum_{y \in \mathcal{V}} P(x,y)(\phi(y) - \phi(x))^2 \right)$$

and if $a \geq \frac{1}{\varepsilon} \sup_{x \notin E} \sum_{y \in \mathcal{V}} P(x,y)(\phi(y) - \phi(x))^2$, then $P\phi_a(x) - \phi_a(x) \geq 0$. We conclude by Corollary 3.3.2. ∎

The next criterion, due to R. Tweedie, uses a submartingale and a direct computation of L^1 convergence.

Theorem 3.3.4 (Tweedie criterion) *Let P be an irreducible transition matrix. If there exists a nonempty subset E of $\mathcal{V}$ and a function ϕ on $\mathcal{V}$ satisfying*

(1) $\sup_E \phi < \sup \phi$, $\mathbb{E}_x(\phi(X_1)) - \phi(x) := P\phi(x) - \phi(x) \geq 0$, $x \notin E$,

(2) $\sup_{x \notin E} \mathbb{E}_x(|\phi(X_1) - \phi(x)|) := \sup_{x \notin E} \sum_{y \in \mathcal{V}} P(x,y)|\phi(y) - \phi(x)| < \infty$,

then P cannot be positive recurrent.

Proof: If P is transient, this is obvious, and we assume that P is recurrent. Hence, $\mathbb{P}_x(S < \infty) = 1$ for $S = S_E = \inf\{n \geq 0 : X_n \in E\}$. For x in $\mathcal{V}$ and $n \geq 1$,

$$\mathbb{E}_x(\phi(X_{S \wedge n})) - \phi(x) = \sum_{k=0}^{n-1} \mathbb{E}_x((\phi(X_{k+1}) - \phi(X_k))\mathbb{1}_{\{S>k\}})$$

$$= \sum_{k=0}^{n-1} \sum_{y \notin E} \mathbb{E}_x((\phi(X_{k+1}) - \phi(X_k))\mathbb{1}_{\{S>k,X_k=y\}})$$

$$= \sum_{k=0}^{n-1} \sum_{y \notin E} \mathbb{P}_x(S > k, X_k = y)\mathbb{E}_y(\phi(X_1) - \phi(y))$$

and hence, $\mathbb{E}_x(\phi(X_{S \wedge n})) \geq \phi(x)$. If $x \notin E$ is s.t. $\phi(x) > \sup_E \phi$ and hence $(\mathbb{E}_x(\phi(X_{S \wedge n})))_{n \geq 0}$ cannot converge to $\mathbb{E}_x(\phi(X_S)) \leq \sup_E \phi$, then

$$\mathbb{E}_x(|\phi(X_S) - \phi(X_{S \wedge n})|) \leq \sum_{k=n}^{\infty} \mathbb{E}_x(|\phi(X_{k+1}) - \phi(X_k)| \mathbb{1}_{\{S > k\}})$$

$$\leq \sum_{k=n}^{\infty} \sum_{y \notin E} \mathbb{P}_x(S > k, X_k = y) \mathbb{E}_y(|\phi(X_1) - \phi(y)|)$$

$$\leq \sup_{y \notin E} \mathbb{E}_y(|\phi(X_1) - \phi(y)|) \sum_{k=n}^{\infty} \mathbb{P}_x(S > k)$$

yields that $\lim_{n \to \infty} \sum_{k=n}^{\infty} \mathbb{P}_x(S > k) \neq 0$, and hence that

$$\mathbb{E}_x(S) = \sum_{k=0}^{\infty} \mathbb{P}_x(S > k) = \infty .$$

As P is irreducible, $\mathbb{P}_x(R_x < S) < 1$, the strong Markov property implies that

$$\mathbb{E}_x(S) = \sum_{k=0}^{\infty} \mathbb{P}_x(R_x < S)^k \mathbb{P}_x(R_x \geq S)(k\mathbb{E}_x(R_x \mid R_x < S) + \mathbb{E}_x(S \mid R_x \geq S))$$

and hence, $\mathbb{E}_x(R_x \mid R_x < S) = \infty$ or $\mathbb{E}_x(R_x \mid R_x \geq S) \geq \mathbb{E}_x(S \mid R_x \geq S) = \infty$, and we conclude that $\mathbb{E}_x(R_x) = \infty$, and P is not positive recurrent. ∎

3.3.2.2 Positive recurrence criteria

Such criteria cannot be based on solid results such as Theorem 3.3.1 and are more delicate. We are back to functions that are basically nonnegative superharmonic and to supermartingales.

Theorem 3.3.5 *Let P be an irreducible transition matrix. If there exists a nonempty finite subset F of $\mathcal{V}$ and a function ϕ on $\mathcal{V}$ satisfying*

$$Card\{x \in \mathcal{V} : \phi(x) \leq K\} < \infty , \qquad \forall K > 0 ,$$

and

$$\mathbb{E}_x(\phi(X_1)) - \phi(x) := P\phi(x) - \phi(x) \leq 0 , \qquad x \notin F ,$$

then P is recurrent.

Proof: We may assume that $\phi \geq 0$ by adding an appropriate constant. Let $S = S_F$ and $Y_n = \phi(X_n) \mathbb{1}_{\{S > n\}}$. Then, $Y_{n+1} \leq \phi(X_{n+1}) \mathbb{1}_{\{S > n\}}$ and

$$\mathbb{E}(\phi(X_{n+1})\mathbb{1}_{\{S>n\}}) = \sum_{x\notin F} \mathbb{E}(\phi(X_{n+1})\mathbb{1}_{\{S>n,X_n=x\}})$$

$$= \sum_{x\notin F} \mathbb{P}(S > n, X_n = x)\mathbb{E}_x(\phi(X_1))$$

$$\leq \sum_{x\notin F} \mathbb{P}(S > n, X_n = x)\phi(x) = \mathbb{E}(\phi(X_n)\mathbb{1}_{\{S>n\}}) = \mathbb{E}(Y_n)$$

and hence, $\mathbb{E}(Y_{n+1}) \leq \mathbb{E}(Y_n)$. The Fatou Lemma (Lemma A.3.3) yields that

$$\mathbb{E}_x\left(\lim_{n\to\infty}\inf Y_n\right) \leq \lim_{n\to\infty}\inf \mathbb{E}_x(Y_n) \leq \mathbb{E}_x(Y_0) \leq \phi(x) < \infty .$$

By contradiction, if $(X_n)_{n\geq 0}$ were transient, then Lemma 3.1.2 would imply that $\mathbb{P}_x(S = \infty) > 0$, and on $\{S = \infty\}$, the chain would visit only a finite number of times each state so that $\lim_{n\to\infty} Y_n = \infty$, and thus $\mathbb{E}_x(\lim\inf_{n\to\infty} Y_n) = \infty$, which is false. Thus, $(X_n)_{n\geq 0}$ is recurrent. ∎

Theorem 3.3.5 in conjunction with Theorem 3.3.4 provides a null recurrence criterion. Under stronger assumptions, a positive recurrence criterion is obtained.

Theorem 3.3.6 (Foster criterion) *Let P be an irreducible transition matrix. If there exists a nonempty finite subset F of $\mathcal{V}$ and a nonnegative function ϕ on $\mathcal{V}$ satisfying for some $\varepsilon > 0$ that*

(1) $\mathbb{E}_x(\phi(X_1)) - \phi(x) := P\phi(x) - \phi(x) \leq -\varepsilon ,$ $x \notin F ,$
(2) $\mathbb{E}_x(\phi(X_1)) := P\phi(x) < \infty ,$ $x \in F ,$

then P is positive recurrent. Moreover,

$$\mathbb{E}_x(S_F) \leq \frac{1}{\varepsilon}\phi(x) , \qquad \forall x \notin F ,$$

and this result remains true under only hypothesis (1) even for infinite F.

Proof: As in the preceding proof,

$$\mathbb{E}(Y_{n+1}) \leq \sum_{x\notin F} \mathbb{P}(S > n, X_n = x)(\phi(x) - \varepsilon) = \mathbb{E}(Y_n) - \varepsilon\mathbb{P}(S > n) .$$

Thus,

$$0 \leq \mathbb{E}(Y_{n+1}) \leq \mathbb{E}(Y_n) - \varepsilon\mathbb{P}(S > n) \leq \cdots \leq \mathbb{E}(Y_0) - \varepsilon\sum_{k=0}^{n} \mathbb{P}(S > k)$$

and hence, for every x in $\mathcal{V}$,

$$\mathbb{E}_x(S) = \sum_{k=0}^{\infty} \mathbb{P}_x(S > k) \leq \frac{1}{\varepsilon}\mathbb{E}_x(Y_0) \leq \frac{1}{\varepsilon}\phi(x) < \infty \ .$$

If x is in F, then the "one step forward" method and $P\phi < \infty$ yield that

$$\mathbb{E}_x(R_F) = 1 + \sum_{y \notin F} P(x,y)\mathbb{E}_y(S_F) \leq 1 + \frac{1}{\varepsilon} \sum_{y \notin F} P(x,y)\phi(y) < \infty \ ,$$

and as F is finite, Lemma 3.2.8 yields that P is positive recurrent.   ∎

3.3.2.3 Queuing application

A system (for instance a processor) processes jobs (such as computations) in synchronized manner. A waiting room (buffer) allows to store jobs before they are processed. The instants of synchronization are numbered $n = 0, 1, 2, \ldots$, and X_n denotes the number of jobs in the system just after time n.

Between time $n - 1$ and time n, a random number A_n of new jobs arrive, and up to a random number D_n of the X_{n-1} jobs already there can be processed, so that

$$X_n = (X_{n-1} - D_n)^+ + A_n \ , \qquad n \geq 1 \ .$$

In a simple special case, there is an integer K s.t. $\mathbb{P}(D_n = K) = 1$, and it is said that the queue has K servers.

The r.v. $(A_n, D_n)_{n \geq 1}$ are assumed to be i.i.d. and independent of X_0, and then Theorem 1.2.3 yields that $(X_n)_{n \geq 0}$ is a Markov chain on $\mathbb{N}$. We assume that

$$\mathbb{P}(A_1 > D_1) > 0 \ , \qquad \mathbb{P}(A_1 = 0, D_1 > 0) > 0 \ ,$$

so that there is a closed irreducible class containing 0, and we consider the chain restricted to this class, which is irreducible. We further assume that A_1 and D_1 are integrable, and will see that then the behavior of $(X_n)_{n \geq 0}$ depends on the joint law of (A_1, D_1) essentially only through $\mathbb{E}(A_1) - \mathbb{E}(D_1) := \mu$.

For $X_0 = x$, we have $X_1 \geq x - D_1 + A_1$ and hence, $\mathbb{E}_x(X_1) - x \geq \mu$, and $\lim_{x \to \infty} X_1 = x - D_1 + A_1$ a.s., and $|X_1 - x| \leq \max(A_1, D_1)$, and thus $\lim_{x \to \infty} \mathbb{E}_x(X_1 - x) = \mu$ by dominated convergence (Theorem A.3.5). We deduce a number of facts from this observation.

- If $\mu < 0$, then there exists $x_0 \geq 0$ s.t. if $x > x_0$ then $\mathbb{E}_x(X_1) - x \leq \mu/2$. Then, the Foster criterion (Theorem 3.3.6) with $F = \{0, 1, \ldots, x_0\}$ and $\phi(x) = x$ yields that the chain is positive recurrent, and $\mathbb{E}_x(S_F) \leq -2x/\mu$.

- If $\mu \geq 0$, then the Tweedie criterion (Theorem 3.3.4) with $E = \{0\}$ and $\phi(x) = x$ yields that the chain cannot be positive recurrent.

- If $\mu > 0$ and A_1 and D_1 are square integrable, then the Lamperti criterion (Theorem 3.3.3) with $E = \{0\}$ and $\phi(x) = x$ yields that the chain is transient. If only D_1 is square integrable, then if a is large enough then

$$\mathbb{E}(a \wedge A_1) - \mathbb{E}_x(D_1) > 0 ,$$

  and the chain $(X_n^a)_{n \geq 0}$ with arrival and potential departures given by $(a \wedge A_n, D_n)_{n \geq 1}$ is transient, and as $X_n \geq X_n^a$ then $(X_n)_{n \geq 0}$ is transient.

- If $\mu = 0$ and $D_1 \leq x_0 \in \mathbb{N}^*$, then Theorem 3.3.5 with $F = \{0, 1, \dots, x_0 - 1\}$ and $\phi(x) = x$ yields that the chain is recurrent, and hence null recurrent.

3.3.2.4   Necessity of jump amplitude control

In Theorems 3.3.3 and 3.3.4, the hypotheses (1) are quite natural, but the hypotheses (2) controlling the jump amplitudes cannot be suppressed (but can be weakened).

We will see this on an example. Consider the queuing system with

$$\mathbb{P}((A_1, D_1) = (1, 0)) = p > 0 , \qquad \mathbb{P}((A_1, D_1) = (0, 1)) = q = 1 - p > 0 .$$

Then, $(X_n)_{n \geq 0}$ is s.t. $P(x, x + 1) = p$ and $P(x + 1, x) = P(0, 0) = q$ for $x \geq 0$ and is a random walk on $\mathbb{N}$ reflected at 0. For $F = \{0\}$ and $\phi(x) = a^x$ with $a > 1$ and $a \geq q/p$, it holds that $\phi(0) = 1$ and $\sup \phi = \infty$ and, for $x \notin F$,

$$\mathbb{E}_x(\phi(X_1)) - \phi(x) = pa^{x+1} + qa^{x-1} - a^x = p(a - 1)(a - q/p)a^{x-1} \geq 0 ,$$

so that hypothesis (1) of Theorem 3.3.4 is satisfied, and if $a > q/p$, then

$$\mathbb{E}_x(\phi(X_1)) - \phi(x) \geq p(a - 1)(a - q/p) > 0 , \qquad x \notin F ,$$

so that hypothesis (1) of Theorem 3.3.3 is satisfied.

Nevertheless, $\mu = p - q = 2p - 1$ and we have seen that $(X_n)_{n \geq 0}$ is positive recurrent for $p < 1/2$, null recurrent for $p = 1/2$, and transient for $p > 1/2$.

The hypothesis in Theorem 3.3.5 and hypothesis (1) in Theorem 3.3.6 are also quite natural. The latter is enough to obtain the bound $\mathbb{E}_x(S_F) \leq \frac{1}{\varepsilon}\phi(x)$ for $x \notin F$, but an assumption such as hypothesis (2) must be made to conclude to positive recurrence. For instance, let $(X_n)_{n \geq 0}$ be a Markov chain with matrix P s.t.

$$\sum_{x \in \mathbb{N}} xP(0, x) = \infty , \qquad P(x, x - 1) = 1 , \quad x \geq 1 .$$

For $F = \{0\}$ and $\phi(x) = x$, it holds that $\mathbb{E}_x(\phi(X_1)) - \phi(x) = -1$ for $x \notin F$, and hypothesis (1) of Theorem 3.3.6 holds, but the chain is null recurrent as

$$\mathbb{E}_0(S_0) = 1 + \sum_{x \in \mathbb{N}} xP(0, x) = \infty .$$

3.3.3   Time reversal, reversibility, and adjoint chain

3.3.3.1   Time reversal in equilibrium

Let $(X_n)_{n \geq 0}$ be a Markov chain, K an integer, and $\tilde{X}_n = X_{K-n}$ for $0 \leq n \leq K$. Then,

$$
\begin{aligned}
\mathbb{P}(\tilde{X}_0 = x_0, \dots, \tilde{X}_K = x_K) &= \mathbb{P}(X_0 = x_K, \dots, X_K = x_0) \\
&= \pi_0(x_K) P(x_K, x_{K-1}) \cdots P(x_1, x_0) \\
&= \pi_K(x_0) \frac{\pi_{K-1}(x_1)}{\pi_K(x_0)} \, P(x_1, x_0) \times \cdots \\
&\quad \times \frac{\pi_0(x_K)}{\pi_1(x_{K-1})} \, P(x_K, x_{K-1})
\end{aligned}
$$

and $(\tilde{X}_n)_{0 \leq n \leq K}$ corresponds to a time-inhomogeneous Markov chain with transition matrices given by

$$
\tilde{P}(n; x, y) = \mathbb{P}(\tilde{X}_{n+1} = y \mid \tilde{X}_n = x) = \frac{\pi_{K-(n+1)}(y)}{\pi_{K-n}(x)} \, P(y, x) \, .
$$

This chain is homogeneous if and only if π_0 is an invariant law, that is, if and only the chain is at equilibrium.

Lemma 3.3.7 *Let $(X_n)_{n \geq 0}$ be an irreducible Markov chain on $\mathcal{V}$ with transition matrix P, having an invariant law π. Then, the transition matrix $\tilde{P} = (\tilde{P}(x, y))_{x,y \in \mathcal{V}}$ of the time reversal of the chain in equilibrium is given by the equations*

$$
\pi(x) \tilde{P}(x, y) = \pi(y) P(y, x) \, ,
$$

That is, by

$$
\tilde{P}(x, y) = \frac{\pi(y)}{\pi(x)} \, P(y, x) \, .
$$

The transition matrix $\tilde{P}$ is irreducible recurrent positive and has invariant law π.

Proof: Straightforward using the previous computations and the invariant law criterion (Theorem 3.2.4). ∎

3.3.3.2   Doubly stationary Markov chain

Let P be a transition matrix with an invariant law π. Let $X_0 = \tilde{X}_0$ have law π, let $(X_n)_{n \geq 0}$ and $(\tilde{X}_n)_{n \geq 0}$ be Markov chains of matrices P and $\tilde{P}$, which are independent conditional on $X_0 = \tilde{X}_0$, and let $X_{-n} = \tilde{X}_n$. Then, $(X_k)_{k \in \mathbb{Z}}$ is a stationary Markov chain in time $\mathbb{Z}$ with transition matrix P, called the doubly stationary Markov chain of matrix P, which can be imagined to be "started" at $-\infty$.

3.3.3.3   Reversible measures

The equality $P = \tilde{P}$ holds if and only if π solves the local balance equations (3.2.5).

Then, the chain and its matrix are said to be *reversible* (in equilibrium), and π to be a *reversible law* for P. The equations (3.2.5) are also called the *reversibility equations* and their nonnegative and nonzero solutions the *reversible measures*.

In equilibrium, the probabilistic evolution of a reversible chain is the same in direct or reverse time, which is natural for many statistical mechanics models such as the Ehrenfest Urn.

Lemma 3.3.8 *Let* $P = (P(x, y))_{x,y \in V}$ *be a transition matrix on* V*. If there exists a reversible measure* μ*, then* $P(x, y) > 0 \iff P(y, x) > 0$ *and, for every* $x_0, x_1, \ldots, x_{k-1}, x_k = x$ *in* V *s.t.* $\prod_{i=1}^{k} P(x_{i-1}, x_i) > 0$,

$$\mu(x) = \mu(x_0) \frac{P(x_0, x_1) \cdots P(x_{k-1}, x)}{P(x, x_{k-1}) \cdots P(x_1, x_0)} = \mu(x_0) \prod_{i=1}^{k} \frac{P(x_{i-1}, x_i)}{P(x_i, x_{i-1})} .$$

A necessary and sufficient condition for the existence of a reversible measure is the Kolmogorov condition: *for every circuit* $x_0, x_1, \ldots, x_{k-1}, x_k = x_0$ *in* V,

$$P(x_0, x_1) \cdots P(x_{k-1}, x_0) = P(x_0, x_{k-1}) \cdots P(x_1, x_0) .$$

If P *is irreducible, then there is uniqueness (up to proportionality) of reversible measures, and if a reversible measure exists, then it can be obtained by the above formula for a particular choice of* x_0 *and* $\mu(x_0)$ *and for every* x *in* V *of* $x_1, \ldots, x_k = x$ *s.t.* $\prod_{i=1}^{k} P(x_{i-1}, x_i) > 0$.

Proof: The formula for $\mu(x)$ is obtained by iteration from 3.2.5, and this necessary form yields uniqueness if P is irreducible. The problem of existence is then a problem of compatibility: in order for all the equations in 3.2.5 to hold, it is easy to check that it is enough that for every x the formulae obtained using two different paths $x_0 = y_0, y_1, \ldots, y_j = x$ and $x_0 = z_0, z_1, \ldots, z_m = x$ satisfying $\prod_{i=1}^{j} P(y_{i-1}, y_i) > 0$ and $\prod_{i=1}^{m} P(z_{i-1}, z_i) > 0$ coincide, that is, satisfy

$$P(x_0, y_1) \cdots P(y_{j-1}, x) P(x, z_{m-1}) \cdots P(z_1, x_0)$$
$$= P(x_0, z_1) \cdots P(z_{m-1}, x) P(x, y_{j-1}) \cdots P(y_1, x_0) ,$$

and this is the Kolmogorov condition for the circuit $x_0 = y_0, y_1, \ldots, y_j = x = z_m, \ldots, z_1, z_0 = x_0$. It is a simple matter to conclude. $\blacksquare$

3.3.3.4   Adjoint chain, superinvariant measures, and superharmonic functions

We gather some simple facts in the following lemma.

Lemma 3.3.9 *Let P be an irreducible transition matrix on $\mathcal{V}$ having an invariant measure μ, and*

$$\tilde{P}(x, y) = \frac{\mu(y)}{\mu(x)} P(y, x) , \qquad x, y \in \mathcal{V} .$$

Then, $\tilde{P}$ is an irreducible transition matrix with invariant measure μ, called the adjoint of P with respect to μ, depends on μ only up to a multiplicative constant,

$$\tilde{\tilde{P}} = P , \qquad \tilde{P}^n := (\tilde{P})^n = \widetilde{P^n} ,$$

and P and $\tilde{P}$ are simultaneously either transient or null recurrent or positive recurrent. Moreover, a measure v is superinvariant for P if and only if the function

$$x \mapsto \frac{v(x)}{\mu(x)} \qquad \text{(density of } v \text{ with respect to } \mu)$$

is superharmonic nonnegative and nonzero for $\tilde{P}$. Conversely, a nonnegative and nonzero function f is superharmonic for P if and only if the measure

$$(f(x)\mu(x))_{x \in \mathcal{V}} \qquad \text{(with density f with respect to } \mu)$$

is superinvariant for $\tilde{P}$.

Proof: Recall that $0 < \mu < \infty$. Thus, $\tilde{P}$ is an irreducible transition matrix as

$$\tilde{P}(x, y) \geq 0 , \qquad \sum_{y \in \mathcal{V}} \tilde{P}(x, y) = \frac{1}{\mu(x)} \sum_{y \in \mathcal{V}} \mu(y) P(y, x) = \frac{\mu(x)}{\mu(x)} = 1 ,$$

and

$$\sum_{x \in \mathcal{V}} \mu(x) \tilde{P}(x, y) = \mu(y) \sum_{x \in \mathcal{V}} P(y, x) = \mu(y) .$$

Clearly, $\tilde{\tilde{P}} = P$ and $\tilde{P}$ depends on μ only up to a multiplicative constant. By a simple recursion,

$$(\tilde{P})^n(x, y) = \sum_{z \in \mathcal{V}} (\tilde{P})^{n-1}(x, z) \tilde{P}(z, y)$$

$$= \frac{\mu(y)}{\mu(x)} \sum_{z \in \mathcal{V}} P^{n-1}(z, x) P(y, z)$$

$$= \frac{\mu(y)}{\mu(x)} P^n(y, x) .$$

As $\tilde{P}(x,x) = P(x,x)$, the Potential matrix criterion (Section 3.1.3) yields that P and $\tilde{P}$ are both transient or recurrent and the invariant law criterion (Theroem 3.2.4) yields that they are both positive recurrent if and only if $\|\mu\|_{\mathrm{var}} < \infty$. Moreover,

$$\sum_{y \in \mathcal{V}} \tilde{P}(x,y)\frac{\mu(y)}{\mu(y)} = \frac{1}{\mu(x)} \sum_{y \in \mathcal{V}} \mu(y)P(y,x)$$

and

$$\sum_{y \in \mathcal{V}} f(y)\mu(y)\tilde{P}(y,x) = \mu(x) \sum_{y \in \mathcal{V}} P(x,y)f(y) \,,$$

and it is a simple matter to conclude.   ∎

This allows to understand the relations between Theorems 3.2.3 and 3.3.1. Let P be an irreducible recurrent transition matrix on $\mathcal{V}$ and μ an arbitrary canonical invariant law.

If v is a superinvariant measure for P, then the function $\frac{v}{\mu}$ is nonnegative and superharmonic for $\tilde{P}$, which is irreducible recurrent, and Theorem 3.3.1 yields that $\frac{v}{\mu}$ is constant, that is, Theorem 3.2.3.

Conversely, if f nonnegative and superharmonic for P, then $(f(x)\mu(x))_{x \in \mathcal{V}}$ is a superharmonic measure for $\tilde{P}$, and Theorem 3.2.3 yields that f is constant, that is, the "only if" part of Theorem 3.3.1.

The following result can be useful in the (rare) cases in which a matrix is not reversible, but its time reversal in equilibrium can be guessed.

Lemma 3.3.10 *Let P be an irreducible transition matrix on $\mathcal{V}$. If $\tilde{P}$ is a transition matrix and μ a nonzero measure s.t.*

$$\mu(x)\tilde{P}(x,y) = \mu(y)P(y,x) \,, \qquad x,y \in \mathcal{V} \,,$$

then μ is an invariant measure for P and $\tilde{P}$ is the adjoint of P with respect to μ.

Proof: As

$$\sum_{y \in \mathcal{V}} \mu(y)P(y,x) = \mu(x) \sum_{y \in \mathcal{V}} \tilde{P}(x,y) = \mu(x) \,,$$

it is a simple matter to conclude.   ∎

3.3.4   Birth-and-death chains

A Markov chain on $\mathbb{Z}$ (or on an interval of $\mathbb{Z}$) s.t.

$$P(x,x-1) + P(x,x) + P(x,x+1) = 1 \,, \qquad x \in \mathbb{Z} \,,$$

is called a birth-and-death chain. All other terms of P are then zero, this matrix is determined by the Birth-and-death probabilities

$$p_x = P(x,x+1) \geq 0 \,, \quad q_x = P(x,x-1) \geq 0 \,, \qquad x \in \mathbb{Z} \,,$$

which satisfy $p_x + q_x \leq 1$, and its graph is given by

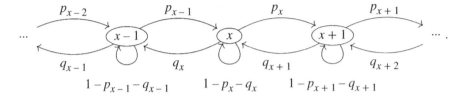

Several of the examples we have examined are birth-and-death chains: Nearest-neighbor random walks on $\mathbb{Z}$, Nearest-neighbor random walks reflected at 0 on $\mathbb{N}$, gambler's ruin, and macroscopic description for the Ehrenfest Urn on $\{0, \ldots, N\}$.

The chain is irreducible on $\mathbb{Z}$ if and only if

$$p_x > 0 , \quad q_x > 0 , \qquad \forall x \in \mathbb{Z} .$$

Birth-and-death on $\mathbb{N}$ or $\{0, \ldots N\}$ Similarly, $\mathbb{N}$ is a closed irreducible class if and only if

$$p_x > 0 , \ \forall x \in \mathbb{N} , \qquad q_0 = 0 , \quad q_x > 0 , \ \forall x \geq 1 ,$$

and $\{0, \ldots, N\}$ is a closed irreducible class if and only if

$$p_x > 0 , \ \forall x \in \{0, \ldots, N-1\} , \quad p_N = q_0 = 0 , \quad q_x > 0 , \ \forall x \in \{1, \ldots, N\} .$$

In these cases, the restriction of the chain is considered. It can be interpreted as describing the evolution of a population that can only increase or decrease by one individual at each step, according to whether there has been a birth or a death of an individual, hence the terminology.

We now give several helpful results, among which a generalization of the gambler's ruin law.

Theorem 3.3.11 *Consider an irreducible birth-and-death chain on $\mathbb{N}$, with birth probabilities $(p_x)_{x \geq 0}$ and death probabilities $(q_x)_{x \geq 1}$.*

1. There exists a unique invariant measure $\mu = (\mu(x))_{x \in \mathbb{N}}$, which is reversible and given if $\mu(0) = 1$ by

$$\mu(x) = \frac{p_0}{q_1} \cdots \frac{p_{x-1}}{q_x} = \frac{p_0 \cdots p_{x-1}}{q_1 \cdots q_x} .$$

2. A necessary and sufficient condition for positive recurrence, and hence for having an invariant law $\pi = (\pi(x))_{x \in \nu}$, is

$$\sum_{x \geq 0} \frac{p_0 \cdots p_{x-1}}{q_1 \cdots q_x} < \infty , \quad \text{and then } \pi(x) = \frac{p_0 \cdots p_{x-1}}{q_1 \cdots q_x} \left(\sum_{x \geq 0} \frac{p_0 \cdots p_{x-1}}{q_1 \cdots q_x} \right)^{-1} .$$

3. *The harmonic functions are constant. If* $D = \mathbb{N}$ *and* $\partial D = \{0\}$ *or* $D = \{0, \ldots, N\}$ *and* $\partial D = \{0, N\}$, *then any function defined on* D, *which is harmonic on* $D - \partial D$ *(s.t.* $(I - P)u = 0$ *on* $D - \partial D$*) is of the form* $u = u(0) + (u(1) - u(0))\phi$ *for*

$$\phi : x \in \mathbb{N} \mapsto \phi(x) = \sum_{y=0}^{x-1} \frac{q_1 \cdots q_y}{p_1 \cdots p_y} \quad \text{(with } \phi(0) = 0 \text{ and } \phi(1) = 1),$$

Notably,

$$\mathbb{P}_x(S_a < S_b) = \frac{\phi(b) - \phi(x)}{\phi(b) - \phi(a)}, \qquad \forall a \le x \le b \in \mathbb{N}.$$

4. *A necessary and sufficient condition for transience is*

$$\lim_{x \to \infty} \phi(x) := \sum_{x \ge 0} \frac{q_1 \cdots q_x}{p_1 \cdots p_x} < \infty.$$

Proof: 1. The global balance equations (3.2.4) are given by

$$\mu(0)p_0 = \mu(1)q_1,$$

$$\mu(x)(p_x + q_x) = \mu(x-1)p_{x-1} + \mu(x+1)q_{x+1}, \quad x \ge 1.$$

Thus, $\mu(1) = \mu(0)p_0/q_1$ and recursively $\mu(x+1)$ can be determined in terms of $\mu(x)$ and $\mu(x-1)$, hence the uniqueness. For $\mu(0) = 1$, the reversibility equations (3.2.5) or Lemma 3.3.8 yield the formula for μ, which is reversible.

2. There exists an invariant law π if and only if $\|\mu\|_{\text{var}} < \infty$, and then $\pi = \mu/\|\mu\|_{\text{var}}$. We conclude with the invariant law criterion (Theorem 3.2.4).

3. The harmonic functions u satisfy $(I - P)u = 0$, that is,

$$p_0 u(0) - p_0 u(1) = 0, \quad -q_x u(x-1) + (p_x + q_x)u(x) - p_x u(x+1) = 0, \quad x \ge 1,$$

and hence

$$u(1) = u(0), \quad p_x(u(x+1) - u(x)) = q_x(u(x) - u(x-1)), \quad x \ge 1.$$

By iteration, $u(x) = u(0)$ for $x \ge 0$. Similarly, if $(I - P)u(x) = 0$ for $x \in D - \partial D$,

$$u(x+1) - u(x) = \frac{q_x}{p_x}(u(x) - u(x-1)) = \cdots = \frac{q_1 \cdots q_x}{p_1 \cdots p_x}(u(1) - u(0))$$

and thus

$$u(x) = u(0) + \sum_{y=0}^{x-1}(u(y+1) - u(y)) := u(0) + (u(1) - u(0))\phi(x), \quad x \in D.$$

For the sequel, we may assume that $a = 0$ and $b = N$. Theorem 2.2.2 yields that $u : x \in \{0, \ldots, N\} \mapsto \mathbb{P}_x(S_0 < S_N)$ satisfies $u(0) = 1$, $u(N) = 0$ and $(I - P)u = 0$

on $\{1, \ldots, N-1\}$. Hence, $u(x) = 1 + (u(1) - 1)\phi(x)$ for $x = 0, 1, \ldots, N$, and $u(N) = 0$ yields that

$$u(x) = 1 - \frac{\phi(x)}{\phi(N)} = \frac{\phi(N) - \phi(x)}{\phi(N) - \phi(0)} .$$

4. Let $w : x \in \mathbb{N} \mapsto \mathbb{P}_x(S_0 < \infty)$. The "one step forward" method yields that

$$\mathbb{P}_0(R_0 < \infty) = 1 - p_0 + p_0 w(1) .$$

Theorem 2.2.2 yields that w is the least nonnegative solution of $w(0) = 1$ and $(I - P)w = 0$ on $\mathbb{N}^*$ and hence that

$$w = 1 + (a-1)\phi$$

with $a = w(1)$ as small as possible compatible with $w \geq 0$. If $\lim \phi = \infty$, then $a = 1$ and thus $\mathbb{P}_0(R_0 < \infty) = 1$ and the chain is recurrent. If $\lim \phi < \infty$, then $a = 1 - (\lim \phi)^{-1} < 1$ (corresponding to $\lim w = 0$) and the chain is transient. ∎

If the birth-and-death chain is irreducible on $\{0, 1, \cdots, N\}$ for a finite N, then it is always positive recurrent, and the invariant law π is obtained by replacing $\sum_{x \geq 0}$ by $\sum_{x=0}^{N}$ in the formula.

Remark 3.3.12 *Parts of this lemma can be generalized to an irreducible birth-and-death chain on $\mathbb{Z}$. Note that the uniqueness results for invariant measures and for harmonic functions are false: we have seen that the nearest-neighbor random walk on $\mathbb{Z}$, for which $p_x = p > 0$ and $q_x = q = 1 - p > 0$ for all x in $\mathbb{Z}$, has two invariant measures if $p \neq 1/2$. For transience and recurrence, the subsets $x \geq 0$ and $x < 0$ must be considered separately; the "one step forward" method yields that*

$$\mathbb{P}_0(R_0 < \infty) = 1 - p_0 - q_0 + p_0 \mathbb{P}_1(S_0 < \infty) + q_0 \mathbb{P}_{-1}(S_0 < \infty) .$$

Exercises

Several exercises of Chapter 1 involve irreducibility and invariant laws.

3.1   Generating functions and potential matrix Let $(X_n)_{n \geq 0}$ be a Markov chain on $\mathcal{V}$ with matrix P. For x and y in $\mathcal{V}$, consider the power series, for $s \geq 0$,

$$G_{x,y}(s) = \sum_{n \in \mathbb{N}} \mathbb{P}_x(R_y = n)s^n , \qquad H_{x,y}(s) = \sum_{n \in \mathbb{N}} P^n(x, y)s^n .$$

a) Prove that $G_{x,y}(s)$ converges for $s \leq 1$ and $H_{x,y}(s)$ for $s < 1$ and that

$$\lim_{s \uparrow 1} G_{x,y}(s) = \mathbb{P}_x(R_y < \infty) , \quad \lim_{s \uparrow 1} H_{x,y}(s) = \mathbb{E}_x(N_y) \in [0, \infty] .$$

b) Prove that $P^n(x, y) = \sum_{k=1}^{n} \mathbb{P}_x(R_y = k)P^{n-k}(y, y)$ for $n \geq 1$. Prove that, I denoting the identity matrix,

$$H_{x,y}(s) = I(x, y) + G_{x,y}(s)H_{y,y}(s), \qquad s < 1.$$

c) Prove that $\mathbb{E}_x(N_x) = \infty \Longleftrightarrow \mathbb{P}_x(R_x < \infty) = 1$ and that if $x \to y$ and $\mathbb{E}_y(N_y) = \infty$, then $\mathbb{E}_x(N_y) = \infty$.

d) We are going to apply these results to the nearest-neighbor random walk on $\mathbb{Z}$, with matrix given by $P(x, x + 1) = p > 0$ and $P(x, x - 1) = 1 - p > 0$. We recall that $(1 - t)^{-1/2} = \sum_{k \geq 0} \binom{2k}{k} (t/4)^k$.

Compute $P^n(x, x)$ for $x \in \mathbb{Z}$ and $n \geq 0$. Compute $H_{x,x}(s)$ and then $\mathbb{E}_x(N_x)$. When is this random walk recurrent?

3.2 Symmetric random Walks with independent coordinates Let P be the transition matrix of the symmetric random walk on $\mathbb{Z}$, given by $P(x, x + 1) = P(x, x - 1) = 1/2$. Use the Potential matrix criterion to prove that P and $P \otimes P$ are recurrent and $P \otimes P \otimes P$ is transient.

3.3 Lemma 3.1.3, alternate proofs Let $(X_n)_{n \geq 0}$ be a Markov chain on $\mathcal{V}$ with matrix P, and $x \neq y$ in $\mathcal{V}$ be s.t. x is recurrent and $x \to y$.

a) Prove that $\mathbb{P}_x(N_x < \infty) \geq \mathbb{P}_x(S_y < \infty)\mathbb{P}_y(N_x < \infty)$. Deduce from this that $\mathbb{P}_y(N_x = \infty) = 1$ and thus that $y \to x$. Prove that y is recurrent, for instance using the Potential matrix criterion. Deduce from all this that $\mathbb{P}_x(N_y = \infty) = 1$.

b) Prove   that $\mathbb{P}_x(S_y = \infty) = \mathbb{P}_x(S_y \geq R_x)\mathbb{P}_x(S_y = \infty)$.   Prove   that $\mathbb{P}_x(S_y \geq R_x) < 1$. Prove that $\mathbb{P}_x(S_y = \infty) = 0$. Prove that $\mathbb{P}_y(S_x = \infty) = 0$. Conclude that y is recurrent and then that $\mathbb{P}_x(N_y = \infty) = 1$.

c) Prove that $\mathbb{P}_x(S_y < R_x) > 0$ and, for $k \geq 1$,

$$\mathbb{P}_x(N_y \geq k) \geq \mathbb{P}_x(S_y \geq R_x)\mathbb{P}_x(N_y \geq k) + \mathbb{P}_x(S_y < R_x)\mathbb{P}_x(N_y \geq k - 1).$$

Deduce from this that $\mathbb{P}_x(N_y \geq k) = \mathbb{P}_x(N_y \geq k - 1)$ and then that $\mathbb{P}_x(N_y \geq k) = 1$. Conclude that $\mathbb{P}_x(N_y = \infty) = 1$.

3.4 Decomposition Find the transient class and the recurrent classes for the Markov chains with graph (every arrow corresponding to a positive transition probability) and matrices given by

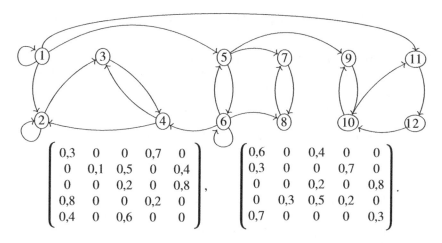

$$\begin{pmatrix} 0,3 & 0 & 0 & 0,7 & 0 \\ 0 & 0,1 & 0,5 & 0 & 0,4 \\ 0 & 0 & 0,2 & 0 & 0,8 \\ 0,8 & 0 & 0 & 0,2 & 0 \\ 0,4 & 0 & 0,6 & 0 & 0 \end{pmatrix}, \quad \begin{pmatrix} 0,6 & 0 & 0,4 & 0 & 0 \\ 0,3 & 0 & 0 & 0,7 & 0 \\ 0 & 0 & 0,2 & 0 & 0,8 \\ 0 & 0,3 & 0,5 & 0,2 & 0 \\ 0,7 & 0 & 0 & 0 & 0,3 \end{pmatrix}.$$

3.5 **Genetic models, see Exercise 1.8** Decompose each state space into its transient class and its recurrent classes. Prove without computations that the population will eventually be composed of individuals having all the same allele, a phenomenon called "allele fixation."

3.6 **Balance equations on a subset** Let P be a transition matrix on $\mathcal{V}$. Prove that for any subset A of $\mathcal{V}$ an invariant measure μ satisfies the balance equation

$$\sum_{x \in A} \mu(x)P(x, \mathcal{V} - A) = \sum_{y \in \mathcal{V} - A} \mu(y)P(y, A) .$$

3.7 **See Lemma 3.1.1** Let $(X_n)_{n \geq 0}$ be a Markov chain on $\mathcal{V}$ having an invariant law π. Prove that $\mathbb{P}_\pi(R_x = \infty) = \mathbb{P}_\pi(S_x = \infty)$. Deduce from this that $\pi(x) = 0$ or $x \in \mathcal{R}$.

3.8 **Induced chain, see Exercise 2.3** Let P be an irreducible transition matrix.

a) Prove that the induced transition matrix Q is recurrent if and only if P is recurrent.

b) If μ is an invariant measure for P, find an invariant measure v for Q. If v is an invariant measure for Q, find an invariant measure μ for P.

c) Prove that if P is positive recurrent, then Q is positive recurrent.

d) Let $0 < p < 1/2$ and $q = 1 - p$. Let P on $\mathbb{N}$ given by $P(x, x - 1) = q(p/q)^x$ and $P(x, x) = 1 - (p/q)^x$ and $P(x, x + 1) = p(p/q)^x$ for $x \geq 1$, and $P(0,1) = 1$. Compute Q, prove that Q is positive recurrent and that P is null recurrent.

3.9   Difficult advance Let $(X_n)_{n\geq 0}$ be the Markov chain on $\mathbb{N}$ with matrix given by $P(x, x + 1) = a$ for $x \geq 0$ and $P(0,0) = P(x, ([x/k] - 1)k) = 1 - a$ for $x \geq 1$, with $0 < a < 1$ and $k \geq 2$.

a) Prove that P is irreducible. Are there any reversible measures?

b) Find the invariant measures. Is there uniqueness for these?

c) Prove that the chain is recurrent positive if and only if $a^k < 1 - a$, and compute then the invariant law. What is the value of $\mathbb{E}_{jk}(R_{jk})$ for $j \geq 0$?

d) Let $S_0 = \inf\{n \geq 0 : X_n \in k\mathbb{N}\}$ and $S_{i+1} = \inf\{n > S_i : X_n \in k\mathbb{N}\}$ and $Y_i = \frac{1}{k}X_{S_i}$, $i \geq 0$. Prove that the S_i are stopping times which are finite, a.s. Prove that $(Y_i)_{i\geq 0}$ is a Markov chain on $\mathbb{N}$ and give its transition matrix Q.

e) Let $(Z_n)_{n\geq 0}$ be the induced chain constituted of the successive distinct states visited by $(Y_n)_{n\geq 0}$. We admit it is a Markov chain, see Exercise 2.3. Give its transition matrix Q'.

f) Use known results to prove that $(Z_n)_{n\geq 0}$ is null recurrent if $a^k = 1 - a$ and transient if $a^k > 1 - a$. Deduce from this that the same property for $(Y_n)_{n\geq 0}$ and for $(X_n)_{n\geq 0}$.

3.10   Queue with K servers, 1 The evolution of a queue with K servers is given by a birth-and-death chain on $\mathbb{N}$ s.t. $p_x = \frac{\lambda}{\lambda + \mu(x \wedge K)}$ and $q_x = \frac{\mu(x \wedge K)}{\lambda + \mu(x \wedge K)}$ for $x \in \mathbb{N}$, with $\lambda, \mu > 0$ and $K \geq 1$. Let $\rho := \lambda/\mu$.

a) Is this chain irreducible?

b) Prove that there is a unique invariant measure, and compute it in terms of ρ and K.

c) Prove that the chain is positive recurrent if and only if $\rho < K$.

d) Prove that the chain is null recurrent if $\rho = K$ and transient if $\rho > K$, first using results on birth-and-death chains, and then using Lyapunov functions.

e) We now considered the generalized case in which $K = \infty$, so that $q_x = \frac{\mu x}{\lambda + \mu x}$. Prove that there is an invariant law π, compute it explicitly, and prove that the chain is positive recurrent. Compute $\mathbb{E}_0(R_0)$.

3.11   ALOHA The ALOHA protocol was established in 1970 in order to manage a star-shaped wireless network, linking a large number of computers through a central hub on two frequencies, one for sending and the other for receiving.

A signal regularly emitted by the hub allows to synchronize emissions by cutting time into timeslots of same duration, sufficient for sending a certain quantity of bits called a packet. If a single computer tries to emit during a timeslot, it is successful and the packet is retransmitted by the hub to all computers. If two or more computers attempt transmission in a timeslot, then the packets interfere and the attempt is unsuccessful; this event is called a *collision*. The only information available to the computers is whether there has been at least one collision or not. When there is a collision, the computers attempt to retransmit after a random duration. For a simple Markovian analysis, we assume that this happens with probability $p > 0$ in each subsequent timeslot, so that the duration is geometric.

Let X_0 be the initial number of packets awaiting retransmission, $(A_n)_{n\geq 1}$ be i.i.d., where A_n is the number of new transmission attempts in timeslot n, and $(R_{n,i})_{n,i\geq 1}$ be i.i.d., where $R_{n,i} = 1$ with probability $p > 0$ if the i-th packet awaiting retransmission after timeslot $n - 1$ undergoes a retransmission attempt in timeslot n, or else $R_{n,i} = 0$. All these r.v. are independent. We assume that

$$\mathbb{P}(A_1 = 0) > 0 \,, \qquad \mathbb{P}(A_1 \geq 2) > 0 \,, \qquad \mathbb{E}(A_1) < \infty \,.$$

a) Prove that the number of packets awaiting retransmission after timeslot n is given by

$$X_n = X_{n-1} + A_n - \mathbb{1}_{\{A_n + \sum_{i=1}^{X_{n-1}} R_{n,i}=1\}} \,,$$

and that $(X_n)_{n\geq 0}$ is an irreducible Markov chain on $\mathbb{N}$.

b) If $\mathbb{E}(A_1^2) < \infty$, use the Lamperti criterion (Theorem 3.3.3) to prove that this chain is transient for every $p > 0$.

c) Conclude to the same result when $\mathbb{E}(A_1^2) = \infty$.

3.12 Queuing by sessions Time is discrete, A_k jobs arrive at time $k \geq 1$ in i.i.d. manner, and $\mathbb{P}(A_1 = 0) > 0$ and $0 < \mathbb{E}(A_1) := \alpha < \infty$. Session n is devoted to servicing exclusively and exhaustively the X_{n-1} jobs that were waiting at its start, and has an integer-valued random duration $T_n \geq 1$, which conditional on $X_{n-1} = x$ is integrable, independent of the rest, and has law not depending on n.

a) Prove that $(X_n)_{n\geq 0}$ is a Markov chain on $\mathbb{N}$.

b) Prove that 0 belongs to a closed irreducible class and that all states outside this class are transient.

c) We consider the restriction of the chain to this closed irreducible class. Prove that if $\limsup_{x\to\infty}(\alpha\mathbb{E}_x(T_1) - x) < 0$ then the chain is positive recurrent.

d) Prove that if $\mathbb{P}_x(T_1 = 1) = \mathbb{P}_x(T_1 \geq 2x/\alpha) = 1/2$, then $(X_n)_{n\geq 0}$ is positive recurrent. Prove that $(X_n)_{n\geq 0}$ satisfies hypothesis (1) of the Lamperti criterion for $\phi : x \mapsto x$ and $\varepsilon = \alpha/2$. Does it satisfy hypothesis (2) of the Lamperti criterion or of the Tweedie criterion?

3.13 Big jumps Let $(X_n)_{n \geq 0}$ be a Markov chain on $\mathbb{N}$ with matrix P with only nonzero terms $P(x, x+1) = P(x, 2x) = \alpha > 0$ and $P(x, \lfloor x/2 \rfloor) = 1 - 2\alpha > 0$.

a) Prove that this chain is irreducible.

b) Prove that if $\alpha < 1/3$, then the chain is positive recurrent and that there exists $\varepsilon > 0$ and a finite subset F of $\mathbb{N}$ s.t. $\mathbb{E}_x(S_F) < \frac{\ln x}{\varepsilon}$.

c) Prove that if $\alpha > 1/3$, then the chain is transient.

3.14 Quick return Let $(X_n)_{n \geq 0}$ be a Markov chain on $\mathcal{V}$ with matrix P. Assume that there exists a nonempty subset E of $\mathcal{V}$ and $\rho < 1$ and $\phi : \mathcal{V} \to \mathbb{R}_+$ s.t. if $x \notin E$, then $\phi(x) \geq 1$ and $\mathbb{E}_x(\phi(X_1)) := P\phi(x) \leq \rho\phi(x)$. (We may assume that ϕ vanishes on E.) Let $T = \inf\{n \geq 0 : X_n \in E\}$.

a) Prove that $\mathbb{P}_x(T > k+1) = \sum_{y \notin E} P(x, y) \mathbb{P}_y(T > k)$ for $x \notin E$ and $k \geq 0$. Deduce from this that $\mathbb{P}_x(T > k) \leq \phi(x)\rho^k$ for $x \notin E$ and $k \geq 0$.

b) Prove that $\mathbb{E}_x(T) \leq 1 + \frac{\rho}{1-\rho}\phi(x)$ and that $\mathbb{E}_x(s^T) \leq s + \frac{\rho s(s-1)}{1-\rho s}\phi(x)$ for $x \notin E$ and $1 < s < \frac{1}{\rho}$.

c) Let $K \geq 1$, and $(A_n)_{n \geq 1}$ be i.i.d., the power series $g(s) := \mathbb{E}(s^{A_1})$ have convergence radius $R > 1$, and $g'(1) = \mathbb{E}(A_1) < K$. Consider X_0 independent of $(A_n)_{n \geq 1}$ and (queue with K servers)

$$X_{n+1} = (X_n - K)^+ + A_{n+1}, \qquad n \geq 0.$$

Prove that there exists $\beta > 1$ and $\rho < 1$ s.t. $g(\beta) \leq \rho\beta^K$. Find a finite subset E of $\mathcal{V}$ and a function ϕ satisfying the above.

d) Let E be a nonempty subset of $\mathcal{V}$ s.t. $\inf_{x \notin E} P(x, E) > 0$. Find ϕ and ρ satisfying the above. For which renewal processes does this hold for $E = \{0\}$?

3.15 Adjoints Let P be the transition matrix of the nearest-neighbor random walk on $\mathbb{Z}$, given by $P(x, x+1) = p > 0$ and $P(x, x-1) = 1 - p = q > 0$. Give the adjoint transition matrix with respect to $\mu = ((p/q)^x)_{x \in \mathbb{Z}}$ and the adjoint transition matrix with respect to the uniform measure.

3.16 Labouchère system, see Exercises 1.4 and 2.10 Let $(X_n)_{n \geq 0}$ be the random walk on $\mathbb{Z}$ with matrix P given by $P(x, x-2) = p > 0$ and $P(x, x+1) = 1 - p > 0$.

a) Are there any reversible measures?

b) Prove that if $p \neq 1/3$, then an invariant measure must be of the form $(\alpha_- \rho_-^x + \alpha_+ \rho_+^x + \beta)_{x \in \mathbb{Z}}$ for $\rho_\pm = -\frac{1}{2} \pm \sqrt{\frac{1}{p} - \frac{3}{4}}$, and if $p = 1/3$ of the form $(\alpha_-(-2)^x + \alpha_+ x + \beta)_{x \in \mathbb{Z}}$.

c) By considering the behaviors for $x \to \infty$ and $x \to -\infty$, prove that if $p \neq 1/3$, then the invariant measures are of the form $(\alpha \rho_+^x + \beta)_{x \in \mathbb{Z}}$ with $\alpha \geq 0$ and $\beta \geq 0$ not both zero and that if $p = 1/3$, then the unique invariant measure is uniform. Deduce from this that if $p \neq 1/3$, then this random walk is transient.

d) Let $(Y_n)_{n \geq 0}$ be the random walk reflected at 0 on $\mathbb{N}$, with matrix Q given by $Q(x, \max(x - 2, 0)) = p > 0$ and $Q(x, x + 1) = 1 - p > 0$ for $x \in \mathbb{N}$, and $S_0 = \inf\{n \geq 0 : Y_n = 0\}$. Write the global balance equations. Prove that if $p \neq 1/3$, then the unique invariant measure is $(\rho_+^x)_{x \geq 0}$, and if $p = 1/3$, then the unique invariant measure is uniform.

e) Prove that Q is positive recurrent if and only if $p > 1/3$, and compute the invariant law π if it exists. Compute $\mathbb{E}_1(S_0)$.

f) Use a Lyapunov function technique to prove that Q is positive recurrent if $p > 1/3$, null recurrent if $p = 1/3$, and transient if $p < 1/3$.

3.17 **Random walk on a graph** A discrete set $\mathcal{V}$ is furnished with a (nonoriented) graph structure as follows. The elements of $\mathcal{V}$ are the nodes of the graph, and the elements of $\mathcal{L} \subset \{\{x, y\} : x, y \in \mathcal{V}\}$ are the links of the graph. The neighborhood of x is given by $V(x) = \{y \in \mathcal{V} : \{x, y\} \in \mathcal{L}\}$, and the degree of x is given by its number of neighbors $d(x) = \mathrm{Card}(V(x))$. It is assumed that $d(x) > 0$ for every x. The random walk on the graph $\mathcal{G}$ is defined as the Markov chain $(X_n)_{n \geq 0}$ on $\mathcal{V}$ s.t. if $X_n = x$, then X_{n+1} is chosen uniformly in $V(x)$.

a) Give an explicit expression for the transition matrix P of $(X_n)_{n \geq 0}$. Give a simple condition for it to be irreducible.

b) Describe the symmetric nearest-neighbor random walk on $\mathbb{Z}^d$ and the microscopic representation of the Ehrenfest Urn as random walks on graphs.

c) Assume that $(X_n)_{n \geq 0}$ is irreducible. Find a reversible measure for this chain. Give a simple necessary and sufficient condition for the chain to be positive recurrent and then give an expression for the invariant law π.

d) Let $\mathcal{V} = \mathbb{Z} \times \mathbb{N}$ and

$$\mathcal{L} = \{\{(k, n), (k + 1, 2n)\}, \{(k, n), (k + 1, 2n + 1)\} : (k, n) \in \mathcal{V}\}.$$

Find two (nonproportional) invariant measures for the random walk. Prove that the random walk is transient.

3.18 **Caricature of TCP** In a caricature of the transmission control protocol (TCP), which manages the window sizes for data transmission in the Internet, any packet is independently received with probability $0 < \theta < 1$, and the consecutive distinct window sizes (in packets) W_n for $n \geq 0$ constitute a Markov chain on $\mathbb{N}$, with matrix P with nonzero terms $P(x, x+1) = \theta^x$ and $P(x, \lfloor x/2 \rfloor) = 1 - \theta^x$.

a) Prove that this Markov chain is irreducible. Prove that there exists a unique invariant law $\pi = (\pi(x))_{x \in \mathbb{N}}$, by using a Lyapunov function technique.

b) Write the global balance equations satisfied by the invariant law π.

c) Prove that, for $x \geq 1$,

$$\pi(x)\theta^x = \sum_{y=x+1}^{2x+1} \pi(y)(1 - \theta^y) \geq \pi(2x)(1 - \theta^{2x}) + \pi(2x+1)(1 - \theta^{2x+1}) .$$

d) Deduce from this, for $x \geq 1$, that $\pi(x) \leq \pi(x-1)\theta^{x-1} + \pi(x)\theta^x$ and then that $\pi(x) \leq \pi(0)\theta^{\frac{x(x-1)}{2}} \prod_{y=1}^{x} (1 - \theta^y)^{-1}$.

e) Prove that $L := \prod_{y=1}^{\infty} (1 - \theta^y)^{-1} < \infty$.

f) Let $\pi(x) = \beta(x)\theta^{\frac{x(x-1)}{2}}$ and $Z = \sum_{x \geq 0} \theta^{\frac{x(x-1)}{2}} < \infty$. Prove that β is nondecreasing, $\pi(0) \leq \beta(x) \leq \pi(0)L$, and $Z^{-1}L^{-1} \leq \pi(0) \leq Z^{-1}$.

4

Long-time behavior

4.1 Path regeneration and convergence

In this section, let $(X_n)_{n\geq 0}$ be an *irreducible recurrent* Markov chain on $\mathcal{V}$. For $n \geq 0$, let the counting measure N_n with integer values, and the empirical measure $\tilde{N}_n$, which is a *probability* measure, be the random measures given by

$$N_n = \sum_{k=0}^{n} \delta_{X_k} \,, \qquad \tilde{N}_n = \frac{1}{n+1} \sum_{k=0}^{n} \delta_{X_k} \,,$$

so that if f is a real function on $\mathcal{V}$, then

$$N_n f = \sum_{k=0}^{n} f(X_k) \,, \qquad \tilde{N}_n f = \frac{1}{n+1} \sum_{k=0}^{n} f(X_k) \,, \tag{4.1.1}$$

and if x is in $\mathcal{V}$, then

$$N_n(x) = \sum_{k=0}^{n} \mathbb{1}_{\{X_k=x\}} \,, \qquad \tilde{N}_n(x) = \frac{1}{n+1} \sum_{k=0}^{n} \mathbb{1}_{\{X_k=x\}} \,.$$

For any state x, Lemma 3.1.3 yields that $\lim_{n\to\infty} N_n(x) = \infty$, and hence that the successive hitting times $0 \leq S_0^x < S_1^x < \cdots$ of x defined in (2.2.2) are stopping times, which are finite, a.s. Moreover, if $N_n(x) \geq 1$, then

$$S_{N_n(x)-1}^x \leq n < S_{N_n(x)}^x \,.$$

As $S_m^x - S_{m-1}^x$ is the first strict future hitting time of x by the shifted chain $(X_{S_{m-1}^x + n})_{n\geq 0}$, the strong Markov property yields that the $(X_k)_{S_{m-1}^x \leq k < S_m^x}$ for $m \geq 1$

are i.i.d. and have same law as $(X_k)_{0 \leq k < R_x}$ when $X_0 = x$. In particular, for any real function f, the

$$Y_m^x f = \sum_{k=S_{m-1}^x}^{S_m^x - 1} f(X_k) , \qquad m \geq 1 , \tag{4.1.2}$$

are i.i.d. and have same law as $\sum_{k=0}^{R_x - 1} f(X_k)$ when $X_0 = x$.

The path followed by the chain can be decomposed into its excursions from x, by setting (with the convention that an empty sum is null)

$$N_n f = \sum_{k=0}^{S_0^x - 1} f(X_k) + \sum_{m=1}^{N_n(x)-1} Y_m^x f + \sum_{k=S_{N_n(x)-1}^x}^{n} f(X_k) . \tag{4.1.3}$$

This will enable to use classic convergence results for i.i.d. sequences.

4.1.1   Pointwise ergodic theorem, extensions

4.1.1.1   Pointwise ergodic theorem

The first result is basically a strong law of large numbers and is often used for positive recurrent chains. The terminology "pointwise" refers to a.s. convergence.

Theorem 4.1.1 (Pointwise ergodic theorem) *Let $(X_n)_{n \geq 0}$ be an irreducible recurrent Markov chain on $\mathcal{V}$, with arbitrary initial law, and μ be its invariant measure. For all functions f and g on $\mathcal{V}$ satisfying $f \geq 0$ or $\mu|f| < \infty$, and $g \geq 0$ or $\mu|g| < \infty$, it holds that*

$$\lim_{n \to \infty} \frac{N_n f}{N_n g} = \frac{\mu f}{\mu g}, \quad a.s.,$$

see (4.1.1), as soon as the r.h.s. has a meaning in $\mathbb{R} \cup \{-\infty, \infty\}$.

If the chain is positive recurrent, and thus has invariant law $\pi := \mu / \|\mu\|_{\mathrm{var}}$, then

$$\lim_{n \to \infty} \tilde{N}_n f = \pi f , \quad a.s.$$

If the chain is null recurrent and $\mu|f| < \infty$, then $\lim_{n \to \infty} \tilde{N}_n f = 0$, a.s

In particular, for x and y in $\mathcal{V}$, a.s.,

$$\lim_{n \to \infty} \frac{N_n(x)}{N_n(y)} = \frac{\mu(x)}{\mu(y)} , \quad \lim_{n \to \infty} \tilde{N}_n(x) = \frac{1}{\mathbb{E}_x(R_x)} = \begin{cases} \pi(x) & \text{if positive recurrent,} \\ 0 & \text{if null recurrent.} \end{cases}$$

Proof: Let us start with $f \geq 0$ and $g = \mathbb{1}_{\{x\}}$ for some state x. With probability 1, for n large enough $N_n(x) \geq 1$, and the decomposition (4.1.3) yields the lower and upper bounds

$$\frac{1}{N_n(x)} \sum_{m=1}^{N_n(x)-1} Y_m^x f \leq \frac{N_n f}{N_n(x)} \leq \frac{1}{N_n(x)} \sum_{k=0}^{S_0^x - 1} f(X_k) + \frac{1}{N_n(x)} \sum_{m=1}^{N_n(x)} Y_m^x f .$$

In this $\lim_{n\to\infty} \frac{1}{N_n(x)} \sum_{k=0}^{S_0^x - 1} f(X_k) = 0$, a.s., and the strong law of large numbers applied to $(Y_m^x f)_{m\geq 1}$ yields that, a.s.,

$$\lim_{n\to\infty} \frac{1}{N_n(x)} \sum_{m=1}^{N_n(x)-1} Y_m^x f = \lim_{n\to\infty} \frac{1}{N_n(x)} \sum_{m=1}^{N_n(x)} Y_m^x f = \mathbb{E}(Y_1^x f) \in [0, \infty] .$$

By definition of the canonical invariant measure μ_x generated at x, the Fubini theorem yields that

$$\mathbb{E}(Y_1^x f) = \mathbb{E}_x \left(\sum_{k=0}^{R_x - 1} f(X_k) \right) = \sum_{y\in\mathcal{V}} \mathbb{E}_x \left(\sum_{k=0}^{R_x - 1} \mathbb{1}_{X_k = y} \right) f(y) = \mu_x f .$$

Thus, $\lim_{n\to\infty} \frac{N_n f}{N_n(x)} = \mu_x f$, a.s., because of the bounds converging. The decomposition $f = f^+ - f^-$ shows that this is also true for a signed f, which is μ-integrable. By the uniqueness of the invariant measure,

$$\lim_{n\to\infty} \frac{N_n f}{N_n g} = \lim_{n\to\infty} \frac{N_n f}{N_n(x)} \frac{N_n(x)}{N_n g} = \frac{\mu_x f}{\mu_x g} = \frac{\mu f}{\mu g}$$

if this has a meaning in $\mathbb{R} \cup \{-\infty, \infty\}$. For $g = 1$, it holds that $N_n 1 = n + 1$ and thus

$$\lim_{n\to\infty} \tilde{N}_n f = \lim_{n\to\infty} \frac{N_n f}{N_n 1} = \frac{\mu f}{\mu 1} = \frac{\mu f}{\|\mu\|_{\text{var}}} .$$

It is a simple matter to conclude. ∎

Note that if x is a transient state for a Markov chain $(X_n)_{n\geq 0}$, then it is visited at most a finite number of times, and $\lim_{n\to\infty} \frac{1}{n+1} \sum_{k=0}^{n} \mathbb{1}_{X_k = x} = 0$, a.s.

Statistical estimation of transition matrix The pointwise ergodic theorem allows to estimate the transition matrix of an irreducible positive recurrent Markov chain.

Indeed, the snake chain $(X_n, X_{n+1})_{n\geq 0}$ is irreducible on its natural state space

$$\{(x, y) \in \mathcal{V} : P(x, y) > 0\} ,$$

and on this space has invariant law with generic term $\pi(x)P(x, y)$, and hence, it is positive recurrent. The pointwise ergodic theorem then yields that

$$\frac{\sum_{k=0}^{n-1} \mathbb{1}_{\{X_k = x, X_{k-1} = y\}}}{\sum_{k=0}^{n-1} \mathbb{1}_{\{X_k = x\}}} \xrightarrow[n\to\infty]{\text{a.s}} P(x, y) .$$

In theory, this could be also used for null recurrent chains (replacing the invariant law by the invariant measure); however, in practice, the convergence is much too slow.

4.1.1.2   Ergodic theorem in probability

Given this proof, it is natural to try to relax the assumptions and consider the case in which $\mathbb{E}(|Y_1^x f|) < \infty$, instead of $\mu_x|f| = \mathbb{E}(Y_1^x|f|) < \infty$. The difficulty is that the last hitting time of x before time n, given by

$$D_x(n) = S^x_{N_n(x)-1} = \sup\{k \in \{0,\ldots,n\} : X_k = x\} \qquad (4.1.4)$$

with the convention $S^x_{-1} = 0$, *is not a stopping time.*

The following lemma is usually proved using some results of convergence of $(\pi_n)_{n\geq 0}$, which we discuss later. We replace these by a coupling argument.

Lemma 4.1.2 *Let $(X_n)_{n\geq 0}$ be an irreducible positive recurrent Markov chain on $\mathcal{V}$, with arbitrary initial law. Then, for any state x and any real function f on $\mathcal{V}$, the sequence $(\sum_{k=D_x(n)}^{n} f(X_k))_{n\geq 0}$ is bounded in probability, in the sense that*

$$\lim_{a\to\infty} \sup_{n\geq 0} \mathbb{P}\left(\left|\sum_{k=D_x(n)}^{n} f(X_k)\right| > a\right) = 0 .$$

Proof: It is enough to consider $f \geq 0$. Let P be the transition matrix, π the invariant law, and $\tilde{P}$ the adjoint transition matrix w.r.t. π, that is, the matrix of the time reversal in equilibrium (see Lemma 3.3.7).

Let $(X'_n)_{n\geq 0}$ be a Markov chain with matrix P in equilibrium, X_0 an independent r.v. with arbitrary law, and $S' = \inf\{n \geq 0 : X'_n = X_0\}$. Then, $\mathbb{P}(S' < \infty) = 1$ as on $\{X_0 = y\}$, it holds that $S' = S'_y = \inf\{n \geq 0 : X'_n = y\} < \infty$, a.s. The strong Markov property yields that $(X_n)_{n\geq 0} = (X'_{S'+n})_{n\geq 0}$ is a Markov chain with matrix P started at X_0.

Let $D'_x(n)$ be the last hitting time of x before time n for $(X'_k)_{k\geq 0}$, analogous to $D_x(n)$ defined in (4.1.4). As $D'_x(n) \leq S' + D_x(n)$,

$$\sum_{k=D_x(n)}^{n} f(X_k) = \sum_{k=S'+D_x(n)}^{S'+n} f(X'_k) \leq \sum_{k=D'_x(n)}^{n} f(X'_k) + \sum_{k=n+1}^{S'+n} f(X'_k) ,$$

and it is enough to prove that each term in the r.h.s. is bounded in probability.

Let $(\tilde{X}_n)_{n\geq 0}$ be a Markov chain with matrix $\tilde{P}$ in equilibrium, and $\tilde{S}_x = \inf\{n \geq 0 : \tilde{X}_n = x\}$. Then, $\sum_{k=D'_x(n)}^{n} f(X'_k)$ has same law as $\sum_{k=0}^{\min(\tilde{S}_x,n)} f(\tilde{X}_k)$, and $\sum_{k=0}^{\min(\tilde{S}_x,n)} f(\tilde{X}_k) \leq \sum_{k=0}^{\tilde{S}_x} f(\tilde{X}_k)$ and this upper bound does not depend on n and hence is bounded in probability. For all $a, b > 0$ with integer b, the stationarity of $(X'_n)_{n\geq 0}$ yields that

$$\mathbb{P}\left(\sum_{k=n+1}^{S'+n} f(X'_k) > a\right) \leq \mathbb{P}(S' > b) + \mathbb{P}\left(\sum_{k=1}^{b} f(X'_k) > a\right),$$

which does not depend on n and converges to 0 when b and then a go to infinity. ∎

The following lemma is technical and is used to clarify statements.

Lemma 4.1.3 *Let $(X_n)_{n\geq 0}$ be an irreducible positive recurrent Markov chain on $\mathcal{V}$, f a function on $\mathcal{V}$, and $a \geq 0$. Then, $\mathbb{E}(|Y_1^x f|^a) = \mathbb{E}_x(|\sum_{k=0}^{R_x-1} f(X_n)|^a) < \infty$ either for all states x or for none.*

Proof: Assume that $a \geq 1$, the case $0 < a < 1$ being similar. Let $(X_n)_{n\geq 0}$ be started at x such that $\mathbb{E}(|Y_1^x f|^a) < \infty$ and $y \neq x$. Let $N = \inf\{n \geq 0 : S_1^y < S_n^x\}$, so that $0 = S_0^x < S_0^y < S_1^y < S_N^x$ are all stopping times, and

$$W = \sum_{m=1}^N Y_m^x f = \sum_{k=0}^{S_0^y-1} f(X_k) + \sum_{k=S_0^y}^{S_1^y-1} f(X_k) + \sum_{k=S_1^y}^{S_N^x-1} f(X_k) , \quad W' = W - Y_1^y f .$$

The strong Markov property yields that $Y_1^y f$ and W' are independent, hence if $\mathbb{E}(|W|^a) < \infty$, then there exists w such that $\mathbb{E}(|Y_1^y f + w|^a) < \infty$ and hence, $\mathbb{E}(|Y_1^y f|^a) < \infty$. Thus, we will be done as soon as we prove that $\mathbb{E}(|W|^a) < \infty$. The sequence $(Y_m^x f, Y_m^x \mathbb{1}_{\{y\}})_{m\geq 1}$ is i.i.d., thus

$$N_1 = \inf\{m \geq 1 : Y_m^x \mathbb{1}_{\{y\}} \geq 1\} , \quad N_2 = \inf\{m \geq 1 : Y_{N_1+m}^x \mathbb{1}_{\{y\}} \geq 1\}$$

are i.i.d., with geometric law given by $\mathbb{P}(N_i > k) = \mathbb{P}(Y_1^x \mathbb{1}_{\{y\}} = 0)^k$ for $k \geq 0$ with $\mathbb{P}(Y_1^x \mathbb{1}_{\{y\}} = 0) < 1$. Thus, $\mathbb{E}(N_i^b) < \infty$ for $b \geq 0$, and as $N \leq N_1 + N_2$,

$$\mathbb{E}(|W|^a) = \mathbb{E}\left(\left|\sum_{m=1}^N Y_m^x f\right|^a\right) \leq \mathbb{E}\left(\left(\sum_{m=1}^{N_1+N_2} |Y_m^x f|\right)^a\right)$$

$$\leq 2^{a-1}\mathbb{E}\left(\left(\sum_{m=1}^{N_1} |Y_m^x f|\right)^a\right) + 2^{a-1}\mathbb{E}\left(\left(\sum_{m=N_1+1}^{N_1+N_2} |Y_m^x f|\right)^a\right)$$

$$\leq 2^a\mathbb{E}\left(\left(\sum_{m=1}^{N_1} |Y_m^x f|\right)^a\right)$$

and the Jensen or Hölder inequality (Lemmas A.3.6 or A.3.8) yields that

$$\mathbb{E}\left(\left(\sum_{m=1}^{N_1} |Y_m^x f|\right)^a\right) \leq \mathbb{E}\left(N_1^{a-1} \sum_{m=1}^{N_1} |Y_m^x f|^a\right)$$

$$\leq \sum_{k=1}^\infty k^{a-1}\mathbb{E}\left(\sum_{m=1}^k |Y_m^x f|^a \,\middle|\, N_1 = k\right) \mathbb{P}(N_1 = k) .$$

Conditional on $N_1 = k$, the $Y_m^x f$ for $1 \leq m < k$ have same law as $Y_1^x f$ conditional on $Y_1^x \mathbb{1}_{\{y\}} = 0$ and $Y_k^x f$ as $Y_1^x f$ conditional on $Y_1^x \mathbb{1}_{\{y\}} \geq 1$. As $\mathbb{E}(|Y_1^x f|^a) < \infty$, it holds that $\mathbb{E}(|Y_1^x f|^a \mid Y_1^x \mathbb{1}_{\{y\}} = 0) < \infty$ and $\mathbb{E}(|Y_1^x f|^a \mid Y_1^x \mathbb{1}_{\{y\}} \geq 1) < \infty$, and denoting by B the largest of these two quantities,

$$\mathbb{E}\left(\left(\sum_{m=1}^{N_1} |Y_m^x f|\right)^a\right) \leq B \sum_{k=1}^{\infty} k^a \mathbb{P}(N_1 = k) = B\mathbb{E}(N_1^a) < \infty$$

and hence $\mathbb{E}(|W|^a) < \infty$, and the proof is done. ∎

Theorem 4.1.4 (Ergodic theorem in probability) *Let* $(X_n)_{n \geq 0}$ *be an irreducible positive recurrent Markov chain on* $\mathcal{V}$, *with arbitrary initial law. Let* π *denote its invariant law. For any real function* f *such that* $\mathbb{E}(|Y_1^x f|) < \infty$ *for some state* x; *hence, for all states (Lemma 4.1.3), the quantity* $m_f = \pi(x)\mathbb{E}(Y_1^x f)$ *does not depend on* x, *and*

$$\lim_{n \to \infty} \tilde{N}_n f = m_f \quad \text{in probability.}$$

Proof: Consider the decomposition (4.1.3). Using Lemma 4.1.2,

$$\lim_{n \to \infty} \frac{1}{n+1} \sum_{k=0}^{S_0^x-1} f(X_k) = 0 \text{ a.s.,} \quad \lim_{n \to \infty} \frac{1}{n+1} \sum_{k=D_x(n)}^{n} f(X_k) = 0 \text{ in probability,}$$

and the strong law of large numbers applied to $(Y_m^x f)_{m \geq 1}$ and the pointwise ergodic theorem (Theorem 4.1.1) yields that

$$\lim_{n \to \infty} \frac{1}{n+1} \sum_{m=1}^{N_n(x)-1} Y_m^x f = \lim_{n \to \infty} \tilde{N}_n(x) \frac{1}{N_n(x)} \sum_{m=1}^{N_n(x)} Y_m^x f = \pi(x)\mathbb{E}(Y_1^x f) \text{ a.s.}$$

It is a simple matter to conclude, and $\pi(x)\mathbb{E}(Y_1^x f)$ does not depend on x by the uniqueness of a limit. ∎

In general, there is not a.s. convergence, see Chung, K.L. (1967), p. 97.

4.1.2   Central limit theorem for Markov chains

Obtaining confidence intervals is essential in practice, notably for elaborating and calibrating Monte Carlo methods (see Section 5.2) or statistical estimations. Note that $Y_1^x m_f = m_f(S_1^x - S_0^x)$ has expectation $\mathbb{E}(Y_1^x f)$. Recall (4.1.1) and (4.1.2).

Theorem 4.1.5 (Central limit theorem) *Let the hypotheses in Theorem 4.1.4 be satisfied. Then,* $m_f = \pi(x)\mathbb{E}(Y_1^x f)$ *does not depend on* $x \in \mathcal{V}$ *and takes value* πf *if* $\pi|f| < \infty$,

$$\sigma_f^2 = \pi(x)\text{Var}(Y_1^x(f - m_f)) = \pi(x)\mathbb{E}((Y_1^x(f - m_f))^2)$$

does not depend on x, and if $\sigma_f^2 < \infty$, then

$$G_n f = \sqrt{n+1}(\tilde{N}_n f - m_f) = \frac{N_n f - (n+1)m_f}{\sqrt{n+1}} , \qquad n \geq 0 ,$$

converges in law to a $\mathcal{N}(0, \sigma_f^2)$ Gaussian r.v.

Proof: Let $\bar{f} = f - m_f$ so that $G_n f = \frac{1}{\sqrt{n+1}} N_n \bar{f}$. On $\{N_n(x) \geq 1\}$, the decomposition (4.1.3) yields that

$$G_n f = \frac{1}{\sqrt{n+1}} \sum_{k=0}^{S_0^x - 1} \bar{f}(X_k) + \frac{1}{\sqrt{n+1}} \sum_{m=1}^{N_n(x)-1} Y_m^x \bar{f} + \frac{1}{\sqrt{n+1}} \sum_{k=D_x(n)}^{n} \bar{f}(X_k) ,$$

Moreover, $\lim_{n \to \infty} N_n(x) = \infty$ and Lemma 4.1.2 yields

$$\lim_{n \to \infty} \frac{1}{\sqrt{n+1}} \sum_{k=0}^{S_0^x - 1} \bar{f}(X_k) = 0 \text{ a.s.}, \quad \lim_{n \to \infty} \frac{1}{\sqrt{n+1}} \sum_{k=D_x(n)}^{n} \bar{f}(X_k) = 0 \text{ in probability},$$

and hence, $G_n f$ will have same limit in law as $\frac{1}{\sqrt{n}} \sum_{m=1}^{N_n(x)-1} Y_m^x \bar{f}$.

The sequence $(Y_m^x \bar{f})_{m \geq 1}$ is i.i.d., and centered as

$$\mathbb{E}(Y_1^x f) - m_f \mathbb{E}(S_1^x - S_0^x) = \mathbb{E}(Y_1^x f) - m_f \mathbb{E}_x(R_x) = 0 .$$

As $\tilde{N}_n(x)$ converges to $\pi(x)$ (see Theorem 4.1.1), for $0 < \varepsilon < 1$, there is $n_0(\varepsilon)$ large enough that, with the notation $n^{\pm}(\varepsilon) = \lfloor \pi(x)n(1 \pm \varepsilon^3) \rfloor$ and

$$A(\varepsilon) = \{\omega : n^-(\varepsilon) < N_n(x, \omega) - 1 < n^+(\varepsilon), \ n \geq n_0(\varepsilon)\} ,$$

we have $\mathbb{P}(A_\varepsilon) > 1 - \varepsilon$ and also that

$$n \geq n_0(\varepsilon) \Rightarrow n^-(\varepsilon) < \lfloor \pi(x)n \rfloor < n^+(\varepsilon) .$$

On $A(\varepsilon)$, if $n \geq n_0(\varepsilon)$, then

$$\left| \frac{1}{\sqrt{n}} \sum_{m=1}^{N_n(x)-1} Y_m^x \bar{f} - \frac{1}{\sqrt{n}} \sum_{m=1}^{\lfloor \pi(x)n \rfloor} Y_m^x \bar{f} \right| \leq \frac{2}{\sqrt{n}} \max_{n^-(\varepsilon) < k < n^+(\varepsilon)} \left| \sum_{m=n^-(\varepsilon)+1}^{k} Y_m^x \bar{f} \right|$$

and the Kolmogorov inequality, see Feller, W. (1968), p. 234, yields that

$$\mathbb{P}\left(\max_{n^-(\varepsilon) < k < n^+(\varepsilon)} \left| \sum_{m=n^-(\varepsilon)+1}^{k} Y_m^x \bar{f} \right| \geq \varepsilon \sqrt{n} \right) \leq \frac{(n^+(\varepsilon) - n^-(\varepsilon) - 1)\sigma_f^2}{\varepsilon^2 n} \leq 2\varepsilon \sigma_f^2$$

and as ε is arbitrary

$$\lim_{n\to\infty}\left|\frac{1}{\sqrt{n}}\sum_{m=1}^{N_n(x)-1}Y_m^x\bar{f}-\frac{1}{\sqrt{n}}\sum_{m=1}^{\lfloor\pi(x)n\rfloor}Y_m^x\bar{f}\right|=0\ \text{ in probability}$$

Hence, $G_n f$ will have same limit in law as $\frac{1}{\sqrt{n}}\sum_{m=1}^{\lfloor\pi(x)n\rfloor}Y_m^x\bar{f}$, which converges in law to a $\mathcal{N}(0,\sigma_f^2)$ Gaussian r.v., by the central limit theorem applied to $(Y_m^x\bar{f})_{m\geq1}$. Lemma 4.1.3 and the uniqueness of a limit allow to conclude. ∎

Theorem 14.7 and its Corollary in Chung, K.L. (1967), p. 88 give unpractical formulae for computing σ_f^2, which can only be useful in very simple cases. Statistical or Monte Carlo methods may yield good approximations for σ_f^2. The most interesting framework is the one for the pointwise ergodic theorem, when $\pi|f| < \infty$ and $m_f = \pi f$. If $\pi f^2 < \infty$ and $\mathbb{E}((Y_1^x f)^2) < \infty$

$$\sigma_f^2 = \pi(x)\mathbb{E}\left((Y_1^x(f-m_f))^2\right) = \pi(x)\mathbb{E}_x\left(\left(\sum_{k=0}^{R_x-1}(f(X_k)-\pi f)\right)^2\right)$$

$$= \pi((f-\pi f)^2) + 2\pi(x)\mathbb{E}_x\left(\sum_{0\leq i<j<R_x}(f(X_i)-\pi f)(f(X_j)-\pi f)\right)$$

$$= \pi((f-\pi f)^2) + \sum_{y,z\in\mathcal{V}}c(y,z)(f(y)-\pi f)(f(z)-\pi f)$$

in which the coefficients $c(y,z) = 2\pi(x)\mathbb{E}_x(\sum_{0\leq i<j<R_x}\mathbb{1}_{\{X_i=y,X_j=z\}})$ depend on the chain evolution in a complex way. Note that $\pi((f-\pi f)^2) = \text{var}_\pi(f(X_0))$.

It is possible to accumulate similar limit theorems using these ideas, such as an iterated logarithm law for Markov chains, see Chung, K.L. (1967), Section I.16.

4.1.3   Detailed examples

4.1.3.1   Nearest-neighbor random walks

Consider the nearest-neighbor random walk reflected at 0 on $\mathbb{N}$, with matrix P given for $x \geq 1$ by

$$P(x,x+1)=p>0\,,\quad P(x,x-1)=q=1-p>0\,,\quad P(0,1)=b>0\,.$$

We have seen that it is positive recurrent if and only if $p < 1/2$, and we have then computed its invariant law π. For instance, the proportion of time before time n spent at the boundary 0 is given by

$$\frac{1}{n+1}\sum_{k=0}^{n}\mathbb{1}_{\{X_k=0\}}\xrightarrow[n\to\infty]{\text{a.s.}}\pi(0)=\frac{q-p}{q-p+b}=\frac{1-2p}{1-2p+b}\,,$$

the limit being given by the pointwise ergodic theorem. We have seen that it is null recurrent for $p = 1/2$, and we have computed its invariant measure μ. For instance, the ratio of the time before time n spent at 0 and that spent at 1 is likewise given by

$$\frac{N_0(n)}{N_1(n)} \xrightarrow[n\to\infty]{\text{a.s.}} \frac{\mu(0)}{\mu(1)} = \frac{q}{b} = \frac{1}{2b} \, .$$

4.1.3.2   Ehrenfest Urn

See Section 1.4.4. Before time $n \geq 0$, the proportion (or fraction) of time in which compartment 1 is empty is given by

$$\frac{1}{n+1} \sum_{k=0}^{n} \mathbb{1}_{\{S_n=0\}} \xrightarrow[n\to\infty]{\text{a.s.}} \beta(0) = 2^{-N} \, ,$$

and that in which the number of molecules in compartment 1 is $\lfloor N/2 \rfloor$ by

$$\frac{1}{n+1} \sum_{k=0}^{n} \mathbb{1}_{\{S_n=\lfloor N/2 \rfloor\}} \xrightarrow[n\to\infty]{\text{a.s.}} \beta(\lfloor N/2 \rfloor) = 2^{-N} \binom{N}{\lfloor N/2 \rfloor} \sim \sqrt{\frac{2}{\pi N}} \, ,$$

see (3.2.6) and thereafter. State $\lfloor N/2 \rfloor$ is hugely more frequent than state 0. Its frequency of order $1/\sqrt{N}$ can be interpreted in terms of the central limit theorem, according to which the mass of the invariant law π is concentrated on a scale of $\sqrt{N}$ around $N/2$. The pointwise ergodic theorem yields that

$$\frac{1}{n+1} \sum_{k=0}^{n} \mathbb{1}_{\{|S_k-\frac{N}{2}|\geq a\sqrt{N}\}} \xrightarrow[n\to\infty]{\text{a.s.}} \mathbb{P}_\beta(|S_0 - \frac{N}{2}| \geq a\sqrt{N}) \simeq 2 \int_{2a}^{\infty} e^{-\frac{x^2}{2}} \frac{dx}{\sqrt{2\pi}} \, ,$$

and for instance for $N = 6 \ 10^{23}$ and $a = 2.25$ this yields

$$\lim_{n\to\infty} \frac{1}{n+1} \sum_{k=0}^{n} \mathbb{1}_{\{|S_k-3 \ 10^{23}|\geq 1,74.10^{12}\}} \simeq 6 \ 10^{-6} \, .$$

4.1.3.3   Word search

See Section 1.4.6. The invariant law π has been computed in a specific occurrence and seen that the Markov chain is irreducible. Theorem 3.2.4 or its Corollary 3.2.7 yields that the chain is positive recurrent.

The frequency of nonoverlapping occurrences of the word GAG in the first n characters of the infinite string is given by

$$\frac{1}{n} \sum_{k=1}^{n} \mathbb{1}_{\{X_k=\text{GAG}\}} \xrightarrow[n\to\infty]{\text{a.s.}} \pi(\text{GAG}) = \frac{p^2 q}{1 + pq} \, .$$

Such results can be used for statistical tests to check whether the character string is indeed generated according to the model we have described.

4.2   Long-time behavior of the instantaneous laws

4.2.1   Period and aperiodic classes

4.2.1.1   Period and aperiodicity

For certain Markov chains, certain states are periodically "forbidden": for instance, the nearest-neighbor random walk on $\mathbb{Z}$, with matrix P given by $P(x, x + 1) = p \geq 0$ and $P(x, x - 1) = 1 - p \geq 0$, alternates between even and odd states and exhibits a period of 2.

Definition 4.2.1 *Let P be a transition matrix on* $\mathcal{V}$. *The period* $d = d(x) \geq 1$ *of a state x satisfying* $x \to x$ *is the greatest common divisor (g.c.d.) of the nonempty set*

$$A(x) = \{n \geq 1 \: : \: P^n(x, x) > 0\} \; .$$

A state with period 1 is said to be aperiodic, with period $d \geq 2$ *to be periodic. If all states of a subset of* $\mathcal{V}$ *have same period or are aperiodic, then the subset is said to have that period, or to be aperiodic.*

If $x \nrightarrow x$, then the period is not defined (it could be said that it is infinity).

Note that $A(x)$ is stable by addition, as $P^{k+n}(x, x) \geq P^k(x, x)P^n(x, x)$ for $k \geq 1$ and $n \geq 1$, and that the period $d(x)$ is the least $k \geq 1$ such that x is aperiodic for the transition matrix P^k. These remarks are used in the proof of the following.

Theorem 4.2.2 *Let P be a transition matrix on* $\mathcal{V}$. *It is equivalent to say that a state x satisfying* $x \to x$ *has period d or that d is the least nonzero integer such that there exists* $j = j(x) \geq 1$ *such that* $P^{dn}(x, x) > 0$ *for all* $n \geq j$, *that is, such that*

$$\{dj, d(j + 1), \dots\} \subset A(x) \; .$$

Proof: By the previous remark, by considering P^d, we may reduce the proof to the case $d = 1$.

If $P^k(x, x) > 0$ for all $k \geq j$, then the period of x is a divisor of j and $j + 1$ and thus is 1.

Conversely, let x be of period 1. The additive subgroup of $\mathbb{Z}$ generated by $A(x)$ is not limited to $\{0\}$, and hence is of the form $m\mathbb{Z}$ in which m is its least positive element. Then, $m \geq 1$ is a divisor of every element of $A(x)$, and hence of its g.c.d., which is 1. Thus, $m = 1$ and, as $A(x)$ is stable by addition, there exists a and b in $A(x) \cup \{0\}$ such that $a - b = 1$. Necessarily $a \geq 1$ and hence $a \in A(x)$.

If $b = 0$ or $b = 1$, then $1 \in A(x)$ and hence, $A(x) = \mathbb{N}^*$ and the proof is finished, with $j(x) = 1$. Else, the two consecutive integers $b \geq 2$ and $b + 1 = a$ are in $A(x)$, so that

$$\{b, b + 1\} + \{b, b + 1\} = \{2b, 2b + 1, 2(b + 1)\}$$

is constituted of three consecutive integers that are in $A(x)$, and so on up to

$$\sum_{i=1}^{b-1}\{b, b+1\} = \{(b-1)b, (b-1)b + 1, \ldots, (b-1)(b+1) = b^2 - 1\},$$

which is constituted of b consecutive integers that are in $A(x)$, and by adding to the latter qb for $q \geq 1$, we see that all the subsequent integers are in $A(x)$, and the proof is terminated, with $j(x) = (b-1)b$.

A more concise but obscure proof can be given as follows: if $n \geq (b-1)b$, then the Euclidian division of $n - (b-1)b \geq 0$ by $b \geq 2$ yields $q \geq 0$ and $0 \leq r < b$ such that

$$n = (b-1)b + qb + r(a-b) = (b-1+q-r)b + ra$$

in which $b - 1 - r + q$ and r cannot both vanish and hence $n \in A(x)$.   ∎

Theorem 4.2.3 *Let P be a transition matrix on $\mathcal{V}$. If $x \leftrightarrow y$, then x and y have same period. Thus, the period of an irreducible class is defined as the period of one of its elements.*

Proof: Let $i \geq 1$ and $j \geq 1$ be such that $P^i(x, y) > 0$ and $P^j(y, x) > 0$. Let $d(x)$ be the period of x, and $n \geq 1$ such that $P^{d(x)n}(x, x) > 0$. Then,

$$P^{d(x)(j+n+i)}(y, y) \geq P^j(y, x)P^{d(x)n}(x, x)P^i(x, y)(P^j(y, x)P^i(x, y))^{d(x)-1} > 0$$

and hence, the period $d(y)$ of y is a divisor of $d(x)$. Similarly, $d(x)$ is a divisor of $d(y)$. Thus, $d(x) = d(y)$.   ∎

Aperiodicity, strong irreducibility, and Doeblin condition

Theorem 4.2.4 *Let P be an irreducible aperiodic transition matrix on $\mathcal{V}$. For all states x and y, there exists $k(x, y) \geq 1$ such that $P^n(x, y) > 0$ for $n \geq k(x, y)$. Moreover, if $\mathcal{V}$ is finite, then there exists $m \geq 1$ such that $P^n > 0$ for all $n \geq m$, and then in particular P is strongly irreducible and satisfies the Doeblin condition (see Corollary 1.3.5), and hence, the conclusions of Theorem 1.3.4 are satisfied with $\pi > 0$ on $\mathcal{V}$.*

Proof: There exists $i(x, y) \geq 1$ such that $P^{i(x,y)}(x, y) > 0$ as P is irreducible, and $j(y)$ such that $P^n(y, y) > 0$ for $n \geq j(y)$ as P is aperiodic (Theorem 4.2.2). Then, we may take $k(x, y) = i(x, y) + j(y)$ as

$$P^n(x, y) \geq P^{i(x,y)}(x, y)P^{n-i(x,y)}(y, y) .$$

If $\mathcal{V}$ is finite, then $P^n > 0$ for all $n \geq m := \max_{x,y \in \mathcal{V}} k(x, y)$.   ∎

4.2.1.2 Aperiodic class decomposition

Note that if P is irreducible and has period $d \geq 2$, then P^d is *not* irreducible.

Lemma 4.2.5 *Let P be a transition matrix on $\mathcal{V}$. Let x be a state with period d. Then, $x \leftrightarrow x$ for P^d, and x is transient, null recurrent, or positive recurrent simultaneously for P and for P^d. If a state y has period d and is such that $x \to y$ for P^d and $y \to x$ for P, then $x \leftrightarrow y$ for P^d.*

Proof: By definition, a state with period d communicates with itself for P^d, and it is clearly simultaneously transient, null recurrent, or positive recurrent simultaneously for P and for P^d. If there exists $i \geq 1$ such that $P^{di}(x, y) > 0$ and $j \geq 1$ such that $P^j(y, x) > 0$, then $P^{di+j}(x, x) \geq P^{di}(x, y)P^j(y, x) > 0$ and d is a divisor of j.   ∎

The statements in the following definition follow from Lemma 4.2.5 or are trivial.

Definition 4.2.6 *Let P be an irreducible transition matrix on $\mathcal{V}$ with period $d \geq 2$. The equivalence relation "x communicates with y for P^d" has d equivalence classes, called the aperiodic classes of P, and P^d restricted to each aperiodic class is irreducible and aperiodic. These classes can be numbered and called $\mathcal{A}_i$ for $0 \leq i \leq d - 1$ in such a way that if x is in $\mathcal{A}_i$, then $P(x, \mathcal{A}_j) = 1$ for $j = i + 1 \pmod{d}$, and then a Markov chain with matrix P goes successively from $\mathcal{A}_i$ to $\mathcal{A}_{i+1}$ if $0 \leq i \leq d - 2$ and from $\mathcal{A}_{d-1}$ to $\mathcal{A}_0$.*

Successive decompositions Given a transition matrix P, the state space $\mathcal{V}$ should first be decomposed in irreducible closed classes, then the restriction of P to each such class should be studied, as in Section 3.1.2. Then, the period of each of these classes should be determined, as well as its aperiodic class decomposition, so as to be able to apply important results for irreducible aperiodic transition matrices, which are soon to be stated and proved.

Period and spectrum For a measure μ on $\mathcal{V}$ and $\mathcal{W} \subset \mathcal{V}$, let $\mu|_{\mathcal{W}}$ denote its restriction to $\mathcal{W}$ as well as the extension of the latter on $\mathcal{V}$ which vanishes on $\mathcal{V} - \mathcal{W}$, the context allowing to make the difference. We do likewise for a function on $\mathcal{V}$.

Theorem 4.2.7 *Let P be an irreducible transition matrix on $\mathcal{V}$ with period d, and $\mathcal{A}_i$ for $0 \leq i \leq d - 1$ its aperiodic classes numbered as in Definition 4.2.6.*

If P has an invariant measure μ, then the $\mu|_{\mathcal{A}_i}$ for $0 \leq i \leq d - 1$ are invariant measures for P^d, are linearly independent, $\mu|_{\mathcal{A}_i} P = \mu|_{\mathcal{A}_j}$ for $j = i + 1 \pmod{d}$, and the operator P^* restricted to the vector space generated by the $\mu|_{\mathcal{A}_i}$ has eigenvalues, which are simple and given by the dth roots of the unit $e^{i2\pi k/d}$ for $0 \leq k \leq d - 1$.

If P is recurrent, then it has a unique invariant measure μ, and the space of invariant measures is the positive cone generated by the $\mu|_{\mathcal{A}_i}$ for $0 \leq i \leq d - 1$.

If P is positive recurrent, that is, has an invariant law π, then the restriction of P^d to $\mathcal{A}_i$ has unique invariant law $d\pi|_{\mathcal{A}_i}$.

If $f \neq 0$ is an harmonic function for P, then the $f|_{\mathcal{A}_i}$ for $0 \leq i \leq d-1$ are harmonic functions for P^d, are linearly independent, $P(f|_{\mathcal{A}_i}) = f|_{\mathcal{A}_j}$ for $j = i - 1 \pmod{d}$, and the operator P restricted to the vector space generated by the $f|_{\mathcal{A}_i}$ has eigenvalues, which are simple and given by the dth roots of the unit.

If P is recurrent, then the harmonic functions for P^d which are upper bounded or lower bounded are constant on each aperiodic class.

Proof: If μ is an invariant measure, then $\mu \geq 0$ and $\mu \neq 0$, and Lemma 1.3.2 or Lemma 3.2.1 shows that $\mu > 0$, so that the $\mu|_{\mathcal{A}_i}$ are linearly independent. If $j = i + 1 \pmod{d}$, then $P(x, \mathcal{A}_j) = 1$ for x in $\mathcal{A}_i$ and $P(x, \mathcal{A}_j) = 0$ for x in $\mathcal{V} - \mathcal{A}_i$, and hence $\mu|_{\mathcal{A}_i} P = \mu|_{\mathcal{A}_j}$ as

$$\sum_{x \in \mathcal{A}_i} \mu(x) P(x, y) = \begin{cases} \sum_{x \in \mathcal{V}} \mu(x) P(x, y) = \mu(y) & \text{if } y \in \mathcal{A}_j , \\ 0 & \text{if } y \notin \mathcal{A}_j . \end{cases}$$

The matrix in the basis $(\mu|_{\mathcal{A}_i})_{0 \leq i \leq d-1}$ of the restriction of P^* is given by

$$\begin{pmatrix} 0 & 0 & \cdots & 0 & 1 \\ 1 & 0 & \cdots & 0 & 0 \\ 0 & 1 & \cdots & 0 & 0 \\ \vdots & \vdots & \ddots & \vdots & \vdots \\ 0 & 0 & \cdots & 1 & 0 \end{pmatrix} = \begin{pmatrix} 0 & 1 & 0 & \cdots & 0 \\ 0 & 0 & 1 & \cdots & 0 \\ \vdots & \vdots & \vdots & \ddots & \vdots \\ 0 & 0 & 0 & 0 & 1 \\ 1 & 0 & 0 & 0 & 0 \end{pmatrix}^*$$

and has characteristic polynomial $X^d - 1$.

If P is recurrent, then so is P^d, and Theorem 3.2.3 can be used on its restriction on each $\mathcal{A}_i$, which is irreducible, and thus has unique invariant measure $\mu|_{\mathcal{A}_i}$. Denoting by $\mathbb{E}^Q$ the expectation for a Markov chain with transition matrix Q, clearly $\mathbb{E}^P_x(R_x) = d \mathbb{E}^{P^d}_x(R_x)$, and thus if π is an invariant law for P then

$$\frac{1}{\mathbb{E}^{P^d}(R_x)} = \frac{d}{\mathbb{E}^P_x(R_x)} = d\pi(x)$$

so that the invariant law for the restriction of P^d on $\mathcal{A}_i$ is $d\pi|_{\mathcal{A}_i}$. The proof for harmonic functions is similar, using Theorem 3.3.1 instead of Theorem 3.2.3. ■

4.2.1.3 Examples

Note that a state x state such that $P(x, x) > 0$ is aperiodic. For instance, the chain corresponding to the word search by an automaton is such that $P(\emptyset, \emptyset) > 0$ and thus is aperiodic as it is irreducible.

An irreducible birth-and-death chain on $\mathbb{Z}$, $\mathbb{N}$, or $\{0, 1, \ldots, N\}$ has period 2 if and only if $P(x, x) = 0$, that is, if and only if $p_x + q_x = 1$ for all states x, and then its two aperiodic classes are constituted of the subsets of the odd states and even states, else it has period 1.

Notably, the random walk on $\mathbb{Z}$ with $P(x, x+1) = p > 0$ and $P(x, x-1) = 1 - p > 0$, and the macroscopic description of the Ehrenfest Urn, see (1.4.4), have period 2, and the random walk on $\mathbb{N}$ with $P(x, x+1) = p > 0$ and $P(x+1, x) = 1 - p > 0$ and $P(0, 1) = 1 - P(0, 0) = b > 0$ has period 2 if $b = 1$ and period 1 if $b < 1$. For gambler's ruin, the absorbing states 0 and N are aperiodic, and the other states $1, \ldots, N - 1$ have period 2.

The microscopic description of the Ehrenfest Urn has period 2 and two aperiodic classes constituted of the subsets of configurations with an even number of 1 and those with an odd number of 1. Similarly, a nearest-neighbor random walk on $\mathbb{Z}^d$ such that $P(0, 0) = 0$ has period 2 and two aperiodic classes constituted of the vectors with even sum of the coordinates and of those with odd sum and is aperiodic if $P(0, 0) > 0$.

The random walk in Figure 1.1 is aperiodic, as its period must be a divisor of 2 and 3.

A snake chain based on an irreducible chain with period d will have period d on its natural state space. Note that an i.i.d. sequence constitutes an aperiodic Markov chain, as is the basic character sequence in word search.

4.2.2   Coupling of Markov chains and convergence in law

4.2.2.1   Product chain coupling

We use the notions that were introduced in Exercise 2.4. Product chains and matrices are defined in Section 1.4.7.

Theorem 4.2.8 (Coupling) *Let P be a transition matrix on $\mathcal{V}$, and μ and v be two laws on $\mathcal{V}$. Let $(X_n^1, X_n^2)_{n \geq 0}$ be a product chain on $\mathcal{V} \times \mathcal{V}$ such that X_0^1 has law μ and X_0^2 has law v (for instance, (X_0^1, X_0^2) could have law $\mu \otimes v$). Let*

$$T = \inf\{n \geq 0 : X_n^1 = X_n^2\}, \qquad Z_n^1 = \begin{cases} X_n^1 & \text{if } n \leq T, \\ X_n^2 & \text{if } n > T. \end{cases}$$

Then, $(Z_n^1)_{n \geq 0}$ is a Markov chain with matrix P and initial law μ, and

$$\|\mu P^n - v P^n\|_{\text{var}} \leq 2\mathbb{P}(Z_n^1 \neq X_n^2) = 2\mathbb{P}(T > n).$$

Proof: For the fact that $(Z_n^1)_{n \geq 0}$ is a Markov chain, we could try to invoke the strong Markov property, but devising a short and convincing proof is not obvious. We leave that as an exercise and use a more direct proof. We will use the fact that T is a stopping time for $(X_n^1, X_n^2)_{n \geq 0}$ and that $(X_n^1, X_n^2)_{n \geq 0}$ and $(X_n^2, X_n^1)_{n \geq 0}$ are both Markov chains with the same matrix $P \otimes P$. For all $x_0, \ldots, x_n$ in $\mathcal{V}$,

$$\mathbb{P}(Z_0^1 = x_0, \ldots, Z_n^1 = x_n)$$

$$= \sum_{k=0}^{n-1} \mathbb{P}(T = k, X_0^1 = x_0, \ldots, X_k^1 = X_k^2 = x_k, X_{k+1}^2 = x_{k+1}, \ldots, X_n^2 = x_n)$$

$$+ \mathbb{P}(T \geq n, X_0^1 = x_0, \ldots, X_n^1 = x_n)$$

and two applications of Theorem 2.1.1 to $(X_n^1, X_n^2)_{n \geq 0}$ yield that

$$\mathbb{P}(T = k, X_0^1 = x_0, \ldots, X_k^1 = X_k^2 = x_k, X_{k+1}^2 = x_{k+1}, \ldots, X_n^2 = x_n)$$

$$= \mathbb{P}(T = k, X_0^1 = x_0, \ldots, X_k^1 = X_k^2 = x_k)\mathbb{P}_{(x_k, x_k)}(X_1^2 = x_{k+1}, \ldots, X_{n-k}^2 = x_n)$$

$$= \mathbb{P}(T = k, X_0^1 = x_0, \ldots, X_k^1 = X_k^2 = x_k)\mathbb{P}_{(x_k, x_k)}(X_1^1 = x_{k+1}, lcdots, X_{n-k}^1 = x_n)$$

$$= \mathbb{P}(T = k, X_0^1 = x_0, \ldots, X_k^1 = X_k^2 = x_k, X_{k+1}^1 = x_{k+1}, \ldots, X_n^1 = x_n)$$

$$= \mathbb{P}(T = k, X_0^1 = x_0, \ldots, X_n^1 = x_n)$$

and by summing up

$$\mathbb{P}(Z_0^1 = x_0, \ldots, Z_n^1 = x_n) = \mathbb{P}(X_0^1 = x_0, \ldots, X_n^1 = x_n)$$

and thus $(Z_n^1)_{n \geq 0}$, like $(X_n^1)_{n \geq 0}$, is a Markov chain with matrix P. Moreover, $Z_0^1 = X_0^1$ has law μ. Hence, $\mathcal{L}(Z_n^1) = \mu P^n$ and $\mathcal{L}(X_n^2) = v P^n$ and (1.2.2) yield that

$$\|\mu P^n - v P^n\|_{\text{var}} = \sup_{\|f\|_\infty \leq 1} \mathbb{E}(f(Z_n^1) - f(X_n^2)) \leq 2\mathbb{P}(Z_n^1 \neq X_n^2) = 2\mathbb{P}(T > n)$$

(see Lemma A.2.2). ∎

Coupling, success The construction of $(Z_n^1)_{n \geq 0}$ and $(X_n^2)_{n \geq 0}$ is an example of what is called a *coupling* of two Markov chains with same matrix P and initial laws μ and v. In this case, if $T < \infty$, then the coupling is said to be successful.

4.2.2.2 Kolmogorov ergodic theorem

Theorem 4.2.9 (Kolmogorov ergodic theorem) *Let P be an irreducible aperiodic positive recurrent transition matrix on $\mathcal{V}$, and π denote its invariant law. Then, for any initial law μ,*

$$\lim_{n \to \infty} \|\mu P^n - \pi\|_{\text{var}} = 0 .$$

In particular, $\mu P^n f = \mathbb{E}_\mu(f(X_n))$ converges to πf uniformly for f in bounded sets in L^∞ and $P^n(x, y)$ converges to $\pi(y)$ for all x uniformly in y in $\mathcal{V}$, so that P^n converges term wise to the rank 1 matrix with all lines equal to π.

Proof: Theorem 4.2.8 with $v = \pi$ yields that

$$\|\mu P^n - \pi\|_{\text{var}} \leq 2\mathbb{P}(T > n) .$$

As, by monotone limit,

$$\lim_{n \to \infty} \downarrow \mathbb{P}(T > n) = \mathbb{P}(T = \infty) ,$$

we need only prove that $\mathbb{P}(T = \infty) = 0$.

Let (x_1, x_2) and (y_1, y_2) be in $\mathcal{V}^2$. The irreducibility of P yields the existence of $i_1 = i(x_1, y_1) \geq 1$ and $i_2 = i(x_2, y_2) \geq 1$ such that

$$P^{i_1}(x_1, y_1) > 0 , \qquad P^{i_2}(x_2, y_2) > 0,$$

and the aperiodicity of P and Theorem 4.2.2 yields the existence of $j_1 = j(y_1)$ and $j_2 = j(y_2)$ such that

$$P^k(y_1, y_1) > 0 , \quad k \geq j_1 , \qquad P^k(y_2, y_2) > 0 , \quad k \geq j_2 .$$

Hence, if $k \geq \max\{i_1 + j_1, i_2 + j_2\}$, then

$$(P \otimes P)^k((x_1, x_2), (y_1, y_2)) = P^k(x_1, y_1)P^k(x_2, y_2)$$
$$\geq P^{i_1}(x_1, y_1)P^{k-i_1}(y_1, y_1)P^{i_2}(x_2, y_2)P^{k-i_2}(y_2, y_2) > 0 ,$$

and thus $(X_n^1, X_n^2)_{n \geq 0}$ is irreducible (and aperiodic, but this will not be used).

The positive recurrent chain $(X_n)_{n \geq 0}$ has an invariant law π, and hence, $\pi \otimes \pi$ is an invariant law for $(X_n^1, X_n^2)_{n \geq 0}$. Thus, the irreducible chain $(X_n^1, X_n^2)_{n \geq 0}$ is positive recurrent by the invariant law criterion (Theorem 3.2.4), and hence, Lemma 3.1.3 yields that $T = \inf_{x \in \mathcal{V}} S_{(x,x)}$ is finite, a.s., so that $\mathbb{P}(T = \infty) = 0$. ■

An important application of this result is the development of Monte Carlo methods for the approximate simulation of draws from probability measures, which are interpreted as invariant laws. These methods will be further described in Section 5.2.

Periodic chains This result for aperiodic Markov chains readily yields results for chains with period other than 1. As in the study of the Ehrenfest Urn in Section 1.4.4, the constraints related to the period are considered using the decomposition in aperiodic classes.

Corollary 4.2.10 *Let P be an irreducible positive recurrent transition matrix on $\mathcal{V}$ with period $d \geq 2$. Let π be its invariant law, and the decomposition in aperiodic classes in Definition 4.2.6 be given by*

$$\mathcal{A}_0 \cup \cdots \cup \mathcal{A}_{d-1} = \mathcal{V} .$$

For every x and y in $\mathcal{V}$, there exists a unique $r = r(x, y)$ in $\{0, 1, \ldots, d-1\}$ such that $(P^{r+dn}(x, y))_{n \geq 1}$ is not null, given if x is in $\mathcal{A}_i$ and y in $\mathcal{A}_j$ by $r = j - i \pmod d$, and then

$$\lim_{n \to \infty} P^{r+dn}(x, y) = d\pi(y) .$$

For any initial law μ and $k \geq 0$,

$$\lim_{n \to \infty} \mu P^{k+dn} = d \sum_{i=1}^{d-1} \mu(\mathcal{A}_i)\pi|_{\mathcal{A}_{i+k \ (mod \ d)}} .$$

Proof: The existence, uniqueness, and formula for $r = r(x, y)$ follow from the aperiodic class decomposition. If $X_0 = x$, then $(X_{r+dn})_{n \geq 0}$ evolves in the aperiodic class of y, its restriction there is irreducible aperiodic positive recurrent, and Theorem 4.2.7 yields that its invariant law is the restriction of $d\pi$, and all this yields that $\lim_{n \to \infty} P^{r+dn}(x, y) = d\pi(y)$ using Theorem 4.2.9. ∎

Remark 4.2.11 *An irreducible aperiodic positive recurrent Markov chain or transition matrix is often said to be ergodic. This terminology does not seem very appropriate, as the pointwise ergodic theorem has nothing to do with the period, and yields results for null recurrent chains.*

4.2.2.3 Null recurrent chains

Theorem 4.2.12 *Let P be a transition matrix on $\mathcal{V}$. A recurrent state x is null recurrent if and only if $\lim_{n \to \infty} P^n(x, x) = 0$. Then $\lim_{n \to \infty} P^n(y, x) = 0$ for all y.*

Proof: By considering the recurrent class of x, the matrix P may be assumed to be irreducible recurrent. If x has period d, by considering the aperiodic class of x and P^d, the matrix P may be assumed to be aperiodic. In the proof of the Kolmogorov ergodic theorem (Theorem 4.2.9), it was shown that the product chain $(X_n^1, X_n^2)_{n \geq 0}$ is irreducible, and positive recurrent if P is positive recurrent.

If $(X_n^1, X_n^2)_{n \geq 0}$ is transient, then P is null recurrent. Moreover, then Lemma 3.1.1 yields that

$$\lim_{n \to \infty} (P \otimes P)^n((x, x), (x, x)) = P^n(x, x)^2 = 0 \, ,$$

and hence, $\lim_{n \to \infty} P^n(x, x) = 0$.

Assume now that $(X_n^1, X_n^2)_{n \geq 0}$ is recurrent. Then, $\mathbb{P}(T = \infty) = 0$, see the very end of proof of Theorem 4.2.9, and in particular, for all states x and y,

$$\lim_{n \to \infty} (P^n(x, y) - P^n(y, y)) = 0 \, . \tag{4.2.5}$$

For any z such that $(P^n(z, z))_{n \geq 0}$ does not converge to 0, there is a subsequence along which $P^n(z, z)$ converges to $\alpha(z) > 0$. A diagonal extraction of subsequences procedure allows to find a subsequence $n_0, n_1, \dots$ such that

$$\lim_{k \to \infty} P^{n_k}(y, y) = \alpha(y) \geq 0 \, , \qquad \forall y \, ,$$

and (4.2.5) yields that $\lim_{k \to \infty} P^{n_k}(x, y) = \alpha(y)$ for all x. For x in $\mathcal{V}$ and a finite subset F of $\mathcal{V}$,

$$\sum_{y \in F} \alpha(y) = \lim_{k \to \infty} \sum_{y \in F} P^{n_k}(x, y) \leq 1$$

and hence, $0 < \alpha(z) \le \sum_{y \in \mathcal{V}} \alpha(y) \le 1$, and

$$\sum_{z \in F} P^{n_k}(x, z) P(z, y) \le P^{n_k+1}(x, y) = \sum_{z \in \mathcal{V}} P(x, z) P^{n_k}(z, y) .$$

Taking first the limit as k goes to infinity of the r.h.s., using dominated convergence (Theorem A.3.5), then the limit as F tends to $\mathcal{V}$, yields

$$\sum_{z \in \mathcal{V}} \alpha(z) P(z, y) \le \sum_{z \in \mathcal{V}} P(x, z) \alpha(y) = \alpha(y) ,$$

so that α is a superinvariant measure for P. Theorem 3.2.3 yields that $\pi = \alpha / \|\alpha\|_{var}$ is an invariant law for P. Then, P is positive recurrent by the invariant law criterion.

Conversely, if P is positive recurrent, then $\lim_{n \to \infty} P^n(z, z) = \pi(z) > 0$. Lemma 3.1.1 allows to conclude. ∎

4.2.2.4   Doeblin ratio limit theorem

This uses a powerful tool, the "taboo probabilities," introduced by Chung, K.L. (1967).

Let P be an irreducible transition matrix on $\mathcal{V}$ and z a "taboo state." In order to construct a Markov chain corresponding to a Markov chain with matrix P "killed" when hitting z, we consider the enlarged state space $\mathcal{V} \cup \{†\}$, where $†$ is a "cemetery state", and the transition matrix

$$_zP = (_zP(x, y))_{x,y \in \mathcal{V} \cup \{†\}}$$

given for x in $\mathcal{V}$ and $y \ne z$ in $\mathcal{V}$ by

$$_zP(x, y) = P(x, y) , \quad _zP(x, z) = 0 , \quad _zP(x, †) = P(x, z) , \quad _zP(†, †) = 1 .$$

The transition matrix $_zP$ on $\mathcal{V} \cup \{†\}$ is determined by its sub-Markovian restriction to $\mathcal{V}$, which can be obtained from P by replacing the column vector $P(\cdot, z)$ by the null column vector. As the restriction of the n-th power of $_zP$ is the nth power of the restriction of $_zP$, we hereafter consider this restriction and denote it again by $_zP$. The main point is that, for $n \ge 0$,

$$_zP^n(x, y) = \mathbb{P}_x(R_z > n, X_n = y) , \qquad x, y \in \mathcal{V} .$$

For x, y, and z in $\mathcal{V}$, the strong Markov property and $y \to z$ yield that

$$\mathbb{E}_x\left(\sum_{n=0}^{R_z-1} \mathbb{1}_{\{X_n=y\}} \right) = \sum_{k \ge 0} \mathbb{P}_x\left(\sum_{n=0}^{R_z-1} \mathbb{1}_{\{X_n=y\}} > k \right)$$

$$= \mathbb{P}_x(S_y < R_z) \sum_{k \ge 0} \mathbb{P}_y(R_y < R_z)^k$$

$$= \frac{\mathbb{P}_x(S_y < R_z)}{\mathbb{P}_y(R_y \ge R_z)} < \infty .$$

Let the measure $_z\mu_x = (_z\mu_x(y))_{y\in V}$ be defined by

$$_z\mu_x(y) = \mathbb{E}_x\left(\sum_{n=0}^{R_z-1} \mathbb{1}_{\{X_n=y\}}\right) = \sum_{n\geq 0} zP^n(x,y) = \frac{\mathbb{P}_x(S_y < R_z)}{\mathbb{P}_y(R_y \geq R_z)} < \infty . \qquad (4.2.6)$$

In particular, $_x\mu_x = \mu_x$, the canonical invariant measure generated at x, and the computations and proof for the latter can be simplified using the notions we have just introduced.

Theorem 4.2.13 *Let P be an irreducible recurrent transition matrix on V. For x and y in V and $n \geq 0$,*

$$0 \leq \sum_{k=0}^n P^k(x,x) - \sum_{k=0}^n P^k(y,x) \leq {}_y\mu_x(x) < \infty ,$$

$$-1 \leq \frac{1}{{}_y\mu_x(x)} \sum_{k=0}^n P^k(x,x) - \frac{1}{{}_x\mu_y(y)} \sum_{k=0}^n P^k(y,y) \leq 1 .$$

Proof: Let $x \neq y$. As $P^k(y,x) = \sum_{i=0}^{k-1} \mathbb{P}_y(S_x = k - i)P^i(x,x)$,

$$\sum_{k=0}^n P^k(y,x) = \sum_{i=0}^{n-1}\sum_{k=i+1}^n \mathbb{P}_y(S_x = k - i)P^i(x,x) \leq \sum_{i=0}^{n-1} P^i(x,x) ,$$

and as $P^k(x,x) = {}_yP^k(x,x) + \sum_{i=0}^{k-1} \mathbb{P}_x(S_y = k - i)P^i(y,x)$, likewise

$$\sum_{k=0}^n P^k(x,x) \leq {}_y\mu_x(x) + \sum_{i=0}^{n-1} P^i(y,x) ,$$

from which the first inequality follows. Moreover, as

$$P^k(y,y) = {}_xP^k(y,y) + \sum_{i=1}^{k-1} P^{k-i}(y,x) \, {}_xP^i(x,y) ,$$

using the first inequality,

$$\sum_{k=0}^n P^k(y,y) \leq {}_x\mu_y(y) + \mu_x(y) \sum_{k=0}^n P^k(y,x) \leq {}_x\mu_y(y) + \mu_x(y) \sum_{k=0}^n P^k(x,x) ,$$

and as $x \neq y$, (4.2.6) yields that

$${}_x\mu_y(y) = \frac{1}{\mathbb{P}_y(R_y \geq R_x)} , \qquad {}_y\mu_x(x) = \frac{1}{\mathbb{P}_x(R_x \geq R_y)} ,$$

$$\mu_x(y) = \frac{\mathbb{P}_x(S_y < R_x)}{\mathbb{P}_y(R_y \geq R_x)} = \frac{{}_x\mu_y(y)}{{}_y\mu_x(x)},$$

and thus the second inequality holds, as x and y play symmetric roles.   ∎

This yields a quick proof for the following.

Theorem 4.2.14 (Doeblin ratio limit theorem) *Let P be an irreducible recurrent transition matrix on* $\mathcal{V}$, *and* μ *be its invariant measure. For* x_0, x, y_0, *and* y *in* $\mathcal{V}$ *and* n *in* $\mathbb{N}$,

$$\lim_{n \to \infty} \frac{\sum_{k=0}^{n} P^k(x_0, x)}{\sum_{k=0}^{n} P^k(y_0, y)} = \frac{\mu(x)}{\mu(y)} .$$

Proof: Then,

$$\frac{\sum_{k=0}^{n} P^k(x_0, x)}{\sum_{k=0}^{n} P^k(y_0, y)} = \frac{\sum_{k=0}^{n} P^k(x_0, x)}{\sum_{k=0}^{n} P^k(x, x)} \times \frac{\sum_{k=0}^{n} P^k(x, x)}{\sum_{k=0}^{n} P^k(y, y)} \times \frac{\sum_{k=0}^{n} P^k(y, y)}{\sum_{k=0}^{n} P^k(y_0, y)} .$$

In the r.h.s., the sums diverge to infinity as the chain is recurrent, and hence, the first inequality in Theorem 4.2.13 yields that the first and last ratios converge to 1 and the second inequality that the middle ratio converges to $\frac{y\mu_x(x)}{x\mu_y(y)} = \mu_y(x) = \frac{\mu(x)}{\mu(y)}$. ∎

For $x_0 = y_0$, this result can be seen as the version "in expectation" of the version of the pointwise ergodic theorem valid for null recurrent chain with $X_0 = x_0 = y_0$, $f = \mathbb{1}_{\{x\}}$ and $g = \mathbb{1}_{\{y\}}$. Conversely, the version of the pointwise ergodic theorem for positive recurrent chains can be seen as a version "in Cesaro mean, a.s." of the Kolmogorov ergodic theorem (Theorem 4.2.9).

4.2.2.5   Eigenvalues of an irreducible transition matrix

Definitions on eigenvalues and such can be found in Section 1.3.

Theorems 1.3.1 and 4.2.7 are now going to be extended, the proof of the Perron-Frobenius theorem (Theorem 1.3.6) is now going to be finished for matrices that are not strongly irreducible, and these results are going to be extended to infinite state spaces.

Theorem 4.2.15 *Let P be an irreducible transition matrix on a discrete state space* $\mathcal{V}$, *with period d.*

If P is positive recurrent, then the eigenvalues with modulus 1 of $P^* : \mu \mapsto \mu P$ *on* $\mathcal{M}$ *and* $P : f \mapsto Pf$ *on* L^∞ *are the dth roots of unity and are simple. For* $P^* : \mu \mapsto \mu P$, *the sum of their eigenspaces is generated by the restrictions of the invariant law* π *to the aperiodic classes, for* $P : f \mapsto Pf$, *it is the space of function, which are constant on each aperiodic class.*

If P is null recurrent or transient, then $P^* : \mu \mapsto \mu P$ on $\mathcal{M}$ has no eigenvalue of modulus 1.

Proof: If P is positive recurrent, then it has an invariant law π, and the restriction of P^d on each aperiodic class $\mathcal{A}_i$ is irreducible aperiodic recurrent positive with invariant law the restriction of $d\pi$. Let μ be a signed measure. The Corollary 4.2.10 of the Kolmogorov ergodic theorem yields that

$$\lim_{n\to\infty} \mu P^{dn} = d \sum_{i=0}^{d-1} \mu(\mathcal{A}_i)\pi|_{\mathcal{A}_i} \in \text{Vect}(\pi|_{\mathcal{A}_i} : 0 \le i \le d-1) := \mathcal{I} .$$

If μ is a left eigenvector for a complex eigenvalue with modulus 1 or more, then $\mu = \lambda^{-dn}\mu P^{dn}$ is in the complex extension of the space $\mathcal{I}$. It was shown in the proof of Theorem 4.2.7 that the restriction of $P^* : \mu \mapsto \mu P$ to the space $\mathcal{I}$, of dimension d, has eigenvalues given by the dth roots of unity. These eigenvalues are semisimple on $\mathcal{M}$, as else there would be a left eigenvector v and $\mu \in \mathcal{M}(\mathcal{V}, \mathbb{C})$ satisfying $\mu P^{dn} = \mu + dn\lambda^{-1}v$, which is impossible. Thus, they are simple.

The proof for $P : f \mapsto Pf$ is similar, Corollary 4.2.10 yielding

$$\lim_{n\to\infty} P^{dn}f = d \sum_{i=0}^{d-1} \pi|_{\mathcal{A}_i} f|_{\mathcal{A}_i},$$

which belongs to the space of functions that are constant on each $\mathcal{A}_i$.

If P is null recurrent or transient, then Theorem 4.2.12 or Lemma 3.1.1 together with dominated convergence (Theorem A.3.5) yields that, by considering the positive and negative parts of $\mu \in \mathcal{M}$ and $x \in \mathcal{V}$,

$$\lim_{n\to\infty} \mu P^n(x) = \lim_{n\to\infty} \sum_{y\in\mathcal{V}} \mu(y)P^n(y,x) = 0 .$$

The conclusion follows easily as earlier   ∎

4.2.3 Detailed examples

The developments on the examples in Section 4.1.3 can be transcribed here, replacing the study of asymptotic frequencies and the use of the pointwise ergodic theorem (Theorem 4.1.1) by the study of asymptotic probabilities and the use of the Kolmogorov ergodic theorem (Theorem 4.2.9).

The problems related to periods 2 or more should be tackled with due care, see, for example, the convergence results for the Ehrenfest Urn, Section 1.4.4, which can be deduced from Corollary 4.2.10 of the Kolmogorov ergodic theorem (without rates of convergence).

This would express what an observer would experience from a statistical perspective, if he arrives on the system long after it has started with arbitrary and usually unknown initial conditions.

4.3   Elements on the rate of convergence for laws

A modern field of study investigates the rates of convergence for the Kolmogorov ergodic theorem. The scope is to help develop and validate Monte Carlo methods for the approximate simulation from laws, which will be described in Section 5.2. One of the objectives of this section is to facilitate the reading of advanced books on the subject, such as Duflo, M. (1996) and Saloff-Coste, L. (1997). The latter book provides many examples of explicit refined bounds of convergence in law.

We are going to describe the functional analysis framework which is at the basis of the simplest aspects of these studies, which is an extension of the concepts in Section 1.3. If $\mathcal{V}$ is finite, then powerful results can be obtained by classic tools of finite dimensional linear algebra, which they thus illustrate. If $\mathcal{V}$ is infinite, extensions of these tools will be introduced, such as those described in the book of Rudin, W. (1991).

4.3.1   The Hilbert space framework

4.3.1.1   Fundamental Hilbert space, adjoints, and time reversal

Let $(X_n)_{n\geq0}$ be an irreducible positive recurrent Markov chain on $\mathcal{V}$ with matrix P, and π be its invariant law. Consider the functional Hilbert space $L^2(\pi) = L^2(\mathcal{V}, \pi)$, in which the scalar product of f and g is given by

$$\langle f, g \rangle_{L^2(\pi)} = \pi(fg) := \sum_{x\in\mathcal{V}} \pi(x)f(x)g(x) .$$

A fundamental probabilistic interpretation is that

$$\langle f, g \rangle_{L^2(\pi)} = \mathbb{E}_\pi(f(X_0)g(X_0)) , \quad \|f - \pi f\|^2_{L^2(\pi)} = \mathrm{var}_\pi(f(X_0)) := \mathrm{var}_\pi(f) .$$

Lemma 4.3.1 *Assume the above. The operator $P : f \mapsto Pf$ on $\in L^\infty$ can be extended to an operator on $L^2(\pi)$ with operator norm 1, still denoted by P.*

Proof: Let f be in L^∞. The Jensen inequality (Lemma A.3.6), or the Cauchy–Schwarz inequality, yields that

$$\|Pf\|^2_{L^2(\pi)} = \sum_{x\in\mathcal{V}} \pi(x)(P(x,\cdot)f)^2 \leq \sum_{x\in\mathcal{V}} \pi(x)P(x,\cdot)f^2 = \pi f^2 = \|f\|^2_{L^2(\pi)} ,$$

with equality if f is constant. The conclusion follows by a density argument.   ■

The matrix $\tilde{P}$ of the time reversal in equilibrium of P satisfies

$$\pi(x)P(x, y) = \pi(y)\tilde{P}(y, x) , \quad \forall x, y \in \mathcal{V} .$$

Hence, $\tilde{P}$ is the adjoint of P in $L^2(\pi)$, and P is reversible for π if and only if P is self-adjoint in $L^2(\pi)$. The link between adjoints and time reversal can be seen in

$$\langle f, Pg \rangle_{L^2(\pi)} = \sum_{x,y \in \mathcal{V}} \pi(x)f(x)P(x,y)g(y) = \mathbb{E}_\pi(f(X_0)g(X_1)) . \tag{4.3.7}$$

4.3.1.2 Duality between measures and functions, identification of measures and their densities

The natural duality bracket between measures and functions

$$(\mu, f) \in \mathcal{M} \times L^\infty \mapsto \mu f = \sum_{x \in \mathcal{V}} \mu(x)f(x)$$

allows to identify the dual space $L^2(\pi)^*$ with a subspace of $\mathcal{M}$. The Cauchy–Schwarz inequality yields that

$$|\mu f| = \left| \sum_{x \in \mathcal{V}} \mu(x)f(x) \right| = \left| \sum_{x \in \mathcal{V}} \frac{\mu(x)}{\sqrt{\pi(x)}} \sqrt{\pi(x)}f(x) \right| \leq \sqrt{\sum_{x \in \mathcal{V}} \frac{\mu(x)^2}{\pi(x)}} \, \|f\|_{L^2(\pi)} ,$$

with equality if f is equal to the density $\frac{\mu}{\pi}$ of μ w.r.t. π. Thus, $L^2(\pi)^*$ is a Hilbert space included in $\mathcal{M}$, with scalar product given, for μ and μ, by

$$\langle \mu, \mu \rangle_{L^2(\pi)^*} = \sum_{x \in \mathcal{V}} \frac{\mu(x)\mu(x)}{\pi(x)} = \sum_{x \in \mathcal{V}} \pi(x) \frac{\mu(x)}{\pi(x)} \frac{\mu(x)}{\pi(x)},$$

which is the scalar product in $L^2(\pi)$ of their densities w.r.t. π. The adjoint of P on $L^2(\pi)$ is the operator $P^* : \mu \mapsto \mu P$ on $L^2(\pi)^*$, which has norm 1, and

$$\frac{\mu P}{\pi}(x) = \sum_{y \in \mathcal{V}} \mu(y) \frac{P(y,x)}{\pi(x)} = \sum_{y \in \mathcal{V}} \mu(y) \frac{\tilde{P}(x,y)}{\pi(y)} = \tilde{P} \frac{\mu}{\pi} .$$

Duality formulae   The following duality formulae should be kept in mind:

$$\langle \mu, \mu \rangle_{L^2(\pi)^*} = \left\langle \frac{\mu}{\pi}, \frac{\mu}{\pi} \right\rangle_{L^2(\pi)} , \qquad \frac{\mu P}{\pi} = \tilde{P} \frac{\mu}{\pi} . \tag{4.3.8}$$

Choice of identifications and duality   These computations become clearer if we use as a reference duality bracket the scalar product of $L^2(\pi)$ itself, which identifies it with its own dual $L^2(\pi)^*$. For $\mu \in \mathcal{M}$ and $f \in L^\infty$,

$$\mu f = \left\langle \frac{\mu}{\pi}, f \right\rangle_{L^2(\pi)} , \qquad \mu P f = \left\langle \frac{\mu}{\pi}, Pf \right\rangle_{L^2(\pi)} = \left\langle \tilde{P} \frac{\mu}{\pi}, f \right\rangle_{L^2(\pi)} ,$$

which identifies $\mu \in L^2(\pi)^*$ to its density $\frac{\mu}{\pi} \in L^2(\pi)$ and $P^* : \mu \mapsto \mu P$ to $\tilde{P} : \frac{\mu}{\pi} \mapsto \tilde{P} \frac{\mu}{\pi}$. Note that the total variation norm can be expressed as the $L^1(\pi)$ norm of the densities.

We elect to use the natural duality already used in Section 1.3 and use the duality formulae (4.3.8), even though the notations $L^2(\pi)$ and $L^2(\pi)^*$ are compatible with both duality frameworks.

This duality could be identified with the natural duality between the sequence spaces ℓ^1 and ℓ^∞, in which case $L^2(\pi)$ and $L^2(\pi)^*$ are identified with the spaces $\ell^2(\pi)$ and $\ell^2(\pi^{-1})$ of square-summable sequences with the corresponding weights, and the space ℓ^2 is the pivot space in the Gelfand triple

$$\ell^2(\pi^{-1}) \subset \ell^2 \subset \ell^2(\pi) \,.$$

All this makes for a better understanding of Section 3.3.3.

4.3.1.3   Simple bounds and orthogonal projections

Lemma 4.3.2 *Let P be an irreducible transition matrix on $\mathcal{V}$ having an invariant law π. For every function f and signed measure μ,*

$$\|f\|_{L^2(\pi)} \leq \|f\|_\infty \leq \sqrt{\frac{1}{\inf \pi}}\, \|f\|_{L^2(\pi)} \,,$$

$$\|\mu\|_{\mathrm{var}} \leq \|\mu\|_{L^2(\pi)^*} \leq \sqrt{\frac{1}{\inf \pi}}\, \|\mu\|_{\mathrm{var}} \,,$$

in which $\inf \pi := \inf_{x \in \mathcal{V}} \pi(x) > 0$ *if and only if $\mathcal{V}$ is finite.*

Proof: The result for f is obvious. The duality formula (4.3.8) yields the last equality in

$$\|\mu\|_{\mathrm{var}} = \left|\frac{\mu}{\pi}\right|_{L^1(\pi)} \leq \left|\frac{\mu}{\pi}\right|_{L^2(\pi)} = \|\mu\|_{L^2(\pi)^*} \,.$$

Moreover, we may assume that $\mu \geq 0$, then by homogeneity that μ is a probability measure, and then all its terms are bounded by 1 and

$$\|\mu\|^2_{L^2(\pi)^*} = \sum_{x \in \mathcal{V}} \frac{\mu(x)^2}{\pi(x)} \leq \frac{1}{\inf \pi} \sum_{x \in \mathcal{V}} \mu(x) = \frac{1}{\inf \pi} \,.$$

As $\pi > 0$ is summable, $\inf_{x \in \mathcal{V}} \pi(x) > 0$ if and only if $\mathcal{V}$ is finite. ∎

Lemma 4.3.3 *Let P be an irreducible transition matrix on $\mathcal{V}$ having an invariant law π. For f in $L^2(\pi)$ and μ in $L^2(\pi)^*$,*

$$\|f - \pi f\|^2_{L^2(\pi)} = \|f\|^2_{L^2(\pi)} - (\pi f)^2 \,, \quad \|\mu - \mu(\mathcal{V})\pi\|^2_{L^2(\pi)^*} = \|\mu\|^2_{L^2(\pi)^*} - \mu(\mathcal{V})^2 \,,$$

hence πf is the orthogonal projection of f in $L^2(\pi)$ on the line of the constant functions, and $\mu(\mathcal{V})\pi$ is the orthogonal projection of μ in $L^2(\pi)^*$ on the line generated *by π. In particular, if $\mu(\mathcal{V}) = 1$ (for instance, if μ is a probability measure), then its orthogonal projection is π and*

$$\|\mu - \pi\|^2_{L^2(\pi)^*} = \|\mu\|^2_{L^2(\pi)^*} - 1 .$$

Proof: Then,

$$\|f - \pi f\|^2_{L^2(\pi)} = \mathrm{Var}_\pi(f) = \pi f^2 - (\pi f)^2 = \|f\|^2_{L^2(\pi)} - (\pi f)^2 .$$

By duality, using (4.3.8) and $\pi\frac{\mu}{\pi} = \mu(\mathcal{V})$, it holds that

$$\|\mu - \mu(\mathcal{V})\pi\|^2_{L^2(\pi)^*} = \|\mu\|^2_{L^2(\pi)^*} - \mu(\mathcal{V})^2 .$$

The statements on the projections follow easily. ∎

4.3.2   Dirichlet form, spectral gap, and exponential bounds

4.3.2.1   Dirichlet form, spectral gap, and Poincaré inequality

Theorem 4.3.4 *Let $(X_n)_{n\geq 0}$ be an irreducible Markov chain on $\mathcal{V}$ with matrix P, having an invariant law π. Its* Dirichlet form *is given by the quadratic form*

$$\mathcal{A}_P : f \in L^2(\pi) \mapsto \mathcal{A}_P(f,f) = \langle (I - P)f, f\rangle_{L^2(\pi)} ,$$

and satisfies

$$0 \leq \mathcal{A}_P(f,f) = \mathcal{A}_{\tilde{P}}(f,f) = \mathcal{A}_{\frac{P+\tilde{P}}{2}}(f,f) \leq 2\mathrm{Var}_\pi(f) = 2\|f - \pi f\|^2_{L^2(\pi)} ,$$

$$\mathcal{A}_P(f,f) = \frac{1}{2}\mathbb{E}_\pi((f(X_1) - f(X_0))^2) = \frac{1}{2}\sum_{x,y\in\mathcal{V}} \pi(x)P(x,y)(f(y) - f(x))^2 ,$$

and its kernel is constituted of the constant functions, which are the only harmonic functions in $L^2(\pi)$. Its spectral gap *is given by*

$$\lambda_P = \inf\left\{ \frac{\mathcal{A}_P(f,f)}{\mathrm{Var}_\pi(f)} : f \in L^2(\pi), \ \mathrm{Var}_\pi(f) \neq 0 \right\}$$

$$= \inf\{\mathcal{A}_P(f,f) : f \in L^2(\pi), \ \pi f = 0, \ \pi f^2 = 1 \}$$

and satisfies $0 \leq \lambda_P = \lambda_{\tilde{P}} = \lambda_{\frac{P+\tilde{P}}{2}} \leq 2$. For $A > 0$,

$$\lambda_P \geq 1/A > 0 \iff \mathrm{Var}_\pi(f) \leq A\mathcal{A}_P(f,f) , \ \forall f \in L^2(\pi) ,$$

and such an inequality is called a Poincaré inequality. *If $\mathcal{V}$ is finite, then $\lambda_P > 0$.*

Proof: As $\langle Pf, f \rangle_{L^2(\pi)} = \langle \tilde{P}f, f \rangle_{L^2(\pi)}$,

$$\langle (I-P)f, f \rangle_{L^2(\pi)} = \langle (I-\tilde{P})f, f \rangle_{L^2(\pi)} ,$$

and hence, $\mathcal{A}_P(f,f) = \mathcal{A}_{\tilde{P}}(f,f) = \mathcal{A}_{\frac{P+\tilde{P}}{2}}(f,f)$. As in equilibrium X_1 has law π, using (4.3.7),

$$\mathcal{A}_P(f,f) = \mathbb{E}_\pi(f(X_0)^2 - f(X_1)f(X_0)) = \frac{1}{2}\mathbb{E}_\pi(f(X_1)^2 + f(X_0)^2 - 2f(X_1)f(X_0)) ,$$

and for $\bar{f} = f - \pi f$,

$$0 \le \mathcal{A}_P(f,f) = \frac{1}{2}\mathbb{E}_\pi((\bar{f}(X_1) - \bar{f}(X_0))^2) \le \mathbb{E}_\pi(\bar{f}(X_1)^2 + \bar{f}(X_0)^2) = 2\mathrm{Var}_\pi(f) .$$

Moreover,

$$\mathcal{A}_P(f,f) = \frac{1}{2}\mathbb{E}_\pi((f(X_1) - f(X_0))^2) = \frac{1}{2}\sum_{x,y \in \mathcal{V}} \pi(x)P(x,y)(f(y) - f(x))^2$$

vanishes if and only if $f(x) = f(y)$ for all x and y such that $P(x,y) > 0$, and by irreducibility this happens if and only if f is constant. If f is harmonic, then

$$\mathcal{A}_P(f,f) = \langle (I-P)f, f \rangle_{L^2(\pi)} = 0$$

and thus f is constant. The equality in the definition of λ_P follows from the fact that both $\mathcal{A}_P(f,f)$ and $\mathrm{Var}_\pi(f)$ are quadratic and do not change if a constant is added to f, and the following inequality follows from the inequality for Dirichlet forms. The equivalence with the Poincaré inequality is a reinterpretation of the definition. If $\mathcal{V}$ is finite, then $L^2(\pi)$ is finite dimensional, and the continuous function $f \mapsto \mathcal{A}_P(f,f)$ is bounded below above 0 on the compact set

$$\{f \in L^2(\pi) : \pi f = 0, \ \pi f^2 = 1\},$$

which contains no constant functions, and must attain its nonnull infimum.   ∎

The notations $\mathcal{A}$ and λ can be used if the context is clear. If $\lambda > 0$, then often it is said that "there is a spectral gap."

4.3.2.2 Exponential bounds for convergence in law

Theorem 4.3.5 *Let P be an irreducible transition matrix on $\mathcal{V}$ having an invariant law π. For $n \ge 1$ and f in $L^2(\pi)$ and μ in $L^2(\pi)^*$,*

$$\|P^n f - \pi f\|_{L^2(\pi)} \le \left(\sqrt{1 - \lambda_{\tilde{P}P}}\right)^n \|f - \pi f\|_{L^2(\pi)} ,$$

$$\|\mu P^n - \mu(\mathcal{V})\pi\|_{L^2(\pi)^*} \le \left(\sqrt{1 - \lambda_{P\tilde{P}}}\right)^n \|\mu - \mu(\mathcal{V})\pi\|_{L^2(\pi)^*} ,$$

and the inequalities for $n = 1$ are optimal. Moreover, $\lambda_{\tilde{P}P} = \lambda_{P\tilde{P}} \le 1$.

Proof: Lemma 4.3.3 and $\langle Pf, Pf \rangle_{L^2(\pi)} = \langle \tilde{P}Pf, f \rangle_{L^2(\pi)}$ yield that

$$\|Pf - \pi f\|_{L^2(\pi)}^2 - \|f - \pi f\|_{L^2(\pi)}^2 = \langle Pf, Pf \rangle_{L^2(\pi)} - \langle f, f \rangle_{L^2(\pi)}$$
$$= -Q_{\tilde{P}P}(f, f) \, ,$$

and using $\|f - \pi f\|_{L^2(\pi)}^2 = \mathrm{Var}_\pi(f)$ this yields, by definition, the optimal inequality

$$\|Pf - \pi f\|_{L^2(\pi)}^2 \leq \left(1 - \frac{Q_{\tilde{P}P}(f,f)}{\mathrm{Var}_\pi(f)} \right) \|f - \pi f\|_{L^2(\pi)}^2 \leq (1 - \lambda_{\tilde{P}P}) \|f - \pi f\|_{L^2(\pi)}^2 \, .$$

The optimal inequality

$$\|\mu P - \mu(\mathcal{V})\pi\|_{L^2(\pi)^*}^2 \leq (1 - \lambda_{P\tilde{P}}) \|\mu - \mu(\mathcal{V})\pi\|_{L^2(\pi)^*}^2$$

is obtained similarly or by duality using (4.3.8) and the first inequality for $\tilde{P}$. These inequalities yield that $\lambda_{\tilde{P}P} \leq 1$ and $\lambda_{P\tilde{P}} \leq 1$, and by iteration the bounds for $\|P^n f - \pi f\|_{L^2(\pi)}$ and $\|\mu P^n - \mu(\mathcal{V})\pi\|_{L^2(\pi)^*}$. Lemma 4.3.3 yields that

$$\|\mu P - \mu(\mathcal{V})\pi\|_{L^2(\pi)^*} = \sup_{\|f\|_{L^2(\pi)} \leq 1} (\mu P - \mu(\mathcal{V})\pi)f$$
$$= \sup_{\|f\|_{L^2(\pi)} \leq 1} (\mu - \mu(\mathcal{V})\pi)(Pf - \pi f)$$
$$\leq \sup_{\|g\|_{L^2(\pi)} \leq \sqrt{1 - \lambda_{\tilde{P}P}}} (\mu P - \mu(\mathcal{V})\pi)g$$
$$= \sqrt{1 - \lambda_{\tilde{P}P}} \, \|\mu - \mu(\mathcal{V})\pi\|_{L^2(\pi)^*} \, ,$$

and the optimality of

$$\|\mu P - \mu(\mathcal{V})\pi\|_{L^2(\pi)^*} \leq \sqrt{1 - \lambda_{P\tilde{P}}} \, \|\mu - \mu(\mathcal{V})\pi\|_{L^2(\pi)^*}$$

yields that $\lambda_{P\tilde{P}} \geq \lambda_{\tilde{P}P}$. By replacing P by $\tilde{P}$, this yields that $\lambda_{\tilde{P}P} \geq \lambda_{P\tilde{P}}$, and thus $\lambda_{\tilde{P}P} = \lambda_{P\tilde{P}}$. ∎

Exponential convergence   This almost tautological result justifies the notions that have been introduced: if $\lambda_{\tilde{P}P} = \lambda_{P\tilde{P}} > 0$, then $\sqrt{1 - \lambda_{P\tilde{P}}} < 1$ provides a geometric convergence rate for $(\pi_n)_{n \geq 0}$ toward π. If $\mathcal{V}$ is finite, it is so as soon as $\tilde{P}P$ is irreducible, and this will be generalized to an arbitrary irreducible aperiodic matrix P in Exercise 4.10.

If P is not irreducible or is not aperiodic, clearly $\tilde{P}P$ and $P\tilde{P}$ are not irreducible, and then $\lambda_{\tilde{P}P} = \lambda_{P\tilde{P}} = 0$. Exercise 4.6 will show that $\tilde{P}P$ or $P\tilde{P}$ may well not be irreducible, even if P is irreducible aperiodic.

Spectral gap bounds   An effective method for finding explicit lower bounds for the spectral gap can be to establish Poincaré inequalities using graph techniques. This will be done in Exercises 4.11 and 4.12. This technique is developed on many examples in Duflo, M. (1996) and Saloff-Coste, L. (1997), Chapter 3.

4.3.2.3   Application to the chain with two states

The Markov chain on $\{1,2\}$ with matrix $P = (\begin{smallmatrix} 1-a & a \\ b & 1-b \end{smallmatrix})$ for $0 \le a, b \le 1$ is irreducible if and only if $ab > 0$ and aperiodic if and only if $ab < 1$.

If $a = b = 0$, then $P = I$, any law is invariant, and $\lambda_P = 0$. Else, the only invariant law is $\pi = (\frac{b}{a+b}, \frac{a}{a+b})$ and P is reversible for π, and $\pi > 0$ if and only if $ab > 0$. Moreover,

$$A_P(f,f) = \frac{ab}{a+b}(f(2) - f(1))^2,$$

and $\pi f = 0$ and $\pi f^2 = 1$ imply that, up to sign for f,

$$f(1) = \sqrt{\frac{a}{b}}, \quad f(2) = -\sqrt{\frac{b}{a}}, \quad A_P(f,f) = \frac{ab}{a+b}\left(\frac{a}{b} + \frac{b}{a} + 2\right)^2 = a + b.$$

Thus, $\lambda_P = a + b$ varies continuously, between 0 when $a = b = 0$ and the chain is not irreducible and 2 when $a = b = 1$ and the chain has period 2.

As $\tilde{P} = P$ and hence $1 - \lambda_{P\tilde{P}} = 1 - \lambda_{\tilde{P}P} = 1 - \lambda_{P^2}$ and

$$P^2 = \begin{pmatrix} 1 - (2a - a^2 - ab) & 2a - a^2 - ab \\ 2b - b^2 - ab & 1 - (2b - b^2 - ab) \end{pmatrix}, \quad 1 - \lambda_{P^2} = (1 - (a+b))^2,$$

and the explicit form for P^n given in Section 1.3.3 shows that the exponential bounds in Theorem 4.3.5 cannot be improved.

4.3.3   Spectral theory for reversible matrices

4.3.3.1   General principles

The spectrum $\sigma(A)$ of a bounded operator A on a Banach space is the set of all $\lambda \in \mathbb{C}$ such that $\lambda I - A$ is not invertible. The spectral radius is given by $\sup |\sigma(A)|$.

If P is a transition matrix having an invariant law π, then $\frac{P+\tilde{P}}{2}$ is reversible w.r.t. π and $\lambda_P = \lambda_{\frac{P+\tilde{P}}{2}}$, and $\tilde{P}P$ and $P\tilde{P}$ are both reversible w.r.t. π and nonnegative: $\sigma(\tilde{P}P) \ge 0$ and $\sigma(P\tilde{P}) \ge 0$. In this way, we can reduce some problems to reversible matrices. These correspond to self-adjoint operators on a Hilbert space, of which the spectral theory is a powerful tool developed, for example, in Rudin, W. (1991), Chapter 12.

If P is an irreducible transition matrix on a state space $\mathcal{V}$, which is reversible w.r.t. a probability measure π, then P is self-adjoint in the Hilbert space $L^2(\pi)$, and hence, its spectrum $\sigma(P)$ is real, see Rudin, W. (1991) Theorem 12.15. Theorem 4.2.15 then yields that the only elements of the spectrum of modulus 1 can be 1 (always a simple eigenvalue) and -1 and that -1 is in the spectrum if and only if P has period 2 and then it is a simple eigenvalue.

4.3.3.2   Finite state spaces

Let P be an irreducible transition matrix, which is reversible w.r.t. a probability measure π, on a *finite* state space $\mathcal{V}$.

Classic results of linear algebra and Theorem 4.2.15 yield that the spectrum $\sigma(P)$ is constituted of $d = \mathrm{Card}(\mathcal{V})$ (possibly repeated) real eigenvalues

$$1 = \beta_1 > \beta_2 \geq \cdots \geq \beta_d \geq -1$$

and that P can be diagonalized in an orthonormal basis of $L^2(\pi)$ constituted of corresponding eigenvectors $\psi_1 = 1, \psi_2, \ldots, \psi_d$. This yields the spectral decomposition

$$Pf = \sum_{i=1}^{d} \beta_i \langle \psi_i, f \rangle_{L^2(\pi)} \psi_i = \pi f + \sum_{\beta \in \sigma(P) - \{1\}} \beta \psi(\beta) f, \quad f \in L^2(\pi), \qquad (4.3.9)$$

in which

$$\psi(\beta) := \sum_{i : \beta_i = \beta} \beta_i \langle \psi_i, \cdot \rangle_{L^2(\pi)} \psi_i$$

denotes the orthogonal projection in $L^2(\pi)$ on the eigenspace of β. The expression in terms of these projections is unique, even though the orthonormal basis is not.

Setting

$$\rho_P = \sup\{|\beta| : \beta \in \sigma(P) - \{1\}\} = \max\{\beta_2, |\beta_d|\},$$

for $n \geq 1$, it holds that

$$\|P^n f - \pi f\|_{L^2(\pi)}^2 = \sum_{\beta \in \sigma(P) - \{1\}} \beta^{2n} \|\psi(\beta) f\|_{L^2(\pi)}^2 \leq \rho_P^{2n} \|f - \pi f\|_{L^2(\pi)}^2 \qquad (4.3.10)$$

with equality if and only if f is in the vector space generated by the eigenvalues with modulus ρ_P. If P has period 2, then -1 is an eigenvalue and hence $\rho_P = 1$, else if P is aperiodic, then $\rho_P < 1$ gives the geometric rate of convergence. The corresponding result for $\|\mu P^n - \mu(\mathcal{V})\pi\|_{L^2(\pi)^*}$ is obtained analogously or by duality.

Ehrenfest Urn   As an example, we refer to the macroscopic description of the Ehrenfest Urn in Section 1.4.4. Its eigenvalues are of the form

$$\beta_{i+1} = \frac{N - 2i}{N}, \quad 0 \leq i \leq N, \qquad \rho_P = \max\left(\frac{N-2}{N}, |-1|\right) = 1,$$

the latter as P has period 2. Moreover, the corresponding eigenvectors were computed.

4.3.3.3   General state spaces

When $\mathcal{V}$ is infinite, the spectral decomposition in terms of orthogonal projections
(4.3.9) can be generalized in integral form into a *resolution of the identity*, using
the spectral theorem given, for example, in Rudin, W. (1991) Theorem 12.23. This
yields results analogous to (4.3.10). We state without further explanations the following important result, which we have just proved when $\mathcal{V}$ is finite and hence,
$L^2(\pi)$ is finite dimensional, and which is still classic in infinite dimensions.

Theorem 4.3.6 *Let P be an irreducible transition matrix which is reversible w.r.t.
a probability measure π, on an arbitrary state space $\mathcal{V}$. Let*

$$\rho_P = \sup\{|\beta| : \beta \in \sigma(P) - \{1\}\} \le 1 \ .$$

For $n \ge 0$ and f in $L^2(\pi)$ and μ in $L^2(\pi)^*$, the following optimal inequalities hold:

$$\|P^n f - \pi f\|_{L^2(\pi)} \le \rho_P^n \|f - \pi f\|_{L^2(\pi)} \ ,$$

$$\|\mu P^n - \mu(\mathcal{V})\pi\|_{L^2(\pi)^*} \le \rho_P^n \|\mu - \mu(\mathcal{V})\pi\|_{L^2(\pi)^*} \ .$$

If $\rho_P < 1$, then $\|P^n f - \pi f\|_{L^2(\pi)}$ and $\|\mu P^n - \mu(\mathcal{V})\pi\|_{L^2(\pi)^*} \ge \|\mu P^n - \mu(\mathcal{V})\pi\|_{\mathrm{var}}$
converge to 0 exponentially as n goes to infinity. Moreover, $\rho_P = 1$ if and only if
1 or -1 are in the closure of $\sigma(P) - \{1\}$, which can happen when $\mathcal{V}$ is infinite
even when P is irreducible aperiodic. In this situation, a modern research topic is
to establish polynomial rates of convergence bounds.

4.3.3.4   Relation with spectral gaps and Dirichlet forms

The advantage of the Dirichlet form techniques is that they can be applied to a
transition matrix P which is not necessarily reversible w.r.t. its invariant law π.
Moreover, it is often difficult to estimate a spectrum, and the following result allows
the estimation of ρ_P from the spectral gap λ_P, and the latter can often be estimated
using, for instance, Poincaré inequalities.

Lemma 4.3.7 *Let P be an irreducible transition matrix on $\mathcal{V}$, which is reversible
w.r.t. a probability measure π. Then, $\lambda_P = \inf(\sigma(I - P) - \{0\})$, if P is nonnegative,
then $\rho_P = 1 - \lambda_P$, and in general $\rho_P = \sqrt{1 - \lambda_{P^2}}$. In particular, if P is an irreducible transition matrix on $\mathcal{V}$ which is not necessarily reversible with respect its
invariant law π, and if $P\tilde{P}$ is irreducible, then $\rho_{P\tilde{P}} = 1 - \lambda_{P\tilde{P}}$.*

Proof: By definition

$$\lambda_P = \inf\{\langle (I - P)f, f \rangle_{L^2(\pi)} : f \in L^2(\pi),\ \pi f = 0 \ ,\ \pi f^2 = 1\} \ .$$

Lemma 4.3.3 yields that $\{f \in L^2(\pi) : \pi f = 0\}$ is the orthogonal space to the space
of constant functions. If $I - P$ is self-adjoint, then λ_P is the infimum of the spectrum

of the restriction of $I - P$ to this orthogonal space, and this spectrum is $\sigma(I - P) - \{0\}$ as Theorem 4.3.4 yields that only the constant functions are in the kernel of $I - P$. If moreover P is nonnegative, then $\sigma(P) \geq 0$ and hence,

$$\rho_P = \sup\{\beta : \beta \in \sigma(P) - \{1\}\} = 1 - \lambda_P ,$$

and in general P^2 is irreducible, reversible w.r.t. π, and nonnegative, and hence,

$$\rho_P^2 = \sup\{\beta^2 : \beta \in \sigma(P) - \{1\}\} = \sup\{\beta : \beta \in \sigma(P^2) - \{1\}\} = 1 - \lambda_{P^2} .$$

The last result follows by applying what we have just proved to the reversible and nonnegative transition matrix $P\tilde{P}$. ∎

4.3.4   Continuous-time Markov chains

The theory of continuous-time Markov chains $(X_t)_{t\in\mathbb{R}_+}$ is simpler than the theory for discrete-time Markov chains for everything concerning the long-time limits of the instantaneous law and their rates of convergence, for instance in terms of Dirichlet forms, spectral gaps, and spectral theory. Notably, the notion of period disappears.

If the generator (or q-matrix) $Q = (Q(x, y))_{x,y\in V}$ is bounded as an operator on L^∞, that is, if

$$\|Q\|_{op} = \sup_{\|f\|_\infty \leq 1} \|Qf\|_\infty = 2\sup_{x\in V}\sum_{y\neq x} Q(x, y) = 2\sup_{x\in V} |Q(x, x)| < \infty ,$$

then the transition semigroup $(P_t)_{t\in\mathbb{R}_+}$ with generic term $P_t(x, y) = \mathbb{P}_x(X_t = y)$ is given by the sum of the exponential series

$$P_t = e^{tQ} := \sum_{k\geq 0} \frac{t^k Q^k}{k!} ,$$

which is normally convergent for the operator norm $\| \cdot \|_{op}$, and the Gronwall Lemma yields that it is the unique solution of the Kolmogorov equations

$$\frac{d}{dt}P_t = QP_t , \qquad \frac{d}{dt}P_t = P_tQ .$$

Moreover, $Q = \frac{1}{2}\|Q\|_{op}(P - I)$ for some transition matrix P, and the evolution of a continuous-time Markov chain with bounded generator Q corresponds to jumping at the instants of a Poisson process of intensity $\frac{1}{2}\|Q\|_{op}$ according to a discrete-time Markov chain with transition matrix P.

Hence, Saloff-Coste, L. (1997, Sections 1.3.1 and 2.1.1) considers the continuous-time Markov chain with generator $Q = P - I$. The first inequality of the proof of Theorem 4.3.5 is replaced if $\pi f = 0$ by

$$\frac{d}{dt}\langle P_t f, P_t f\rangle_{L^2(\pi)} = 2\langle QP_t f, P_t f\rangle_{L^2(\pi)}$$

$$= -2A_P(P_t f, P_t f)$$

$$\leq -2\lambda_P\langle P_t f, P_t f\rangle_{L^2(\pi)},$$

obtained by the Kolmogorov equations and the differentiation of bilinear forms, yielding the inequalities

$$\|P_t f - \pi f\|_{L^2(\pi)} \leq e^{-\lambda_P t}\|f - \pi f\|_{L^2(\pi)},$$

$$\|\mu P_t - \mu(\mathcal{V})\pi\|_{L^2(\pi)^*} \leq e^{-\lambda_P t}\|\mu - \mu(\mathcal{V})\pi\|_{L^2(\pi)^*},$$

in which the spectral gap λ_P appears directly. If P and hence Q are reversible, formulae such as (4.3.10) become

$$\|P_t f - \pi f\|^2_{L^2(\pi)} = \sum_{\beta\in\sigma(Q)-\{0\}} e^{2\beta}\|\psi(\beta)f\|^2_{L^2(\pi)} \leq e^{-2\lambda_P}\|f - \pi f\|^2_{L^2(\pi)}$$

in which $\lambda_P = -\sup(\sigma(Q) - \{0\}) = \inf(\sigma(I - P) - \{0\})$ is again the spectral gap, which explains this denomination.

Exercises

4.1 **The space station, the mouse, and three-card Monte** See Exercises 1.1, 1.2, and 1.5.

a) What is the asymptotics for the proportion of time in which the astronaut is in the central unit? in which the mouse is in room 1? in which the three cards are in their initial positions?

b) Find the periods of the chains.

c) After a very long period of time, an asteroid hits one of the peripheral units of the space station. Estimate the probability that the astronaut happens to be in the unit when this happens.

d) The entrance door of the apartment is in room 5, and after a very long period of time, the tenant comes back home. Give an approximation for the probability that he does so precisely when the mouse is in room 5.

e) For three-card Monte, give an approximation of the probability that after 1000 steps the three cards are in their initial positions. Same question after 1001 steps.

f) For three-card Monte, an on-looker waits for a large number of card exchanges and then designates the middle card as the ace of spades. He bets 50 dollars, which he will loose if he is wrong and which will be given back to him with an additional 50 dollars if he is right. Give an approximation for the expectation of the sum he will end up with. If $p = 1/2$ and this large number is 10, give an order of magnitude in the error made in this approximation.

4.2 Difficult advance, see Exercise 3.9 For $a^k < 1 - a$, give the long-time limit of the probability that the chain is in $k\mathbb{Z}$. For $a^k = 1 - a$, give for $j \geq 0$ the long-time limit of the ratio between the time spent in jk and the time spent in $\{jk + 1, \ldots, jk + k - 1\}$.

4.3 Queue with K servers, see Exercise 3.10 When $\rho < K$, give the long-time limit of the fraction of time that all K servers are operating simultaneously.

4.4 Centered random walk, 1-d   On $\mathbb{Z}$, let $(\xi_i)_{i\geq 1}$ be a sequence of i.i.d. integrable r.v., X_0 be an independent r.v., and $X_n = X_0 + \xi_1 + \cdots + \xi_n$ be the corresponding random walk. Let $\mathbb{E}(\xi_1) = 0$, recall that then $(X_n)_{n\geq 0}$ is null recurrent, and let $\mathbb{P}(\xi_1 = 0) \neq 1$ and $\mathbb{P}(\xi_1 < -1) = 0$. Let $U_0 = 0$ and $U_{k+1} = \inf\{n > U_k : X_n \geq X_{U_k}\}$ for $k \geq 0$.

a) Prove that the Markov chain is irreducible on $\mathbb{Z}$, that the uniform measure is the unique invariant measure, and that the U_k are stopping times.

b) Prove that if $x \leq 0$, then $\sum_{n=0}^{\infty} \mathbb{P}_0(U_1 > n, X_n = x) = 1$.

c) Prove that if $y \leq 0$, then

$$\mathbb{P}_0(X_{U_1} = y) = \sum_{n=0}^{\infty} \mathbb{P}_0(U_1 > n, X_{n+1} = y) = \mathbb{P}(\xi_1 \geq y) .$$

d) Prove that the $(X_{U_k} - X_{U_{k-1}})_{k\geq 1}$ are i.i.d. and have same law as X_{U_1} when $X_0 = 0$.

e) Let now ξ_1 be square integrable, and $\mathbb{E}(\xi_1^2) = \mathrm{Var}(\xi_1) := \sigma^2$. Prove that $\frac{1}{n}X_{U_n}$ converges a.s. to $\sigma^2/2$.

4.5 Period and adjoint Let P be an irreducible transition matrix with period $d \geq 2$ having an invariant measure μ, and $\tilde{P}$ the adjoint of P w.r.t. μ. Prove that $\tilde{P}$ has period d and describe its decomposition in aperiodic classes. Are the matrices $\tilde{P}P$ and $P\tilde{P}$ irreducible?

4.6 Renewal process Let $m \geq 1$, the renewal process $(X_n)_{n\geq 1}$ with $p_0 = \cdots = p_{m-1} = 1$ and $p_m = p_{m+1} = \cdots = 1/2$, and its matrix P.

a) Prove that P is irreducible, aperiodic, and positive recurrent.

b) Give the limit for large n of $\frac{1}{n+1}\sum_{n=0}^{n} X_n$.

c) Give the long-time limit of the instantaneous probability that the age of the component is greater than or equal to m. Give a bound for the distance between this limit probability and the probability at time $11(m+1)$.

d) Determine for $k \geq 1$ the recurrent classes of the matrices $P^k \tilde{P}^k$ and $\tilde{P}^k P^k$. For which k are these matrices irreducible?

4.7 Distance to equilibrium Let $(X_n)_{n\geq 0}$ be an irreducible Markov chain on $\mathcal{V}$ with matrix P having an invariant law π.

a) Prove that, for f in $L^2(\pi)$ and μ in $L^2(\pi)^*$ and $0 \leq k \leq n$,

$$|\mu P^n f - \mu(\mathcal{V})\pi f| \leq \|\mu P^k - \mu(\mathcal{V})\pi\|_{L^2(\pi)^*} \times \|P^{n-k}f - \pi f\|_{L^2(\pi)} .$$

b) Prove that if $X_0 = x$, then $\|\pi_n - \pi\|^2_{L^2(\pi)^*} = \frac{P^n \tilde{P}^n(x,x)}{\pi(x)} - 1$.

4.8 Dirichlet form Let P be an irreducible transition matrix and μ an invariant measure. As a generalization of the Dirichlet form, for f in $L^2(\mu)$, let $\mathcal{A}(f,f) = \frac{1}{2}\sum_{x,y\in\mathcal{V}}\mu(x)P(x,y)(f(y)-f(x))^2$.

a) Let $f \in L^2(\mu)$. Prove that $0 \leq \mathcal{A}(f,f) \leq \|f\|^2_{L^2(\pi)} < \infty$, that $\mathcal{A}(f,f) = 0$ if and only if f is constant, and that

$$\mathcal{A}(f,f) = \sum_{x\in\mathcal{V}} \mu(x)(f(x) - Pf(x))f(x) .$$

b) Prove that any nonnegative subharmonic function of $L^2(\pi)$ is constant. Generalize to lower-bounded subharmonic functions.

c) Let f be harmonic and in $L^2(\pi)$. Prove that f^+ and f^- are subharmonic. Conclude that f is constant.

4.9 Spectral gap bounds Let P be an irreducible transition matrix on $\mathcal{V}$ having an invariant law π, and $c = \inf_{x\in\mathcal{V}}P(x,x)$.

a) Prove that $(1 - \lambda_P)^2 \leq 1 - \lambda_{\tilde{P}P}$.

b) Prove that if $c > 0$, then $R := (1 - c)^{-1}(P - cI)$ is an irreducible transition matrix, and that $\mathcal{A}_{\tilde{P}P} = (1 - c)^2\mathcal{A}_{\tilde{R}R} + 2c\mathcal{A}_{\frac{P+\tilde{P}}{2}}$. Deduce from this that $1 - \lambda_{\tilde{P}P} \leq 1 - 2c\lambda_P$.

c) Prove that if $c \geq 1/2$, then $\sigma(P)$ is included in the complex half-plane of nonnegative real parts.

4.10 Exponential bounds Let P be an irreducible transition matrix on $\mathcal{V}$ having an invariant law π.

a) Prove that, for $n \geq 1$ and $k \geq 1$ and f in $L^2(\pi)$ and μ in $L^2(\pi)^*$,

$$\|P^n f - \pi f\|_{L^2(\pi)} \leq \left(\sqrt{1 - \lambda_{\tilde{P}^k P^k}}\right)^{\lfloor n/k \rfloor} \|f - \pi f\|^2_{L^2(\pi)} ,$$

$$\|\mu P^n - \mu(\mathcal{V})\pi\|_{L^2(\pi)^*} \leq \left(\sqrt{1 - \lambda_{P^k \tilde{P}^k}}\right)^{\lfloor n/k \rfloor} \|\mu - \mu(\mathcal{V})\pi\|_{L^2(\pi)^*} .$$

b) Prove that if $\mathcal{V}$ is finite and P is aperiodic, then there exists $k \geq 1$ such that $P^k \tilde{P}^k$ is strongly irreducible. Deduce from this some exponential convergence bounds.

4.11 **Poincaré inequality and graphs** Let P be an irreducible transition matrix on $\mathcal{V}$ which is reversible w.r.t. a law π, and

$$G = \{(x, y) \in \mathcal{V}^2 : x \neq y, \; P(x, y) > 0\} \; .$$

Choose a length function $L : (x, y) \in G \mapsto L(x, y) \in \mathbb{R}_+^*$, for every $x \neq y$ a simple path

$$\gamma(x, y) = \{(x, x_1), (x_1, x_2), \ldots, (x_{n-1}, y)\}$$

in which $(x, x_1), (x_1, x_2), \ldots, (x_{n-1}, y)$ are distinct elements of G, and let

$$\Gamma = \{\gamma(x, y) : x \neq y \in \mathcal{V}\}, \qquad |\gamma(x, y)|_L = \sum_{(x', y') \in \gamma(x,y)} L(x', y') ,$$

$$A_{L,\Gamma} = \sup_{(x,y) \in G} \left(\frac{1}{\pi(x)P(x, y)L(x, y)} \sum_{x' \neq y' : (x,y) \in \gamma(x',y')} \pi(x')\pi(y')|\gamma(x', y')|_L \right) .$$

a) Prove that $\mathrm{Var}_\pi(f) = \frac{1}{2}\sum_{x,y \in \mathcal{V}} \pi(x)\pi(y)(f(y) - f(x))^2$. Of which Markov chain, is this the Dirichlet form?

b) Prove that, for $x \neq y$ and f in $L^2(\pi)$,

$$(f(y) - f(x))^2 \leq |\gamma(x, y)|_L \sum_{(x',y') \in \gamma(x,y)} \frac{(f(y') - f(x'))^2}{L(x', y')} \; .$$

Deduce from this that

$$\mathrm{Var}_\pi(f) \leq \frac{1}{2} \sum_{(x,y) \in G} \frac{(f(y) - f(x))^2}{L(x, y)} \sum_{x' \neq y' : (x,y) \in \gamma(x',y')} \pi(x')\pi(y')|\gamma(x', y')|_L \; .$$

c) Prove that $\lambda_P \geq 1/A_{L,\Gamma}$.

4.12 **Spectral gap of the M/M/1 queue** The results of Exercise 4.11 are now going to be applied to the random walk with reflection at 0 on $\mathbb{N}$ with matrix P given by $P(x, x + 1) = p \in]0, 1/2[$ and $P(x, (x - 1)^+) = 1 - p$. Let $r = \frac{p}{1-p}$ and $\pi(x) = (1 - r)r^x$ for x in $\mathbb{N}$.

a) Check that P is irreducible, aperiodic, and reversible for π. Check that any $x \neq y$ in $\mathbb{N}$ are linked by a simple path $\gamma(x, y)$, which goes from neighbor to neighbor. Check that if the length function satisfies $L(x, x + 1) = L(x + 1, x)$, then

$$A_{L,\Gamma} = \frac{1 - r^2}{r} \sup_{x \in \mathbb{N}} \left(\frac{1}{r^x L(x, x + 1)} \sum_{y \leq x < z} r^{y+z} \sum_{y \leq a < z} L(a, a + 1) \right) .$$

b) Prove that the choice $L(x, x+1) = L(x+1, x) = r^{-x/2}$ for x in $\mathbb{N}$ yields

$$A_{L,\Gamma} = \frac{1+r}{\left(1 - \sqrt{r}\right)^2} = \frac{1}{\left(\sqrt{p} - \sqrt{1-p}\right)^2} , \qquad \lambda_p \geq \left(\sqrt{p} - \sqrt{1-p}\right)^2 .$$

Actually, this bound is the exact value for the spectral gap, which is well known as $P - I$ is the generator of the continuous-time Markov chain of the number of jobs in an $M/M/1$ queue.

4.13   Entropy Let $h : x \in \mathbb{R}_+ \mapsto x \log x$ with $h(0) = 0$. For laws $\pi > 0$ and μ on $\mathcal{V}$, the relative entropy of μ w.r.t. π is defined by

$$H(\mu \mid \pi) = \pi h\left(\frac{\mu}{\pi}\right) = \sum_{x \in \mathcal{V}} \pi(x) h\left(\frac{\mu(x)}{\pi(x)}\right) .$$

a) Prove that h is strictly convex and continuous on $\mathbb{R}_+$ and that $\mu \mapsto H(\mu \mid \pi)$ is strictly convex and continuous with values $[0, \infty]$ and vanishes if and only if $\mu = \pi$.

b) Let P be an irreducible transition matrix on $\mathcal{V}$ with an invariant law π. Prove that $H(\mu P \mid \pi) \leq H(\mu \mid \pi)$ and that if there exists y such that $P(x, y) > 0$ for all x, then $H(\mu P \mid \pi) = H(\mu \mid \pi)$ if and only if $\mu = \pi$. The adjoint $\tilde{P}$ can be used for this.

c) Let P hereafter be aperiodic, and the state space $\mathcal{V}$ be finite. Prove that there exists $k \geq 1$ such that $P^k > 0$ and that $(\mu P^n)_{n \geq 0}$ is relatively compact.

d) Prove that the limits $\lim_{n \to \infty} H(\mu P^n \mid \pi)$ and $\lim_{n \to \infty} H(\mu P^n P^k \mid \pi)$ exist and are equal. For any accumulation point μ^* of $(\mu P^n)_{n \geq 0}$, prove that $H(\mu^* \mid \pi) = H(\mu^* P^k \mid \pi)$, and deduce from this that $\mu^* = \pi$. Conclude that $(\mu P^n)_{n \geq 0}$ converges to π.

5

Monte Carlo methods

Monte Carlo methods based on Markov chains are often called MCMC methods, the acronym "MCMC" standing for "Markov Chain Monte Carlo." These are often the only effective methods for the approximate computation of highly combinatorial quantities of interest and may be introduced even in situations in which the basic model is deterministic.

The corresponding research field is at the crossroads of disciplines such as statistics, stochastic processes, and computer science, as well as of various applied sciences that use it as a computation tool. It is the subject of a vast modern literature.

We are going to explain the main bases for these methods and illustrate them on some classic examples. We hope that we thus shall help the readers appreciate their adaptability and efficiency. This chapter thus provides better understanding on the practical importance of Markov chains, as well as some of the problematics they introduce.

5.1 Approximate solution of the Dirichlet problem

5.1.1 General principles

The central idea is to use Theorem 2.2.2, not any longer for computing

$$u(x) = \mathbb{E}_x(f(X_{S_E}) \mathbb{1}_{\{S_E < \infty\}}), \qquad x \in \mathcal{V},$$

by solving the equation as we have done, but on the contrary to use this probabilistic representation in order to approximate the solution of the Dirichlet problem. If

$$\mathbb{E}_x(|f(X_{S_E})| \mathbb{1}_{\{S_E < \infty\}}) < \infty,$$

then the strong law of large numbers yields a Monte Carlo method for this.

The corresponding MCMC method consists in simulating N independent chains of matrix P starting at x, denoted by $(X_k^i)_{k \geq 0}$ for $1 \leq i \leq N$, each until its first exit

---

Markov Chains: Analytic and Monte Carlo Computations, First Edition. Carl Graham.
© 2014 John Wiley & Sons, Ltd. Published 2014 by John Wiley & Sons, Ltd.

time S^i from $D = \mathcal{V} - E$, and use that

$$u^N(x) = \frac{1}{N}(f(X^1_{S^1})\mathbb{1}_{\{S^1<\infty\}} + \cdots + f(X^N_{S^N})\mathbb{1}_{\{S^N<\infty\}}) \xrightarrow[N\to\infty]{\text{a.s.}} u(x) .\qquad(5.1.1)$$

In order to obtain confidence intervals for this method, one can use for instance

- the central limit theorem, which requires information on the variance

$$\text{Var}(f(X_T)\mathbb{1}_{\{S_E<\infty\}}) = \mathbb{E}_x(f(X_{S_E})^2\mathbb{1}_{\{S_E<\infty\}}) - u(x)^2,$$

- exponential Markov inequalities, or a large deviation principle, which requires information on the exponential moments of $f(X_T)\mathbb{1}_{\{S_E<\infty\}}$.

The central limit theorem or a large deviation principle actually only yield asymptotic confidence intervals.

Naturally, this requires that $\mathbb{P}_x(S_E < \infty) = 1$, and even that $\mathbb{E}_x(S_E) < \infty$ as another application of the strong law of large numbers yields that the total number of time steps for N simulations is

$$N\mathbb{E}_x(S_E) + o(N) , \quad \text{a.s.}$$

Hence, the obtention of exact or approximate solutions of the equations in Theorems 2.2.5 and 2.2.6 would allow to handle some essential problems, such as

- the estimation of the mean number of steps

$$\mathbb{E}_x(S_E) = m_E(x)$$

required for each simulation,

- the establishment of confidence intervals on the duration of the simulation of one of the Markov chains and on the total duration

$$N\mathbb{E}_x(S_E) + o(N)$$

of the N simulations: this can be obtained using $\mathbb{E}_x(S_E)$ and $\text{Var}(S_E)$ and the central limit theorem, or using $g_E(x, \cdot)$ and exponential Markov inequalities if the convergence radius is greater than 1,

- the development of criteria for abandoning simulations, which are too long, and for compensating the bias induced by these censored data.

5.1.2   Heat equation in equilibrium

5.1.2.1   Dirichlet problem

The Dirichlet problem for the stationary heat equation on a domain D of $\mathbb{R}^d$ with boundary condition f is given, denoting the Laplacian by $\Delta = \sum_{i=1}^d \partial_i^2$, by

$$u = f \text{ on } \partial D , \qquad \Delta u = 0 \text{ in } D .$$

It corresponds to the physical problem of describing the temperature distribution in equilibrium of a homogeneous solid occupying D, which is heated at its boundary ∂D according to a temperature distribution given by f. Physically, $f \geq 0$ w.r.t. the absolute zero, and the solution is the minimal solution.

5.1.2.2   Discretization

Let us discretize space by a regular grid with mesh of size $h > 0$. A natural approximation of $\Delta u(x)$ is given by

$$\frac{1}{h^2} \sum_{i=1}^{d} (u(x + he_i) + u(x - he_i) - 2u(x)) = \frac{2d}{h^2} \sum_{i=1}^{d} \sum_{y = x \pm he_i} \frac{1}{2d} (u(y) - u(x)) \,.$$

Considering the transition matrix P^h of the symmetric nearest-neighbor random walk $(X_n^h)_{n \geq 0}$ on the grid, an approximation D^h of D on the grid, the set ∂D^h of gridpoints outside D^h, which are at a distance h, and an appropriate approximation f^h of f on ∂D^h, yields the discretized problem

$$u^h = f^h \text{ on } \partial D^h \,, \qquad (I - P^h)u^h = 0 \text{ in } D^h \,.$$

The quality of the approximation of the initial problem by the discretized problem can be treated by analytic methods or by probabilistic methods using the approximation of Brownian motion by random walks. Note that if D is bounded with "characteristic length" L then the number of points in D^h is of order $(L/h)^d$.

5.1.2.3   Monte Carlo approximation

Let S^h be the exit time of $(X_n^h)_{n \geq 0}$ from D^h. Theorem 2.2.2 yields that if f^h is nonnegative then the least nonnegative solution of the discretized problem is given by the probabilistic representation

$$u^h : x \in \bar{D}^h \mapsto u^h(x) = \mathbb{E}_x(f^h(X_{S^h}^h) \mathbb{1}_{S^h < \infty}) \,,$$

and that if f^h is bounded and moreover $\mathbb{P}_x(S^h < \infty) = 1$ for every $x \in D^h$ then u^h is the unique bounded solution. For $x \in D^h$, a Monte Carlo approximation $u^{h,N}(x)$ of $u^h(x)$ is obtained as in (5.1.1).

Note that $\mathbb{P}_x(S^h < \infty) = 1$ as soon as D^h is bounded in at least one principal direction: the effective displacement along this axis will form a symmetric nearest-neighbor random walk on $h\mathbb{Z}$, which always eventually reaches a unilateral boundary, as we have seen.

When D is bounded, the result (2.3.10) on the exit time from a box provides a bound on the expectation of the duration of a simulation. In particular, if D has "characteristic length" L then the mean number of steps for obtaining the Monte Carlo approximation is of order

$$\frac{Nd}{4} \frac{L^2}{h^2}$$

5.1.3 Heat equation out of equilibrium

5.1.3.1 Dirichlet problem

There is an initial heat distribution u_0 on $\bar{D}$, and for $t \in \mathbb{R}_+^*$, a heat distribution $f(t, \cdot)$ is applied to ∂D. The Monte Carlo method is well adapted to this nonequilibrium situation, in which time will be considered as a supplementary spatial dimension, and the initial condition as a boundary condition.

With an adequate choice of the units of time and space in terms of the thermal conductance of the medium,

$$
\begin{cases}
u(0, x) = u_0(x) & \text{for } x \in \bar{D} & \text{(initial condition)}, \\
u(t, x) = f(t, x) & \text{for } t > 0 \text{ and } x \in \partial D & \text{(boundary condition)}, \\
\partial_t u(t, x) - \Delta_x u(t, x) = 0 & \text{for } t > 0 \text{ and } x \in D & \text{(heat equation)}.
\end{cases}
$$

5.1.3.2 Discretization

Discretizing by a spatial mesh of size $h > 0$ and a temporal mesh of size $\varepsilon > 0$ yields

$$
\frac{1}{\varepsilon}(u(t, x) - u(t - \varepsilon, x)) - \sum_{i=1}^{d} \sum_{y=x \pm he_i} \frac{1}{h^2}(u(t, y) - u(t, x)) = 0,
$$

which writes $(I - P^{\varepsilon,h})u(t, x) = 0$ for the transition matrix $P^{\varepsilon,h}$ given by

$$
P^{\varepsilon,h}u(t, x) = \left(1 + 2d\frac{\varepsilon}{h^2}\right)^{-1}\left(u(t - \varepsilon, x) + \sum_{i=1}^{d} \sum_{y=x \pm he_i} \frac{\varepsilon}{h^2}u(t, y)\right).
$$

Hence, ε and h^2 should be of the same order for this diffusive phenomenon. By approximating adequately u_0 by u_0^h on $\bar{D}^h = D^h \cup \partial D^h$, the discretized problem writes

$$
\begin{cases}
u^{\varepsilon,h}(0, x) = u_0^h(x) & \text{for } x \in \bar{D}^h, \\
u^{\varepsilon,h}(t, x) = f^h(t, x) & \text{for } t = \varepsilon, 2\varepsilon, \dots \text{ and } x \in \partial D^h, \\
(I - P^{\varepsilon,h})u^{\varepsilon,h}(t, x) = 0 & \text{for } t = \varepsilon, 2\varepsilon, \dots \text{ and } x \in D^h.
\end{cases}
$$

This constitutes a discretization scheme, implicit in time, for the heat equation, which is unconditionally stable: there is no need for a CFL condition. This scheme has many advantages over an explicit scheme.

The quality of the approximation of the initial problem by the discretized problem can be treated by analytic methods or by probabilistic methods using the approximation of Brownian motion by random walks.

Curse of dimensionality The deterministic numerical solution of this scheme requires to solve a linear system at each time step. If the domain D is bounded with "characteristic length" L, then D^h has cardinality of order $(L/h)^d$. The deterministic methods hence become quickly untractable as d increases even to moderate sizes.

Such difficulties are referred to by the terminology "the curse of dimensionality," due to Bellman.

5.1.3.3   Monte Carlo approximation

Monte Carlo methods provide very good alternative techniques bypassing this curse. Let $S^{\varepsilon,h}$ be the hitting time of

$$(\{0\} \times \bar{D}^h) \cup (\{\varepsilon, 2\varepsilon, \cdots\} \times \partial D^h)$$

by the chain $(X_n^{\varepsilon,h})_{n\geq 0}$ with matrix $P^{\varepsilon,h}$. For (t,x) in $\{0,\varepsilon,2\varepsilon,\ldots\} \times \bar{D}^h$, Theorem 2.2.2 yields that if f is nonnegative (resp. bounded) then the least nonnegative solution (resp. the unique bounded solution) of the above discretization scheme is given by the probabilistic representation

$$u^{\varepsilon,h}(t,x)$$

$$= \mathbb{E}_{t,x}\left(u_0(X_{S^{\varepsilon,h}}^{\varepsilon,h})\mathbb{1}_{\{X_{S^{\varepsilon,h}}^{\varepsilon,h}\in\{0\}\times\bar{D}^h\}} + f^h(X_{S^{\varepsilon,h}}^{\varepsilon,h})\mathbb{1}_{\{X_{S^{\varepsilon,h}}^{\varepsilon,h}\in\{\varepsilon,2h,\ldots\}\times\partial D^\varepsilon\}}\right) .$$

At each time step, the first coordinate of $(X_n^{\varepsilon,h})_{n\geq 0}$ advances of $-\varepsilon$ with probability $p = (1 + 2d\varepsilon/h^2)^{-1} > 0$, and hence $S^{\varepsilon,h} \leq T$ where $T = G_1 + \cdots + G_{\lceil t/\varepsilon\rceil}$ for G_i which are i.i.d., and have geometric law given by $\mathbb{P}(G_i = k) = (1-p)^{k-1}p$ for $k \geq 1$. Simple computations show that T is finite, a.s., has some finite exponential moments, and

$$\mathbb{E}(T) \leq \lceil t/\varepsilon\rceil p^{-1} = \lceil t/\varepsilon\rceil(1 + 2d\varepsilon/h^2) ,$$

$$\text{Var}(G) \leq \lceil t/\varepsilon\rceil(1-p)p^{-2} = \lceil t/\varepsilon\rceil((1 + 2d\varepsilon/h^2)^2 - (1 + 2d\varepsilon/h^2)) ,$$

so that we have bounds linear in the dimension d for the expectation and the standard deviation.

Then, $u^{\varepsilon,h}$ can be approximated using the Monte Carlo method by $u^{\varepsilon,h,N}(x)$ given in (5.1.1). The central limit theorem yields that the precision is of order $1/\sqrt{N}$ with a constant that depends on the variance and hence very reasonably on the dimension d.

This is a huge advantage of Monte Carlo methods over deterministic methods, which suffer of the curse of dimensionally and of which the complexity increases much faster with the dimension.

5.1.4   Parabolic partial differential equations

5.1.4.1   Dirichlet problem

More general parabolic problems may be considered. Let a time-dependent elliptic operator acting on g in $C^2(\mathbb{R}^d)$ be given by

$$\mathcal{L}g(t,x) = \sum_{i,j=1}^{d} a_{ij}(t,x)\partial_{ij}^2 g(x) + \sum_{i=1}^{d} b_i(t,x)\partial_i g(x) ,$$

where the matrix $A = (a_{ij})_{1 \leq i,j \leq d}$ is symmetric nonnegative. Consider the Dirichlet problem

$$
\begin{cases}
u(0, x) = u_0(x) & \text{for } x \in \bar{D} & \text{(initial condition)}, \\
u(t, x) = f(t, x) & \text{for } t > 0 \text{ and } x \in \partial D & \text{(boundary condition)}, \\
\partial_t u(t, x) - \mathcal{L}_x u(t, x) = 0 & \text{for } t > 0 \text{ and } x \in D & \text{(parabolic PDE)}.
\end{cases}
$$

5.1.4.2   Discretization: diagonal case

Let us first assume that A is diagonal. It is always possible to find an orthonormal basis in which is locally the case, corresponding to the principal axes of A, and it is a good idea to find them if A is constant. By discretizing the ith coordinate by a step of h_i and time by a step of ε,

$$
\frac{1}{\varepsilon}(u(t, x) - u(t - \varepsilon, x))
$$

$$
- \sum_{i=1}^{d} \left[\left(\frac{a_{ii}(t, x)}{h_i^2} + \frac{b_i(t, x)}{h_i} \mathbb{1}_{b_i(t,x)>0} \right)(u(t, x + h_i e_i) - u(t, x)) \right.
$$

$$
\left. + \left(\frac{a_{ii}(t, x)}{h_i^2} - \frac{b_i(t, x)}{h_i} \mathbb{1}_{b_i(t,x)<0} \right)(u(t, x - h_i e_i) - u(t, x)) \right] = 0 .
$$

This writes $(I - P^{\varepsilon, h_1, \ldots, h_d})u(t, x) = 0$ for the transition matrix $P^{\varepsilon, h_1, \ldots, h_d}$ equal to

$$
c(t, x) \left(u(t - \varepsilon, x) + \sum_{i=1}^{d} \left[\left(\frac{\varepsilon}{h_i^2} a_{ii}(t, x) + \frac{\varepsilon}{h_i} b_i(t, x) \mathbb{1}_{b_i(t,x)>0} \right) u(t, x + h_i e_i) \right. \right.
$$

$$
\left. \left. + \left(\frac{\varepsilon}{h_i^2} a_{ii}(t, x) - \frac{\varepsilon}{h_i} b_i(t, x) \mathbb{1}_{b_i(t,x)<0} \right) u(t, x - h_i e_i) \right] \right)
$$

in which the normalizing constant is given by

$$
c(t, x)^{-1} = 1 + \sum_{i=1}^{d} \left(2 a_{ii}(t, x) \frac{\varepsilon}{h_i^2} + |b_i(t, x)| \frac{\varepsilon}{h_i} \right).
$$

This is again a discretization scheme for the PDE, which is implicit in time and unconditionally stable. Then, h_i should be of the same order than $\sqrt{\varepsilon}$ if $a_{ii} \neq 0$ and of same order as ε if $a_{ii} \equiv 0$. The relative choices for the h_i should be influenced by the sizes of the $a_{ii} \neq 0$ or the b_i, and in particular if A is constant in (t, x), then a good choice would be $h_i^2 = \varepsilon a_{ii}$ if $a_{ii} \neq 0$ and $h_i = \varepsilon |b_i|$ if $a_{ii} = 0$.

5.1.4.3   Monte Carlo approximation

The matrix $P^{\varepsilon, h_1, \ldots, h_d}$ is a transition matrix, and the previous probabilistic representations and Monte Carlo methods are readily generalized to this situation.

5.1.4.4   Discretization: nondiagonal case

In general, a natural discretization of the Dirichlet problem is given by

$$\frac{1}{\varepsilon}(u(t,x) - u(t - \varepsilon, x))$$

$$- \sum_{1 \le i < j \le d} \left[\frac{a_{ij}(t,x)}{h_i h_j} \mathbb{1}_{a_{ij}(t,x)>0}(u(t, x + h_i e_i + h_j e_j) - u(t,x)) \right.$$

$$+ \frac{a_{ij}(t,x)}{h_i h_j} \mathbb{1}_{a_{ij}(t,x)>0}(u(t, x - h_i e_i - h_j e_j) - u(t,x))$$

$$+ \frac{|a_{ij}(t,x)|}{h_i h_j} \mathbb{1}_{a_{ij}(t,x)<0}(u(t, x + h_i e_i - h_j e_j) - u(t,x))$$

$$\left. + \frac{|a_{ij}(t,x)|}{h_i h_j} \mathbb{1}_{a_{ij}(t,x)<0}(u(t, x - h_i e_i + h_j e_j) - u(t,x)) \right]$$

$$+ \sum_{i=1}^{d} \left[\left(\frac{a_{ii}(t,x)}{h_i^2} - \sum_{j \ne i} \frac{|a_{ij}(t,x)|}{h_i h_j} + \frac{b_i(t,x)}{h_i} \mathbb{1}_{b_i(t,x)>0} \right) \right.$$

$$\times (u(t, x + h_i e_i) - u(t,x))$$

$$+ \left(\frac{a_{ii}(t,x)}{h_i^2} - \sum_{j \ne i} \frac{|a_{ij}(t,x)|}{h_i h_j} - \frac{b_i(t,x)}{h_i} \mathbb{1}_{b_i(t,x)<0} \right)$$

$$\left. \times (u(t, x - h_i e_i) - u(t,x)) \right]$$

$$= 0,$$

which features "diagonal" jumps, and thus extended ∂D^h and $\bar{D}^h = D^h \cup \partial D^h$ w.r.t. the previous situation.

5.1.4.5   Monte Carlo approximation

In order to have an interpretation in terms of a transition matrix, it is necessary and sufficient that

$$a_{ii} - h_i \sum_{j \ne i}^{d} \frac{|a_{ij}|}{h_j} = \left\langle A e_i, e_i - h_i \sum_{j \ne i} \frac{\text{sign}(a_{ij})}{h_j} e_j \right\rangle \ge 0, \qquad 1 \le i \le d,$$

which is not always true when $d \ge 3$, as can be seen with the matrix in which all terms are 1. It this holds for all (t,x), then the previous probabilistic representations and Monte Carlo methods can easily be extended to this situation. Else other methods will be used, which do not involve a spatial grid discretization as this causes a strong anisotropy.

5.2   Invariant law simulation

In this section, we mostly assume that $\mathcal{V}$ is finite and set $d := \text{Card}(\mathcal{V})$.

5.2.1   Monte Carlo methods and ergodic theorems

Let $\mathcal{V}$ be a state space, and π a probability measure on $\mathcal{V}$, which can be interpreted as the invariant law for an irreducible transition matrix P. Finding an explicit expression for π may raise the following difficulties.

1. The invariant measure μ is a nonnegative nonzero solution of the linear system $\mu = \mu P$. This develops into a system of d equations for d unknowns, usually highly coupled and as such difficult to solve, see the global balance equations (3.2.4). If P is reversible, then the local balance equations (3.2.5) can be used, yielding for μ the semiexplicit expression in Lemma 3.3.8.

2. The invariant law π has total mass 1, and thus the following normalization problem arises. Once an invariant law μ is obtained, then $\|\mu\|_{\text{Var}} = \sum_{x \in \mathcal{V}} \mu(x)$ must be computed in order to make $\pi = \mu/\|\mu\|_{\text{Var}}$ explicit. In general, such a summation is an NP-complete problem in terms of d.

These difficulties are elegantly bypassed by Monte Carlo approximation methods. These replace the resolution of an equation, which implicitly defines an invariant measure, and the normalization of the solution, by the simulation of a Markov chain, which is explicitly given by its transition matrix.

5.2.1.1   Pointwise ergodic theorem

Assume that a real function f on $\mathcal{V}$ is given and that the mean of f w.r.t. π is sought. We then are interested in

$$\pi f = \sum_{x \in \mathcal{V}} \pi(x) f(x) .$$

Even if π were explicitly known, such a summation is again, in general, an NP-complete problem. Recall that π is usually at best known only up to a normalizing constant, that is, as an invariant measure μ.

If a term $\pi(x)$ of π or the probability $\pi(A)$ of some subset of $\mathcal{V}$ is sought, then this corresponds to taking $f = \mathbb{1}_{\{x\}}$ or $f = \mathbb{1}_A$. Note that if an invariant law μ is known and a term $\pi(x)$ has been computed, then

$$\pi = \frac{\pi(x)}{\mu(x)} \mu .$$

The pointwise ergodic theorem (Theorem 4.1.1) yields a Monte Carlo method for approximating πf. Indeed, if $(X_n)_{n \geq 0}$ is a Markov chain with matrix P and arbitrary initial condition, then

$$\frac{1}{n+1} \sum_{k=0}^{n} f(X_k) \xrightarrow[n \to \infty]{\text{a.s.}} \pi f .$$

The Markov chain central limit theorem (Theorem 4.1.5) yields confidence intervals for this convergence.

5.2.1.2   Kolmogorov's ergodic theorem

Some situations require to simulate random variables of law π, for instance in view of integrating them in simulations of more complex systems. Even if π were explicitly known, this is usually a very difficult task if d is large; moreover, π is usually at best known only up to a normalizing constant.

If P is aperiodic, then the Kolmogorov ergodic theorem yields a Monte Carlo method for obtaining samples drawn from an approximation of π. Indeed, if $(X_n)_{n \geq 0}$ is a Markov chain with matrix P and arbitrary initial condition, then

$$\|\pi_n - \pi\|_{\mathrm{Var}} \underset{n \to \infty}{\longrightarrow} 0 .$$

Exponential convergence bounds are given by Theorem 1.3.4, in terms of the Doeblin condition, which is satisfied, but usually the bounds are very poor. Much better exponential bounds are given in various results and developments around the Dirichlet forms, spectral gap estimations, and Poincaré inequalities, which have been introduced in Section 4.3.

5.2.2   Metropolis algorithm, Gibbs law, and simulated annealing

5.2.2.1   Metropolis algorithm

Let π be a law of interest, determined at least up to a normalizing constant. It may be given for instance by statistical mechanics considerations. The first step is to interpret π as the reversible law (and thus the invariant law) of an explicit transition matrix P.

Let Q be an irreducible transition matrix Q on $\mathcal{V}$, called the selection matrix, s.t. $Q(x, y) > 0 \Rightarrow Q(y, x) > 0$ for x, y in $\mathcal{V}$, and a function

$$h : u \in \mathbb{R}_+^* \mapsto h(u) \in [0, 1] \text{ s.t. } h(u) = u h(1/u) ,$$

for instance $h(u) = \min(u, 1)$ or $h(u) = \frac{u}{1+u}$. Let

$$R(x, y) = h\left(\frac{\pi(y)Q(y, x)}{\pi(x)Q(x, y)} \right) , \qquad x \neq y \in \mathcal{V} , \ Q(x, y) \neq 0 ,$$

which depends on π only up to a normalizing constant. Let $P = (P(x, y))_{x, y \in \mathcal{V}}$ be defined by

$$P(x, y) = R(x, y)Q(x, y) , \quad x \neq y \in \mathcal{V} , \qquad P(x, x) = 1 - \sum_{y \neq x} P(x, y) ,$$

with $P(x, y) = 0$ if $x \neq y$ and $Q(x, y) = 0$.

It is a simple matter to check that P is an irreducible transition matrix, which is reversible w.r.t. π, and that P is aperiodic if Q is aperiodic or if $h < 1$. Hence, the above-mentioned Monte Carlo methods for approximating quantities related to π can be implemented using P.

The Metropolis algorithm is a method for sequentially drawing a sample $(x_n)_{n \geq 0}$ of a Markov chain $(X_n)_{n \geq 0}$ of transition matrix P, using directly Q and h:

- Step 0 (initialization): draw x_0 according to the arbitrary initial law.

- Step $n \geq 1$ (from $n - 1 \geq 0$ to n): draw $y \in \mathcal{V}$ according to the law $Q(x_{n-1}, \cdot)$:

  - with probability $R(x_{n-1}, y)$ set $x_n = y$

  - else set $x_n = x_{n-1}$.

This is an acceptance–rejection method: the choice of a candidate for a new state y is made according to Q and is accepted with probability $R(X_{n-1}, y)$ and else rejected. The actual evolution is thus made in accordance with P, which has invariant law π, instead of Q.

A classic case uses h given by $h(u) = \min(u, 1)$. Then, y is accepted systematically if

$$\pi(y)Q(y, x) \geq \pi(x)Q(x, y)$$

and else with probability

$$\frac{\pi(y)Q(y, x)}{\pi(x)Q(x, y)} < 1 \ .$$

5.2.2.2 Gibbs laws and Ising model

Let $V : \mathcal{V} \to \mathbb{R}$ be a function and $T > 0$ a real number. The Gibbs law with energy function V and temperature T is given by

$$g_{V,T}(x) = \frac{1}{Z_T} \exp\left(-\frac{V(x)}{T}\right), \quad x \in \mathcal{V}; \qquad Z_T = \sum_{y \in \mathcal{V}} \exp\left(-\frac{V(y)}{T}\right) \ .$$

This law characterizes the thermodynamical equilibrium of a system in which an energy $V(x)$ corresponds to each configuration x. More precisely, it is well known that among all laws π for which the mean energy $\pi V = \sum_{x \in \mathcal{V}} \pi(x)V(x)$ has a given value E, the physical entropy

$$-\sum_{x \in \mathcal{V}} \pi(x) \log \pi(x)$$

is maximal for a Gibbs law $g_{V,T}(x)$ for a well-defined $T := T(E)$.

Owing to the normalizing constant, $g_{V,T}(x)$ does not change if a constant is added to V, and hence

$$\lim_{T \to 0} g_{V,T}(x) = \lim_{T \to 0} g_{(V - \min V), T}(x) = \frac{\mathbb{1}_{\arg \min V}}{\mathrm{Card}(\arg \min V)}, \tag{5.2.2}$$

the uniform law on the set of minima of V.

Ising model A classic example is given by the Ising model. This is a magnetic model, in which $\mathcal{V} = \{-1, +1\}^N$ and the energy of configuration $x = (x^1, \ldots, x^N)$ is given by

$$V(x) = -\frac{1}{2} \sum_{i,j=1}^{N} J_{ij} x^i x^j + h_i x^i \,,$$

in which the matrix $(J_{ij})_{1 \le i,j \le N}$ can be taken symmetric and with a null diagonal, and $x^i = -1$ and $x^i = +1$ are the two possible orientations if a magnetic element at site i. The term J_{ij} quantifies the interaction between the two sites $i \ne j$, for instance $J_{ij} > 0$ for a ferromagnetic interaction in which configurations $x^i = x^j$ are favored. The term h_i corresponds to the effect of an external magnetic field on site i.

The state space $\mathcal{V}$ has cardinal 2^N, so that the computation or even estimation of the normalizing constant Z_T, called the partition function, is very difficult as soon as N is not quite small.

Simulation The simulation by the Metropolis algorithm is straightforward. Let Q be an irreducible matrix on $\mathcal{V}$, for instance corresponding to choosing uniformly i in $\{1, \ldots, N\}$ and changing x^i into $-x^i$. The algorithm proceeds as follows:

- Step 0 (initialization): draw x_0 according to the arbitrary initial law.

- Step $n \ge 1$ (from $n - 1 \ge 0$ to n): draw $y \in \mathcal{V}$ according to the law $Q(x_{n-1}, \cdot)$:

  - set $x_n = y$ with probability

$$h\left(\exp\left(-\frac{V(y) - V(X_{n-1})}{T} \right) \frac{Q(y, x)}{Q(x, y)} \right) \,,$$

  - else set $x_n = x_{n-1}$.

If moreover $h(u) = \min(u, 1)$ and $Q(y, x) = Q(x, y)$, then y is accepted systematically if

$$V(y) \le V(X_{n-1})$$

and else with probability

$$\exp\left(-\frac{V(y) - V(X_{n-1})}{T} \right) .$$

5.2.2.3 Global optimization and simulated annealing

A deterministic algorithm for the minimization of V would only accept states that actually decrease the energy function V. They allow to find a local minimum for V, which may be far from the global minimum.

The Metropolis algorithms also accepts certain states that increase the energy, with a probability, which decreases sharply if the energy increase is large of the temperature T is low. This allows it to escape from local minima and explore the state space, and its instantaneous laws converge to the Gibbs $g_{V,T}$, which distributes its mass mainly close to the global minima. We recall (5.2.2): the limit as T goes to zero of $g_{V,T}$ is the uniform law on the minima of V.

Let $(T_n)_{n\geq 0}$ be a sequence of temperatures, and $(X_n)_{n\geq 0}$ the inhomogeneous Markov chain for which

$$\mathbb{P}(X_{n+1} = y \mid X_n = 0) = P(n; x, y), \qquad n \geq 0,$$

where the transition matrix $(P(n; x, y))_{x,y \in \mathcal{V}}$ has invariant law g_{V,T_n}, for instance corresponds to the Metropolis algorithm of energy function V and temperature T_n.

A natural question is whether it is possible to choose $(T_n)_{n\geq 0}$ decreasing to 0 in such a way that the law of X_n converges to the uniform law on the minima of V. This will yield a stochastic optimization algorithm for V, in which the chain $(X_n)_{n\geq 0}$ will be simulated for a sufficient length of time (also to be estimated), and then the instantaneous values should converge to a global minimum of V.

From a physical viewpoint, this is similar to annealing techniques in metallurgy, in which an alloy is heated and then its temperature let decrease sufficiently slowly that the final state is close to a energy minimum. In quenched techniques, on the contrary, the alloy is plunged into a cold bath in order to obtain a desirable state close to a local minimum.

Clearly, the temperature should be let decrease sufficiently slowly. The resulting theoretic results, see for example Duflo, M. (1996, Section 6.4, p. 264), are for logarithmic temperature decrease, for instance of the form $T_n = C/\log n$ or $T_n = 1/k \; \mathrm{e}^{(k-1)C} \leq n \leq \mathrm{e}^{kC}$ for large enough $C = C(V)$.

This is much too slow in practice, and temperatures are let decrease much faster than that in actual algorithms. These nevertheless provide good results, in particular high-quality suboptimal values in combinatorial optimization problems.

5.2.3 Exact simulation and backward recursion

5.2.3.1 Identities in law and backward recursion

The random recursion in Theorem 1.2.3, given by

$$X_n = F_n(X_{n-1}) = F_n \circ \cdots \circ F_1(X_0),$$

where $(F_k)_{k\geq 1}$ is a sequence of i.i.d. random functions from $\mathcal{V}$ to $\mathcal{V}$, which is independent of the r.v. X_0 of π_0, allows to construct a Markov chain $(X_n)_{n\geq 0}$ on $\mathcal{V}$ with matrix P of generic term $P(x, y) = \mathbb{P}(F_1(x) = y)$. In particular, X_n has law $\pi_n = \pi_0 P^n$.

As the $(F_k)_{k\geq 1}$ are i.i.d., the random functions $F_1 \circ \cdots \circ F_n$ and $F_n \circ \cdots \circ F_1$ have same law, and if Z_0 has law π_0 and is independent of $(F_k)_{k\geq 1}$ then

$$Z_n = F_1 \circ \cdots \circ F_n(Z_0)$$

has also law $\pi_n = \pi_0 P^n$ for $n \geq 0$, as X_n.

Far past start interpretation A possible description of $(Z_n)_{n \geq 0}$ is as follows. If F_n is interpreted as a draw taking the state at time $-n$ to time $-n+1$ of $\mathbb{Z}_-$, then Z_n can be seen as the instantaneous value at time 0 of a Markov chain of matrix P started in the past at time $-n$ at Z_0. Letting n go to infinity then corresponds to letting the initial instant tend to $-\infty$, the randomness at the different following instants remaining fixed.

5.2.3.2   Coalescence and invariant law

On the event in which the random function $F_1 \circ \cdots \circ F_n$ is constant on $\mathcal{V}$, for any $k \geq 0$, it holds that

$$F_1 \circ \cdots \circ F_{n+k} = F_1 \circ \cdots \circ F_n \circ (F_{n+1} \circ \cdots \circ F_{n+k}) = F_1 \circ \cdots \circ F_n \,.$$

Let $Z_n^x = F_1 \circ \cdots \circ F_n(x)$ for $x \in \mathcal{V}$, and their first coalescence time

$$C := \inf\{n \geq 0 : F_1 \circ \cdots \circ F_n \text{ is constant}\} := \inf\{n \geq 0 : Z_n^x = Z_n^y, \; \forall x, y \in \mathcal{V}\} \,. \tag{5.2.3}$$

This is a stopping time, after which all $(Z_n^x)_{n \geq 0}$ are constant and equal. On $\{C = n\}$ let Z_∞ be defined by the constant value taken by $F_1 \circ \cdots \circ F_n$, so that $Z_\infty = Z_{n+k}^x$ for all $k \geq 0$ and $x \in \mathcal{V}$. If $C < \infty$, then for any initial r.v. Z_0, it holds that

$$Z_\infty = Z_{C+k}^{Z_0} := F_1 \circ \cdots \circ F_n(Z_0) \,, \qquad k \geq 0 \,.$$

Theorem 5.2.1 *Let P be an irreducible transition matrix on $\mathcal{V}$, and $(F_k)_{k \geq 1}$ be i.i.d. random functions from $\mathcal{V}$ to $\mathcal{V}$ s.t. $P(x, y) = \mathbb{P}(F_1(x) = y)$. If the coalescence time C defined in 5.2.3 is finite, a.s., then the law π of the r.v. Z_∞ taking the constant value of $F_1 \circ \cdots \circ F_C$ is the invariant law for P; moreover, P is aperiodic.*

Proof: As in the proof for Theorem 4.2.8 or Lemma A.2.2, for Z_0 of law π_0,

$$\|\pi_n - \pi\|_{\mathrm{Var}} \leq 2\mathbb{P}(Z_n^{Z_0} \neq Z_\infty) \leq 2\mathbb{P}(C > n) \underset{n \to \infty}{\longrightarrow} 0 \,,$$

and passing to the limit in the recursion $\pi_n = \pi_{n-1}P$ yields that $\pi = \pi P$. By contradiction, if P were of period $d \geq 2$ then $(\mathbb{P}(Z_n^z = x))_{n \geq 0}$ would vanish infinitely often, which is impossible as it converges to $\pi(x) > 0$. ∎

5.2.3.3   Propp and Wilson algorithm

The Propp and Wilson algorithm uses this for the exact simulation of draws from the invariant law π of P, using the random functions $(F_k)_{k \geq 1}$. The naive idea is the following.

- Start with Φ_0 being the identical mapping on $\mathcal{V}$,

- for $n \geq 1$: draw F_n and compute $\Phi_n = \Phi_{n-1} \circ F_n$, that is, $\Phi_n = F_1 \circ \cdots \circ F_n$: $x \in \mathcal{V} \mapsto Z_n^x$, and then test for coalescence:

  – if Φ_n is constant, then the algorithm terminates issuing this constant value,

  – else increment n by 1 and continue.

If the law of the F_k is s.t. $\mathbb{P}(C < \infty) = 1$, then the algorithm terminates after a random number of iterations and issues a draw from π.

5.2.3.4   Criteria on the coalescence time

It is important to obtain verifiable criteria for having $\mathbb{P}(C < \infty) = 1$ and good estimates on C.

One criteria is to choose the law of F_1 so that there exists $p > 0$ s.t.

$$A \subset \mathcal{V}, \; \text{Card}(A) > 1 \Rightarrow \mathbb{P}(\text{Card}(F_1(A)) < \text{Card}(A)) \geq p \; .$$

Then $C \leq \sum_{i=1}^{d-1} D_i$, where the D_i are i.i.d. with geometric law $\mathbb{P}(D_i = k) = (1-p)^{k-1}p$ for $k \geq 1$. This implies that $\mathbb{P}(C < \infty) = 1$ and $\mathbb{E}(C) < (d-1)/p$ and yields exponential bounds on the duration of the simulation.

If P satisfies the Doeblin condition in Theorem 1.3.4 for $k = 1$ and $\varepsilon > 0$ and a law $\hat{\pi}$, then we may consider independent r.v. $(Y_i)_{i \geq 1}$ of law $\check{\pi}$, random functions $(D_i)_{i \geq 1}$ from $\mathcal{V}$ to $\mathcal{V}$ satisfying

$$\mathbb{P}(D_1(x) = y) = \frac{P(x, y) - \varepsilon \hat{\pi}(y)}{1 - \varepsilon},$$

which may be constructed as in Section 1.2.3, and r.v. $(I_i)_{i \geq 1}$ s.t. $\mathbb{P}(I_i = 1) = \varepsilon$ and $\mathbb{P}(I_i = 0) = 1 - \varepsilon$. Then, for $i \geq 1$,

$$F_i(x) = I_i Y_i + (1 - I_i) D_i(x) \; , \qquad x \in \mathcal{V} \; ,$$

define i.i.d. random functions s.t. $\mathbb{P}(F_1(x) = y) = P(x, y)$, for which C has geometric law $\mathbb{P}(C = n) = (1 - \varepsilon)^{n-1} \varepsilon$ for $n \geq 1$.

Note that the algorithm can be applied to some well-chosen power P^k of P, which has same invariant law π and is more likely to satisfy one of the above-mentioned criteria.

5.2.3.5   Monotone systems

An important problem with the algorithm is that in order to check for coalescence it is in general necessary to simulate the $(Z_n^x)_{n \geq 0}$ for all x in $\mathcal{V}$, which is untractable.

Some systems are monotone, in the sense that there exists a partial ordering on $\mathcal{V}$ denoted by $\preceq$ s.t. one can choose random mappings $(F_n)_{n \geq 0}$ for the random recursion satisfying, a.s.,

$$x \preceq y \Rightarrow F_1(x) \preceq F_1(y) \; .$$

If there exist a least element m and a greatest M in $\mathcal{V}$, hence s.t.

$$m \leq x \leq M, \qquad x \in \mathcal{V},$$

then clearly $C = \inf\{n \geq 0 : Z_n^m = Z_n^M\}$, and in order to check for coalescence, it is enough to simulate Z_n^m and Z_n^M.

Many queuing models are of this kind, for instance the queue with capacity $C \geq 1$ given by

$$X_n = \min((X_{n-1} - S_n)^+ + A_n, C), \qquad n \geq 1,$$

is monotone, with $m = 0$ and $M = C$.

Ising model Another interesting example is given by the ferromagnetic Ising model, in which adjacent spins tend to have the same orientations $+1$ or -1. The natural partial order is that $x \leq y$ if and only if $x_i \leq y_i$ for all sites i. The least state m is the configuration constituted of all -1, and the greatest state M the configuration constituted of all $+1$. Many natural dynamics preserve the partial order, for instance the Metropolis algorithm we have described earlier. This is the scope of the original study in Propp, J.G. and Wilson, D.B. (1996).

Appendix A

Complements

A.1 Basic probabilistic notions

A.1.1 Discrete random variable, expectation, and generating function

A.1.1.1 General random variable and its law

A probability space $(\Omega, \mathcal{F}, \mathbb{P})$ will be considered throughout. In general, an r.v. with values in a measurable state space $\mathcal{V}$ is a measurable function

$$X : \omega \in \Omega \mapsto X(\omega) \in \mathcal{V}.$$

Then, for every measurable subset A of $\mathcal{V}$, it holds that

$$\{X \in A\} := \{\omega \in \Omega : X(\omega) \in A\} := X^{-1}(A) \in \mathcal{F}.$$

The law of X is the probability measure defined on $\mathcal{V}$ by $\mathbb{P}_X := \mathbb{P} \circ X^{-1}$ and, more concretely, for measurable subsets A, by

$$\mathbb{P}_X(A) := \mathbb{P}(X \in A) := \mathbb{P}(\{X \in A\}).$$

A.1.1.2 Random variable with discrete state space

Laws and expectations In this appendix, $\mathcal{V}$ will be assumed to be discrete (finite or countably infinite), with measurable structure (σ-field) given by the collection of all subsets. Then, X is an r.v. if and only if

$$\{X = x\} := \{\omega \in \Omega : X(\omega) = x\} \in \mathcal{F}, \qquad \forall x \in \mathcal{V}.$$

Markov Chains: Analytic and Monte Carlo Computations, First Edition. Carl Graham.
© 2014 John Wiley & Sons, Ltd. Published 2014 by John Wiley & Sons, Ltd.

In the sense of nonnegative or absolutely convergent series,

$$\mathbb{P}_X(A) := \mathbb{P}(A \in E) = \sum_{x \in A} \mathbb{P}(X = x), \quad \mathbb{E}(f(X)) := \sum_{x \in \mathcal{V}} f(x)\mathbb{P}(X = x),$$

for $A \subset \mathcal{V}$ and functions $f : \mathcal{V} \to \mathbb{R}^d$, which are nonnegative (and then $\mathbb{E}(f(X))$ is in $[0, \infty] := \mathbb{R}_+ \cup \{\infty\}$) or satisfy $\mathbb{E}(|f(X)|) < \infty$ (and then $\mathbb{E}(f(X))$ is in $\mathbb{R}^d$).

Thus, the law $\mathbb{P}_X$ of X can be identified with the collection $(\mathbb{P}(X = x))_{x \in \mathcal{V}}$ of nonnegative real numbers with sum 1.

More generally, a (nonnegative) measure μ on a discrete space $\mathcal{V}$ can be identified with a collection $(\mu(x))_{x \in \mathcal{V}}$ of nonnegative real numbers, and

$$\mu : A \subset \mathcal{V} \mapsto \sum_{x \in A} \mu(x) \in [0, \infty], \qquad \int f d\mu = \sum_{x \in \mathcal{V}} f(x)\mu(x),$$

where $f : \mathcal{V} \to \mathbb{R}^d$ is nonnegative or satisfies $\sum_{x \in \mathcal{V}} |f(x)|\mu(x) < \infty$. Then, $\mu(\{x\}) = \mu(x)$ and $\mu = \sum_{x \in \mathcal{V}} \mu(x)\delta_x$.

Note that the sum of a nonnegative or an absolutely converging series does not depend on the order of summation.

Integer valued random variables, possibly infinite The natural state space of some random variables is $\mathbb{N} \cup \{\infty\}$, for instance when they are defined as an infimum of a possibly empty subset of $\mathbb{N}$ or as a possibly infinite sum of integers. The first step in their study is to try to determine whether $\mathbb{P}(X = \infty) > 0$, and if yes to compute this quantity.

Distribution tails For this, the formula

$$\mathbb{P}(X = \infty) = \lim_{k \in \mathbb{N}} \downarrow \mathbb{P}(X > k)$$

is often more practical than $\mathbb{P}(X < \infty) = \sum_{n \in \mathbb{N}} \mathbb{P}(X = n)$. We give a related formula for $\mathbb{E}(X)$. Recall that if $\mathbb{P}(X = \infty) > 0$, then $\mathbb{E}(X) = \infty$, and that

$$\mathbb{E}(X \mathbb{1}_{\{X < \infty\}}) = \sum_{n \in \mathbb{N}} n\mathbb{P}(X = n) \in [0, \infty].$$

Lemma A.1.1 *If X is an r.v. with values in $\mathbb{N} \cup \{\infty\}$, then*

$$\mathbb{E}(X) = \sum_{k \in \mathbb{N}} \mathbb{P}(X > k) \in [0, \infty].$$

Proof: If $\mathbb{P}(X = \infty) > 0$, then clearly $\sum_{k \in \mathbb{N}} \mathbb{P}(X > k) = \infty = \mathbb{E}(X)$. Else,

$$\mathbb{E}(X) = \sum_{n \in \mathbb{N}} \sum_{0 \le k \le n-1} \mathbb{P}(X = n) = \sum_{k \in \mathbb{N}} \sum_{n \ge k+1} \mathbb{P}(X = n) = \sum_{k \in \mathbb{N}} \mathbb{P}(X > k)$$

using the Fubini theorem. It is also possible to use the Abel rule. ∎

This formula is a particular instance of the following integration by parts formula: if X is an $\mathbb{R}_+$-valued r.v., and $f : \mathbb{R}_+ \to \mathbb{R}_+$ is absolutely continuous and has nonnegative density f', then by the Fubini theorem

$$\mathbb{E}(f(X)) = \int_0^\infty \left(f(0) + \int_0^x f'(t)\,dt \right) \mathbb{P}_X(dx) = f(0) + \int_0^\infty f'(t)\mathbb{P}(X > t)\,dt.$$

Generating functions The generating function for (the law of) an r.v. X with values in $\mathbb{N} \cup \{\infty\}$ is denoted by $g := g_X$ and is given by the power series

$$g(s) = \mathbb{E}\left(s^X \mathbb{1}_{\{X<\infty\}} \right) = \sum_{n\in\mathbb{N}} \mathbb{P}(X = n)s^n, \qquad s \in \mathbb{R}_+,$$

possibly extended to $\mathbb{C}$. If $\mathbb{P}(X = \infty) = 0$, then $g(s) = \mathbb{E}(s^X)$, and this formula can be used (with proper care) for $|s| < 1$ with the convention $s^\infty = 0$. If $|s| \le 1$, then

$$\sum_{n\in\mathbb{N}} \mathbb{P}(X = n)|s|^n \le \sum_{n\in\mathbb{N}} \mathbb{P}(X = n) \le 1$$

and thus the convergence radius is greater than or equal to 1. Hence, g is finite and continuous on $[0, 1]$ and has derivatives of all orders $k \ge 1$ at $s \in [0, 1[$ given by

$$g^{(k)}(s) = \mathbb{E}\left(X(X-1)\cdots(X-k+1)s^X \mathbb{1}_{\{X<\infty\}} \right)$$
$$= \sum_{n\ge k} \mathbb{P}(X = n)n(n-1)\cdots(n-k+1)s^{n-k}.$$

The function g is determined by its restriction on $]0, 1[$, and for some computations, it is easier to work on this restriction.

Inversion formula A Taylor series expansion of g at 0 shows that

$$\mathbb{P}(X = n) = \frac{1}{n!}g^{(n)}(0), \qquad n \in \mathbb{N},$$

which provides a theoretical inversion method yielding the law from the generating function. In practice, algebraic series expansions should be preferred.

Moments Using the monotone convergence theorem (Theorem A.3.2), in $[0, \infty]$,

$$\mathbb{P}(X < \infty) = \lim_{s\to 1-} g(s) = g(1),$$

$$\mathbb{E}\left(X(X-1)\cdots(X-k+1)\mathbb{1}_{\{X<\infty\}} \right) = \lim_{s\to 1-} g^{(k)}(s),$$

and the moments can be obtained from the second formula. If $g^{(2)}(1) < \infty$, then $\mathbb{E}(X \mid X < \infty)$ and $\mathrm{Var}(X \mid X < \infty)$ can be computed using the Taylor expansion of order 2 for g at 1, given by

$$g(1 + \varepsilon) = \mathbb{P}(X < \infty) + \mathbb{E}(X \mathbb{1}_{\{X<\infty\}})\varepsilon$$

$$+ (\mathbb{E}(X^2 \mathbb{1}_{\{X<\infty\}}) - \mathbb{E}(X \mathbb{1}_{\{X<\infty\}}))\frac{\varepsilon^2}{2} + o(\varepsilon^2),$$

$$\log g(1 + \varepsilon) = \log \mathbb{P}(X < \infty) + \mathbb{E}(X \mid X < \infty)\varepsilon$$

$$+ (\mathrm{Var}(X \mid X < \infty) - \mathbb{E}(X \mid X < \infty))\frac{\varepsilon^2}{2} + o(\varepsilon^2). \qquad (A.1.1)$$

If $g^{(2)}(1) = \infty$ and $g'(1) < \infty$, then these Taylor expansions limited to order 1 yield $\mathbb{E}(X \mid X < \infty)$.

Sums of independent random variables   The following result is one of the reasons that generating functions are important. The converse of this result uses multivariate generating functions.

Theorem A.1.2 *If $X_1, \dots, X_n$ are independent random variables with values in $\mathbb{N} \cup \{\infty\}$, then*

$$g_{X_1 + \cdots + X_n} = g_{X_1} \cdots g_{X_n}.$$

In particular, if these r.v. are i.i.d. with generating function g, then

$$g_{X_1 + \cdots + X_n} = g^n.$$

Proof: It is sufficient to consider $0 < s < 1$ with the convention $s^\infty = 0$, and

$$g_{X_1 + \cdots + X_n}(s) := \mathbb{E}(s^{X_1 + \cdots + X_n}) = \mathbb{E}(s^{X_1}) \cdots \mathbb{E}(s^{X_n}) := g_{X_1}(s) \cdots g_{X_n}(s)$$

using independence for the middle equality.   ∎

Tail distributions   The following result is an application of the ideas of Lemma A.1.1 to generating functions.

Lemma A.1.3 *Let X be an r.v. with values in $\mathbb{N} \cup \{\infty\}$. Then,*

$$g_X(s) = 1 + (s - 1) \sum_{n \in \mathbb{N}} \mathbb{P}(X > n)s^n$$

for any $s \geq 0$ s.t. the power series on the l.h.s. converges, which is the case if $s < 1$.

Proof: As $\mathbb{P}(X = n + 1)s^n \leq \mathbb{P}(X > n)s^n \leq s^n$, the power series defining g converges if the power series on the l.h.s. converges, and this is the case if $s < 1$. Moreover,

$$\sum_{n \in \mathbb{N}} \mathbb{P}(X = n)s^n = \sum_{n \in \mathbb{N}} (\mathbb{P}(X > n - 1) - \mathbb{P}(X > n))s^n$$

$$= 1 + (s - 1) \sum_{n \in \mathbb{N}} \mathbb{P}(X > n)s^n$$

(this is the Abel rule).   ∎

A.1.2   Conditional probabilities and independence

A.1.2.1   Conditioning and total probability formula

If A and B are two events, $A \cap B$ can be denoted in some circumstances by A, B, and it is said "A and B."

If $\mathbb{P}(B) \neq 0$, then we define the probability of A conditional on B by

$$\mathbb{P}(A \mid B) = \frac{\mathbb{P}(A, B)}{\mathbb{P}(B)},$$

and $\mathbb{P}^B : A \mapsto \mathbb{P}(A \mid B)$ is a probability measure on Ω.

If X is a nonnegative or integrable r.v. for $\mathbb{P}$, then its expectation or variance conditional to B s.t. $\mathbb{P}(B) \neq 0$ is defined as its expectation or variance for $\mathbb{P}^B$, that is,

$$\mathbb{E}(X \mid B) = \mathbb{E}^B(X) = \frac{\mathbb{E}(X \mathbb{1}_B)}{\mathbb{P}(B)},$$

$$\mathrm{Var}(X \mid B) = \mathbb{E}^B((X - \mathbb{E}^B(X))^2) = \mathbb{E}^B(X^2) - \mathbb{E}^B(X)^2.$$

Iterated conditioning   If C is an event s.t. $\mathbb{P}^B(C) \neq 0$, or equivalently $\mathbb{P}(B, C) \neq 0$, then

$$\mathbb{P}^B(A \mid C) := \frac{\mathbb{P}^B(A, C)}{\mathbb{P}^B(C)} := \frac{\frac{\mathbb{P}(A,B,C)}{\mathbb{P}(B)}}{\frac{\mathbb{P}(B,C)}{\mathbb{P}(B)}} = \frac{\mathbb{P}(A, B, C)}{\mathbb{P}(B, C)} := \mathbb{P}(A \mid B, C).$$

Total probability   If $(B_i)_{i \in I}$ is a countable collection of events s.t.

$$\mathbb{P}(\cup_{i \in I} B_i) = 1, \qquad \mathbb{P}(B_i \cap B_j) = 0, \ \ \forall i \neq j,$$

then the probability of any event A or the expectation of any nonnegative or integrable r.v. X can be obtained as

$$\mathbb{P}(A) = \sum_{i \in I} \mathbb{P}(A, B_i) = \sum_{i \in I} \mathbb{P}(A \mid B_i) \mathbb{P}(B_i),$$

$$\mathbb{E}(X) = \sum_{i \in I} \mathbb{E}(X \mathbb{1}_{B_i}) = \sum_{i \in I} \mathbb{E}(X \mid B_i) \mathbb{P}(B_i),$$

with the natural convention $\mathbb{P}(A \mid B) \mathbb{P}(B) = \mathbb{E}(X \mid B) \mathbb{P}(B) = 0$ for $\mathbb{P}(B) = 0$.

A.1.2.2   Independence and conditional independence

Independent events   Two events A and B are independent if and only if

$$\mathbb{P}(A, B) = \mathbb{P}(A) \mathbb{P}(B),$$

and if $\mathbb{P}(B) \neq 0$ this is equivalent to $\mathbb{P}(A \mid B) = \mathbb{P}(A)$. For an arbitrary index set I, the events $(A_i)_{i \in I}$ are independent if and only if

$$\mathbb{P}(A_{i_1}, \dots, A_{i_k}) = \mathbb{P}(A_{i_1}) \cdots \mathbb{P}(A_{i_k}), \qquad \forall k \geq 2, \forall \{i_1, \dots, i_k\} \subset I.$$

Independent random variables, i.i.d. family Two random variables X and Y are independent if and only if, for all measurable E and F in the respective state spaces, $\{X \in E\}$ and $\{Y \in F\}$ are independent, that is,

$$\mathbb{P}(X \in E, Y \in F) := \mathbb{P}_{(X,Y)}(E \times F) = \mathbb{P}(X \in E)\mathbb{P}(Y \in F).$$

This expresses that the joint law $\mathbb{P}_{(X,Y)}$ is the product law $\mathbb{P}_X \otimes \mathbb{P}_Y$. Hence, the Fubini theorem yields that X and Y are independent if and only if, for any f and g, which are nonnegative or satisfy that $f(X)g(Y)$ be L^1,

$$\mathbb{E}(f(X)g(Y)) = \mathbb{E}(f(X))\mathbb{E}(g(X)).$$

If X and Y have discrete state spaces, it is sufficient that $\mathbb{P}(X = x, Y = y) = \mathbb{P}(X = x)\mathbb{P}(Y = y)$ for every x and y in the respective state spaces.

Remark A.1.4 *It is enough to consider all f and g in appropriately "rich" collections of functions. It is not enough to check that $\mathbb{E}(XY) = \mathbb{E}(X)\mathbb{E}(Y)$, in which case the r.v. are said to be decorrelated.*

For an arbitrary index set I, the random variables $(X_i)_{i \in I}$ are independent if and only if for any $k \geq 2$ and $\{i_1, \ldots, i_k\} \subset I$ and measurable $E_1, \ldots, E_k$ included in the respective state spaces

$$\mathbb{P}(X_{i_1} \in E_1, \ldots, X_{i_k} \in E_k) = \mathbb{P}(X_{i_1} \in E_1) \cdots \mathbb{P}(X_{i_k} \in E_k),$$

that is, if and only if the joint laws are given by the product of the marginals

$$\mathbb{P}_{(X_{i_1}, \ldots, X_{i_k})} = \mathbb{P}_{X_{i_1}} \otimes \cdots \otimes \mathbb{P}_{X_{i_k}},$$

and the Fubini theorem can be used as earlier. If I is finite, then it suffices to check this property for $\{i_1, \ldots, i_k\} = I$.

The random variables $(X_i)_{i \in I}$ are independent and identically distributed, i.i.d. for short, if they are independent and all have same law.

Independent σ-fields The most general independence notion is as follows. The sub-σ-fields $(\mathcal{G}_i)_{i \in I}$ of $\mathcal{F}$ (see Section A.3) are independent if and only if, for all $k \geq 2$ and $\{i_1, \ldots, i_k\} \subset I$ and $A_{i_1} \in \mathcal{G}_{i_1}, \ldots, A_{i_k} \in \mathcal{G}_{i_k}$, it holds that

$$\mathbb{P}(A_{i_1}, \ldots, A_{i_k}) = \mathbb{P}(A_{i_1}) \cdots \mathbb{P}(A_{i_k}).$$

If I is finite, then it suffices to check this property for $\{i_1, \ldots, i_k\} = I$.

Note that the random variables $(X_i)_{i \in I}$ are independent if and only if the generated sub-σ-fields $(\sigma(X_i))_{i \in I}$ are independent and that the events $(A_i)_{i \in I}$ are independent if and only if the random variables $(\mathbb{1}_{A_i})_{i \in I}$ are independent.

Conditional independence If B is an event s.t. $\mathbb{P}(B) \neq 0$, all these independence notions can be applied to the conditional probability measure $\mathbb{P}^B := \mathbb{P}(\cdot \mid B)$, and the terminology "independent conditional to B" is then used. In particular, A and C are independent conditional to B if and only if

$$\mathbb{P}(A, C \mid B) = \mathbb{P}(A \mid B)\mathbb{P}(C \mid B),$$

or, equivalently,

$$\mathbb{P}(A, B, C) = \mathbb{P}(A, B)\mathbb{P}(C \mid B) \ \text{ or } \ \mathbb{P}(C \mid A, B) = \mathbb{P}(C \mid B).$$

A.1.2.3 Basic limit theorems for i.i.d. random variables

The notion of a Markov chain is a generalization of the notion of a sequence of i.i.d. random variables. We recall two basic limit theorems for the latter. The first result shows that in a certain scale randomness tends to disappear, and the second quantifies precisely the residual randomness in the appropriate scale.

Theorem A.1.5 (Strong law of large numbers) *Let $(X_i)_{i \geq 1}$ be a sequence of i.i.d. integrable random vectors, and $m := \mathbb{E}(X_1)$ denote their common expectation. Then,*

$$\frac{X_1 + \cdots + X_n}{n} \xrightarrow[n \to \infty]{\text{a.s.}} m.$$

This convergence also holds in L^1.

Theorem A.1.6 (Central limit theorem) *Let $(X_i)_{i \geq 1}$ be a sequence of i.i.d. square-integrable random vectors, and $m := \mathbb{E}(X_1)$ and $K = \mathrm{Cov}(X_1)$ denote their common expectation and covariance matrix. Then,*

$$\sqrt{n}\left(\frac{X_1 + \cdots + X_n}{n} - m\right) \xrightarrow[n \to \infty]{\text{in law}} \mathcal{N}(0, K),$$

that is, converges in law to a centered Gaussian vector with covariance K.

These two results have been adapted to recurrent Markov chains using regenerative techniques in Section 4.1. This has yielded notably the pointwise ergodic theorem and the Markov chain central limit theorem.

A.2 Discrete measure convergence

A.2.1 Total variation norm and maximal coupling

A.2.1.1 Total variation norm and duality

The state space $\mathcal{V}$ will here be discrete, and we develop the notions in Section 1.2.2, see Section A.3.2 for some extensions to general measurable state spaces.

The space $\mathcal{M} := \mathcal{M}(\mathcal{V})$ of signed measures with the total variation norm can be identified to the separable (with dense countable subset) Banach space (complete normed space) $\ell^1 = \ell^1(\mathcal{V})$ of summable real sequences indexed by $\mathcal{V}$ with its natural norm. Its dual space L^∞ of bounded functions with the supremum norm can be identified with the space $\ell^\infty = \ell^\infty(\mathcal{V})$ of bounded sequences.

These vector spaces are of finite dimension if and only if $\mathcal{V}$ is finite. Recall that a vector space is of finite dimensions if and only if all norms are equivalent.

The subset $\mathcal{M}_+ := \mathcal{M}_+(\mathcal{V})$ of finite nonnegative measures is a closed (for the norm) cone (stable by nonnegative linear combinations) of $\mathcal{M}$. The set $\mathcal{M}_+^1 = \mathcal{M}_+^1(\mathcal{V})$ of probability measures is the intersection of $\mathcal{M}_+$ with the unit sphere. Thus, $\mathcal{M}_+^1$ is a closed convex subset of $\mathcal{M}$ and is complete for the distance induced by the norm.

Lemma A.2.1 *On a discrete state space $\mathcal{V}$, if μ is in $\mathcal{M}$, then*

$$\|\mu\|_{\text{var}} = \sum_{x \in \mathcal{V}} |\mu(x)| = \max_{\|f\|_\infty \le 1} \mu f = \max_{A \subset \mathcal{V}} \{\mu(A) - \mu(\mathcal{V} - A)\}.$$

Notably, the total variation norm $\|\cdot\|_{\text{var}}$ is the strong dual norm of the supremum norm on $L^\infty(\mathcal{V})$. Moreover, if μ and v are in $\mathcal{M}_+^1$, then

$$\|\mu - v\|_{\text{var}} = 2 \max_{A \subset \mathcal{V}} \{\mu(A) - v(A)\} \le 2,$$

and if X and Y are random variables with values in $\mathcal{V}$, then

$$\|\mathcal{L}(X) - \mathcal{L}(Y)\|_{\text{var}} = \sum_{x \in \mathcal{V}} |\mathbb{P}(X = x) - \mathbb{P}(Y = x)|$$

$$= \max_{\|f\|_\infty \le 1} \mathbb{E}(f(X) - f(Y))$$

$$= 2 \max_{A \subset \mathcal{V}} \{\mathbb{P}(X \in A) - \mathbb{P}(Y \in A)\} \le 2.$$

Proof: Let $A^+ = \{x \in \mathcal{V} : \mu(x) > 0\}$. For $f : \mathcal{V} \to \mathbb{R}$ s.t. $\|f\|_\infty \le 1$ and $A \subset \mathcal{V}$,

$$\mu f := \sum_{x \in \mathcal{V}} \mu(x) f(x) \le \sum_{x \in \mathcal{V}} |\mu(x)| = \mu(\mathbb{1}_{A^+} - \mathbb{1}_{\mathcal{V} - A^+}),$$

$$\mu(A) - \mu(\mathcal{V} - A) \le \mu(A^+) - \mu(\mathcal{V} - A^+) = \sum_{x \in \mathcal{V}} |\mu(x)|,$$

hence the equalities for $\|\mu\|_{\text{var}}$. Further, $\|\mu\|_{\text{var}} = \sup_{\|f\|_\infty \le 1} \mu f$ expresses the total variation norm as the dual of the supremum norm. Moreover, if μ and v are probability measures, then

$$\mu(\mathcal{V} - A) = 1 - \mu(A), \qquad v(\mathcal{V} - A) = 1 - v(A),$$

and hence

$$\|\mu - v\|_{\mathrm{var}} = 2 \max_{A \subset \mathcal{V}} \{\mu(A) - v(A)\}.$$

The rest is obvious. ∎

A.2.1.2   Total variation norm and maximal coupling

Lemma A.2.2 (Maximal coupling) *On a discrete state space* $\mathcal{V}$*, let* μ *and* v *be in* $\mathcal{M}_+^1$*. If any two random variables X and Y have laws* μ *and* v*, then*

$$\|\mu - v\|_{\mathrm{var}} \leq 2\mathbb{P}(X \neq Y).$$

Moreover, there exists two random variables X and Y with laws μ *and* v *satisfying*

$$\|\mu - v\|_{\mathrm{var}} = 2\mathbb{P}(X \neq Y),$$

and then (X, Y) *is said to be a maximal coupling of* μ *and* v*. Hence,*

$$\|\mu - v\|_{\mathrm{var}} = 2\min\{\mathbb{P}(X \neq Y) : \mathcal{L}(X) = \mu, \ \mathcal{L}(Y) = v\}.$$

In addition,

$$\|\mu - v\|_{\mathrm{var}} = \sum_{x \in \mathcal{V}} (\mu(x) - v(x))^+ = \sum_{x \in \mathcal{V}} (\mu(x) - v(x))^-.$$

Proof: As $\mathbb{E}(f(X) - f(Y)) \leq 2\|f\|_\infty \mathbb{P}(X \neq Y)$, it is obvious that

$$\|\mu - v\|_{\mathrm{var}} = \max_{\|f\|_\infty \leq 1} \mathbb{E}(f(X) - f(Y)) \leq 2\mathbb{P}(X \neq Y).$$

Moreover, for z in $\mathcal{V}$, let $p(z, z) = \min\{\mu(z), v(z)\}$. If $\sum_{z \in \mathcal{V}} p(z, z) = 1$, then $\mu = v$, and it is enough to take $X = Y$ of law μ. Else, for $x \neq y$ in $\mathcal{V}$ let

$$p(x, y) = \frac{(\mu(x) - p(x, x))(v(y) - p(y, y))}{1 - \sum_{z \in \mathcal{V}} p(z, z)} \geq 0.$$

As $p(x, x)$ has value either $\mu(x)$ or $v(x)$,

$$\sum_{y \in \mathcal{V}} p(x, y) = p(x, x) + \sum_{y \neq x} \frac{(\mu(x) - p(x, x))(v(y) - p(y, y))}{1 - \sum_{z \in \mathcal{V}} p(z, z)}$$

$$= p(x, x) + \sum_{y \in \mathcal{V}} \frac{(\mu(x) - p(x, x))(v(y) - p(y, y))}{1 - \sum_{z \in \mathcal{V}} p(z, z)}$$

$$= p(x, x) + \mu(x) - p(x, x)$$

$$= \mu(x),$$

$$\sum_{x \in \mathcal{V}} p(x, y) = v(y), \qquad \text{(by symmetry)}$$

and hence $(p(x, y))_{(x,y)\in\mathcal{V}\times\mathcal{V}}$ defines a law on $\mathcal{V} \times \mathcal{V}$ with marginals μ and ν, and in particular $\sum_{x,y} p(x, y) = \sum_x \mu(x) = 1$. If (X, Y) is taken with this law, then

$$\mathbb{P}(X \neq Y) = 1 - \sum_{x\in\mathcal{V}} p(x, x)$$

$$= \sum_{x\in\mathcal{V}} (\mu(x) - \min\{\mu(x), \nu(x)\})$$

$$= \sum_{x\in\mathcal{V}} (\mu(x) - \nu(x))^+,$$

$$\mathbb{P}(X \neq Y) = \sum_{x\in\mathcal{V}} (\mu(x) - \nu(x))^-, \qquad \text{(by symmetry)}$$

and by summing these two identities

$$2\mathbb{P}(X \neq Y) = \sum_{x\in\mathcal{V}} |\mu(x) - \nu(x)| = \|\mu - \nu\|_{\text{var}}.$$

The conclusion is then obvious. ∎

Remark A.2.3 *Some authors define the total variation distance between probability measures as the half of the definition given here, so as to get rid of factors 2 in some formulae. The definition here is more natural from a functional analytic point of view, as it preserves the strong dual norm formulation, the natural identification with sequence spaces, the fact that a probability measure is on the unit sphere, and so on.*

A.2.2   Duality between measures and functions

A.2.2.1   Dual Banach space and strong dual norm

Let $\mathcal{A}$ be a Banach space. Its dual $\mathcal{A}^*$ is the space of all continuous (for the norm) linear forms (real linear mappings) on $\mathcal{A}$. The action of ϕ in $\mathcal{A}^*$ on E is denoted by duality brackets as

$$\phi : v \in \mathcal{A} \mapsto \langle \phi, v \rangle \in \mathbb{R}.$$

The strong dual norm on $\mathcal{A}^*$ is given by the operator norm

$$\|\phi\|_{\mathcal{A}^*} = \sup_{v\in\mathcal{A}:\|v\|_{\mathcal{A}}\leq 1} \langle \phi, v \rangle = \inf\{B \geq 0 : \langle \phi, v \rangle \leq B\|v\|_{\mathcal{A}}\}, \qquad \phi \in \mathcal{A}^*,$$

and for this norm $\mathcal{A}^*$ is a Banach space.

A.2.2.2   Discrete signed measures and classic sequence spaces

Let $\mathcal{V}$ be a discrete state space. For $1 \leq p < \infty$, let $\ell^p = \ell^p(\mathcal{V})$ and $\ell^\infty = \ell^\infty(\mathcal{V})$ denote the spaces of real sequences $u = (u(x))_{x\in\mathcal{V}}$ s.t., respectively,

$$\|u\|_p = \left(\sum_{x\in\mathcal{V}} |u(x)|^p\right)^{1/p} < \infty, \qquad \|u\|_\infty = \sup_{x\in\mathcal{V}} |u(x)| < \infty.$$

If $\mathcal{V}$ is finite, then all these finite sequence spaces can be identified to elements of $\mathbb{R}^{\mathrm{Card}(\mathcal{V})}$, and all these norms are equivalent. The main focus is on infinite $\mathcal{V}$, and these spaces are isomorphic to the classic spaces of sequences indexed by $\mathbb{N}$.

The Banach space $\mathcal{M} = \mathcal{M}(\mathcal{V})$ of signed measures on $\mathcal{V}$ with the total variation norm can be identified with the separable space ℓ^1, and its dual $\mathcal{M}^* = L^\infty$ with ℓ^∞ by identifying f in ℓ^∞ with the linear form

$$\mu \in \ell^1 \mapsto \langle f, \mu \rangle = \sum_{x \in \mathcal{V}} \mu(x) f(x),$$

and the norms are in duality with this duality bracket.

The Banach space $c_0 = c_0(\mathcal{V})$ is the subspace of ℓ^∞ of the sequences that converge to 0: for all $\varepsilon > 0$, there exists a finite subset F of $\mathcal{V}$ s.t. $|u(x| < \varepsilon$ for x in $\mathcal{V} - F$. Then, with continuous injections,

$$1 \leq p \leq q < \infty \Rightarrow \ell^1 \subset \ell^p \subset \ell^q \subset c_0 \subset \ell^\infty.$$

The countable space of sequences with finite support is dense in c_0 and in ℓ^p for $1 \leq p < \infty$, and these Banach spaces hence are separable.

On the contrary, ℓ^∞ is not separable for in finite $\mathcal{V}$, and its dual contains strictly ℓ^1. Indeed, let $(f_n)_{n \geq 0} = ((f_n(x))_{x \in \mathcal{V}})_{n \geq 0}$ be a sequence with values in ℓ^∞, $(x_k)_{k \geq 0}$ an enumeration of $\mathcal{V}$, and $g(x_n) = 1$ if $f_n(x_n) < 0$ and else $g(x_n) = -1$. Then, $g = (g(x))_{x \in \mathcal{V}}$ is in ℓ^∞ and

$$\|g - f_n\|_\infty \geq |g(x_n) - f_n(x_n)| \geq 1,$$

and thus $(f_n)_{n \geq 0}$ cannot be dense in ℓ^∞.

The dual space of c_0 can be identified with ℓ^1, with duality bracket for μ in ℓ^1 and f in c_0 again given by

$$\langle \mu, f \rangle = \sum_{x \in \mathcal{V}} \mu(x) f(x) = \langle f, \mu \rangle.$$

For μ in ℓ^1, for all $\varepsilon > 0$, there exists a finite subset F of $\mathcal{V}$ s.t. $\sum_{x \notin F} |\mu(x)| < \varepsilon$, which readily yields using Lemma A.2.1 that

$$\|\mu\|_{\mathrm{var}} = \|\mu\|_1 = \sup_{f \in \ell^\infty : \|f\|_\infty \leq 1} \langle \mu, f \rangle = \sup_{f \in c_0 : \|f\|_\infty \leq 1} \langle \mu, f \rangle,$$

so that the total variation norm (or the ℓ_1 norm) is the strong dual norm both considering $\mathcal{M}$ (or ℓ^1) as the dual of c_0 or as a subspace of the dual of ℓ^∞.

A.2.2.3 Weak topologies

The Banach space $\mathcal{M}$ can be given the weak topology

$$\mu_n \xrightarrow[n \to \infty]{\text{weak}} \mu \iff \langle \mu_n, f \rangle \xrightarrow[n \to \infty]{} \langle \mu, f \rangle, \ \forall f \in \ell^\infty,$$

also denoted by $\sigma(\mathcal{M}, \ell^\infty)$. It can also be considered as the dual space of c_0, and given the weak-$*$ topology

$$\mu_n \overset{\text{weak}-*}{\underset{n\to\infty}{\longrightarrow}} \mu \Rightarrow \langle \mu_n, f \rangle \underset{n\to\infty}{\longrightarrow} \langle \mu, f \rangle, \ \forall f \in c_0,$$

also denoted by $\sigma(\mathcal{M}, c_0)$. Recall that in infinite dimension the dual space of ℓ^∞ is much larger than $\mathcal{M}$.

A simple fact is that a sequence $(\mu_n)_{n\geq 0}$ converges for $\sigma(\mathcal{M}, c_0)$ if and only if it is bounded (for the norm) and converges termwise. A diagonal subsequence extraction procedure then shows that a subset of $\mathcal{M}$ is relatively compact for $\sigma(\mathcal{M}, c_0)$ if and only if it is bounded.

Let $\mathcal{V}$ be infinite and identified with $\mathbb{N}$. Then, the sequence $(\delta_n)_{n\geq 0}$ of $\mathcal{M}_+^1$ clearly converges to 0 for $\sigma(\mathcal{M}, c_0)$, and hence $\mathcal{M}_+^1$ is not closed for this topology. Moreover, this sequence cannot have an accumulation point for $\sigma(\mathcal{M}, \ell^\infty)$, as this could only be 0 as per the above-mentioned conditions, whereas $\langle \mu_n, 1 \rangle \equiv 1$. Hence, the bounded set $\mathcal{M}_+^1$ is not relatively compact for $\sigma(\mathcal{M}, \ell^\infty)$ nor for the (stronger) topology of the total variation norm.

These are instances of far more general facts. Recall that a normed vector space is of finite dimension if and only if its unit sphere is compact and that the unit sphere is always compact for the weak-$*$ topology (but not necessarily for the weak topology), which helps explain its popularity, see the Banach–Alaoglu theorem (Rudin, W. (1991), Theorem 3.15).

A.2.3   Weak convergence of laws and convergence in law

Let us now assume that the above-mentioned notions are *restricted* to the space of probability measures $\mathcal{M}_+^1$, that is, that both the sequence $(\mu_n)_{n\geq 0}$ *and* its limit μ are probability measures.

Then, not only the $\sigma(\mathcal{M}, \ell^\infty)$ and $\sigma(\mathcal{M}, c_0)$ topologies coincide (a fact which extends to general state spaces), but as $\mathcal{V}$ is discrete, they also coincide with both the topology of the termwise convergence (product topology) and the topology of the complete metric space given by the (trace of the) total variation norm.

The resulting topology is called the topology of weak convergence of probability measures. The convergence in law of random variables is defined as the weak convergence of their laws.

Indeed, clearly on $\mathcal{M}$, the weakest topology is that of termwise convergence, and the strongest is that of total variation. Let μ_n for $n \geq 0$ and μ be in $\mathcal{M}_+^1$, and $\lim_{n\to\infty} \mu_n(x) = \mu(x)$ for every x in $\mathcal{V}$. Let $\varepsilon > 0$ be arbitrary. It is possible to choose a finite subset F of $\mathcal{V}$ and then $N \geq 0$ s.t.

$$\sum_{x\notin F} \mu(x) < \varepsilon, \qquad n \geq N \Rightarrow \sum_{x\in F} |\mu_n(x) - \mu(x)| < \varepsilon.$$

As these are probability measures, if $n \geq N$, then

$$\left| \sum_{x \notin F} \mu_n(x) - \sum_{x \notin F} \mu(x) \right| = \left| \sum_{x \in F} \mu_n(x) - \sum_{x \in F} \mu(x) \right| < \varepsilon$$

and thus, $\sum_{x \notin F} \mu_n(x) < 2\varepsilon$, and hence,

$$\|\mu_n - \mu\|_{\text{var}} \leq \sum_{x \in F} |\mu_n(x) - \mu(x)| + \sum_{x \notin F} \mu_n(x) + \sum_{x \notin F} \mu(x) < 4\varepsilon.$$

A.2.3.1   Relative compactness and tightness

The fact that $\mathcal{M}_+^1$ is weak-$*$ relatively compact in $\mathcal{M}$, and computations quite similar to that shown earlier, show that a subset C of $\mathcal{M}_+^1$ is relatively compact for the weak convergence of probability measures if and only if C is tight, in the following sense: for every $\varepsilon > 0$, there exists a finite subset F of $\mathcal{V}$ s.t.

$$\sum_{x \notin F} \mu(x) < \varepsilon, \qquad \forall \mu \in C.$$

A.3   Measure-theoretic framework

This appendix introduces without proofs the main notions and results in measure and integration theory, which allow to treat the subject of Markov chains in a mathematically rigorous way.

A.3.1   Probability spaces

A probability space $(\Omega, \mathcal{F}, \mathbb{P})$ is given by

1. a set Ω encoding all possible random outcomes,

2. a σ-field $\mathcal{F}$, which is a set constituted of certain subsets of Ω, and satisfies

   - the set Ω is in $\mathcal{F}$,

   - if A is in $\mathcal{F}$, then its complement $A^c = \Omega - A$ is in $\mathcal{F}$,

   - if A_n for n in $\mathbb{N}$ is in $\mathcal{F}$, then $\bigcup_{n \in \mathbb{N}} A_n$ is in $\mathcal{F}$,

3. a probability measure $\mathbb{P}$, which is a mapping $\mathbb{P} : \mathcal{F} \to \mathbb{R}_+$ satisfying

   - it holds that $\mathbb{P}(\Omega) = 1$,

   - the σ-additivity property: if A_n for n in $\mathbb{N}$ are *pairwise disjoint* sets of $\mathcal{F}$, then

$$\mathbb{P}\left(\bigcup_{n \in \mathbb{N}} A_n \right) = \sum_{n \in \mathbb{N}} \mathbb{P}(A_n).$$

The elements of $\mathcal{F}$ are called events, and regroup certain random outcomes leading to situations of interest in such a way that these can be attributed a "likelihood measure" using $\mathbb{P}$.

Clearly, $\emptyset := \Omega^c \in \mathcal{F}$, and $\mathbb{P}(\emptyset) = 0$ and more generally $\mathbb{P}(A^c) = 1 - \mathbb{P}(A)$, and if A_n for n in $\mathbb{N}$ is in $\mathcal{F}$, then $\bigcap_{n \in \mathbb{N}} A_n \in \mathcal{F}$. Note that in order to consider finite unions or intersections it suffices to use $\emptyset$ or Ω where necessary.

The trivial σ-field $\{\emptyset, \Omega\}$ is included in any σ-field, which is in turn included in the σ-field of all subsets of Ω. The latter is often the one of choice when possible, and notably if Ω is countable, but it is often too large to define an appropriate probability measure $\mathbb{P}$ on it. Moreover, the notion of sub-σ-field is used to encode partial information available in a probabilistic model.

The following important property is in fact equivalent to σ-additivity, using the fact that $\mathbb{P}$ is a finite measure.

Lemma A.3.1 (Monotone limit) *If $(A_n)_{n \geq 0}$ is a nondecreasing sequence of events, then*

$$\mathbb{P}\left(\bigcup_{n \in \mathbb{N}} \uparrow A_n\right) = \lim_{n \to \infty} \uparrow \mathbb{P}(A_n).$$

If $(A_n)_{n \geq 0}$ is a nonincreasing sequence of events, then

$$\mathbb{P}\left(\bigcap_{n \in \mathbb{N}} \downarrow B_n\right) = \lim_{n \to \infty} \downarrow \mathbb{P}(B_n).$$

Proof: Let $A_{-1} = \emptyset$. As $A_{n-1} \subset A_n$, the events $A_n - A_{n-1}$ are pairwise disjoint for $n \geq 0$ and $\mathbb{P}(A_n - A_{n-1}) = \mathbb{P}(A_n) - \mathbb{P}(A_{n-1})$. Hence, the σ-additivity yields that

$$\mathbb{P}\left(\bigcup_{n \in \mathbb{N}} A_n\right) = \mathbb{P}\left(\bigcup_{n \in \mathbb{N}} (A_n - A_{n-1})\right)$$

$$= \sum_{n \in \mathbb{N}} (\mathbb{P}(A_n) - \mathbb{P}(A_{n-1}))$$

$$= \lim_{n \to \infty} \uparrow \mathbb{P}(A_n)$$

and we see that in fact this is an equivalence. The second result is obtained from the first by complementation, using the fact that $\mathbb{P}(\Omega) = 1$. ∎

A.3.1.1   Generated σ-field and information

An arbitrary intersection of σ-fields is a σ-field, and the set of all subsets of Ω is a σ-field. This allows to define the σ-field generated by a set C of subsets of Ω as the intersection of all σ-fields containing C, and thus it is the least σ-field containing C. This σ-field is denoted by $\sigma(C)$ and encodes the probabilistic information available by observing C.

A.3.1.2   Almost sure (a.s.) and negligible

A subset of Ω containing an event of probability 1 is said to be almost sure, a subset of Ω included in an event of probability 0 is said to be negligible, and these are two complementary notions. The σ-additivity property yields that a countable union of negligible events is negligible. By complementation, a countable intersection of almost sure sets is almost sure. A property is almost sure, or holds a.s., if the set of all ω in Ω that satisfy it is almost sure. The classical abbreviation for almost sure is "a.s." and is often left implicit, but care needs to be taken if a uncountable number of operations are performed.

A.3.2   Measurable spaces and functions: signed and nonnegative

A set $\mathcal{V}$ furnished with a σ-field $\mathcal{A}$ is said to be measurable. A mapping f from a measurable set $\mathcal{V}$ with σ-field $\mathcal{A}$ to another measurable set $\mathcal{W}$ with σ-field $\mathcal{B}$ is said to be measurable if and only if

$$\forall B \in \mathcal{B}, \qquad f^{-1}(B) := \{x \in \mathcal{V} : f(x) \in B\} \in \mathcal{A}.$$

A (nonnegative) measure μ on a measurable set $\mathcal{V}$ with σ-field $\mathcal{A}$ is a σ-additive mapping $\mu : \mathcal{A} \to [0, \infty] := \mathbb{R}_+ \cup \{\infty\}$.

By σ-additivity, if A and B are in $\mathcal{A}$ and $A \subset B$, then $0 \leq \mu(A) \leq \mu(B) \leq \mu(\mathcal{V}) \leq \infty$. The measure μ is said to be finite if $\mu(\mathcal{V}) < \infty$, and then $\mu : \mathcal{A} \to \mathbb{R}_+$, and to be a probability measure or a law if $\mu(\mathcal{V}) = 1$, and then $\mu : \mathcal{A} \to [0, 1]$.

Many results for probability spaces can be extended in this framework (which is usually introduced first) using the classical computation conventions in $[0, \infty]$.

For instance, $\mu(A^c) = \mu(\mathcal{V}) - \mu(A)$ if this quantity has a meaning. As in Lemma A.3.1, the σ-additivity property is equivalent to the fact that if $(A_n)_{n \geq 0}$ is a nondecreasing sequence of events in $\mathcal{A}$, then $\mu(\bigcup_{n \in \mathbb{N}} A_n) = \lim_{n \to \infty} \uparrow \mu(A_n)$. Moreover, by complementation, if $(B_n)_{n \geq 0}$ is a nonincreasing sequence of events in $\mathcal{A}$ s.t. $\mu(B_k) < \infty$ for some k, then $\mu(\bigcap_{n \in \mathbb{N}} B_n) = \lim_{n \to \infty} \downarrow \mu(B_n)$.

A further extension is given by signed measures μ, which are σ-additive mappings $\mu : \mathcal{A} \to \mathbb{R}$. The Hahn–Banach decomposition yields an essentially unique decomposition of a signed measure μ into a difference of nonnegative finite measures, under the form $\mu = \mu^+ - \mu^-$, in which the supports A^+ and A^- of μ^+ and μ^- are disjoint. The finite nonnegative measure $|\mu| = \mu^+ + \mu^-$ is called the total variation measure of μ, and its total mass $\|\mu\|_{var} = |\mu|(\mathcal{V})$ is called the total variation norm of μ.

The space $\mathcal{M}$ of all signed measures is a Banach space for this norm, which can be identified with a closed subspace of the strong dual of the functional space L^∞.

For every (nonnegative, possible infinite) reference measure λ, the Banach space $\mathcal{M}$ contains a subspace that can be identified with $L^1(\lambda)$ by identifying any measure μ, which is absolutely continuous w.r.t. λ with its Radon–Nikodym derivative $\frac{d\mu}{d\lambda}$. If $\mathcal{V}$ is discrete, then a natural and universal choice for λ is the counting measure, and thus μ can be identified with the collection $(\mu(x))_{x \in \mathcal{V}}$ and $\mathcal{M}$ with $\ell^1(\mathcal{V})$.

A.3.3   Random variables, their laws, and expectations

A.3.3.1   Random variables and their laws

A probability space $(\Omega, \mathcal{F}, \mathbb{P})$ is given. A random variable (r.v.) with values in a measurable set $\mathcal{V}$ with σ-field $\mathcal{A}$ is a measurable function $X : \Omega \mapsto \mathcal{V}$, which satisfies

$$\forall A \in \mathcal{A}, \qquad X^{-1}(A) := \{\omega \in \Omega : X(\omega) \in E\} := \{X \in E\} \in \mathcal{F}.$$

For an arbitrary mapping $X : \Omega \mapsto \mathcal{V}$, the set

$$\sigma(X) = \{\{X \in E\} : E \in \mathcal{A}\}$$

is a σ-field, called the σ-field generated by X, encoding the information available on Ω by observing X. Notably, X is measurable if and only if $\sigma(X) \subset \mathcal{F}$.

The probability space $(\Omega, \mathcal{F}, \mathbb{P})$ is often only assumed to be fixed without further precision and represents some kind of ideal probabilistic knowledge. Only the properties of certain random variables are precisely given. These often represent indirect observations or effects of the random outcomes, and it is natural to focus on them to get useful information.

The law of the r.v. X is the probability measure $\mathbb{P} \circ X^{-1}$ on $\mathcal{V}$, which is well defined as X is measurable. It is denoted by $\mathcal{L}(X)$ or $\mathbb{P}_X$ and is given by

$$\mathbb{P}_X(A) := \mathbb{P}(X \in A), \qquad A \in \mathcal{A}.$$

Then, $(\mathcal{V}, \mathcal{A}, \mathbb{P}_X)$ is a probability space which encodes the probabilistic information available on the outcomes of X.

A.3.3.2   Expectation for $[0, \infty]$-valued random variables

The expectation $\mathbb{E}$ will be defined as a monotone linear extension of the probability measure $\mathbb{P}$, first for random variables taking a finite number of values in $[0, \infty]$, then for general random variables with values in $[0, \infty]$, and finally for real random variables satisfying an integrability condition. The notation $\mathbb{E}^{\mathbb{P}}$ is sometimes used to stress $\mathbb{P}$.

This procedure allows to define the integral $\int f \, d\mu$ of a measurable function f, from Ω with σ-field $\mathcal{F}$ to $\mathcal{V}$ with σ-field $\mathcal{A}$, by a measure $\mu : \mathcal{F} \to [0, \infty]$, but we restrict this to probability measures for the sake of concision.

The classic structure of $\mathbb{R}$ is extended to $[0, \infty] = \mathbb{R}_+ \cup \{\infty\}$ by setting

$$x + \infty = \infty, \ x \in [0, \infty], \qquad x \times \infty = \infty, \ x > 0, \qquad 0 \times \infty = 0.$$

Finite number of values   If X is an r.v. taking a finite number of values in $[0, \infty]$, then

$$\mathbb{E}(X) := \int_\Omega X(\omega) \mathbb{P}(d\omega) := \sum_{x \in X(\Omega)} x \mathbb{P}(X = x) \in [0, \infty].$$

In particular,

$$\mathbb{E}(\mathbb{1}_A) = \mathbb{P}(A), \qquad A \in \mathcal{F}.$$

For such random variables, this defines a monotone operator, in the sense that

$$X \geq Y \Rightarrow \mathbb{E}(X) \geq \mathbb{E}(Y),$$

which moreover is nonnegative linear, in the sense that

$$a, b \in [0, \infty] \Rightarrow \mathbb{E}(aX + bY) = a\mathbb{E}(X) + b\mathbb{E}(Y) \in [0, \infty].$$

Extension by supremum For an r.v. X with values in $[0, \infty]$, let

$$S(X) := \{Y \text{ r.v., s.t. } Y \leq X \text{ and } Y(\Omega) \subset [0, \infty] \text{ is finite}\},$$

and

$$\mathbb{E}(X) := \int_{\Omega} X(\omega)\mathbb{P}(d\omega) := \sup_{Y \in S(X)} \mathbb{E}(Y) \in [0, \infty].$$

This extension of $\mathbb{E}$ is still monotone and nonnegative, from which we deduce the following extension of the monotone limit lemma (Lemma A.3.1). This is where the fact that X is measurable becomes crucial.

Theorem A.3.2 (Monotone convergence theorem) *In* $[0, \infty]$, *if* $(X_n)_{n\geq 0}$ *is a nondecreasing sequence of random variables, then*

$$\mathbb{E}\left(\lim_{n\to\infty} \uparrow X_n\right) = \lim_{n\to\infty} \uparrow \mathbb{E}(X_n).$$

Proof: Note that, in $[0, \infty]$,

$$\lim_{n\to\infty} \uparrow X_n = \sup_{n\geq 0} X_n, \qquad \lim_{n\to\infty} \uparrow \mathbb{E}(X_n) = \sup_{n\geq 0} \mathbb{E}(X_n).$$

Monotonicity yields that

$$\mathbb{E}\left(\sup_{n\geq 0} X_n\right) \geq \sup_{n\geq 0} \mathbb{E}(X_n).$$

For every $Y \in S(\sup_{n\geq 0} X_n)$ and $0 < c < 1$, monotonicity again yields that

$$\mathbb{E}(X_n) \geq \mathbb{E}(cY\mathbb{1}_{\{X_n \geq cY\}}) = \sum_{y \in Y(\Omega)} cy\mathbb{P}(Y = y, X_n \geq cy).$$

As $Y(\Omega)$ is finite, by monotone limit (Lemma A.3.1)

$$\lim_{n\to\infty} \uparrow \sum_{y \in Y(\Omega)} cy\mathbb{P}(Y = y, X_n \geq cy) = c \sum_{y \in Y(\Omega)} y\mathbb{P}(Y = y) := c\mathbb{E}(Y).$$

Hence $\sup_{n\geq 0}\mathbb{E}(X_n) \geq \mathbb{E}(\sup_{n\geq 0} X_n)$, and these two quantities are equal. ∎

Nonnegative linearity   This theorem allows to prove that $\mathbb{E}$ is nonnegative linear, by replacing the supremum in the definition by the limit of an adequate nonde-creasing sequence. If Z is a $[0, \infty]$-valued r.v., then we define for $n \geq 1$ the dyadic approximation $[Z]_n$ satisfying

$$[Z]_n = n\mathbb{1}_{\{Z \geq n\}} + \sum_{k=0}^{n2^n - 1} \frac{k}{2^n} \mathbb{1}_{\left\{ \frac{k}{2^n} \leq Z < \frac{k+1}{2^n} \right\}} \in S(Z), \qquad \lim_{n \to \infty} \uparrow [Z]_n = Z.$$

If X and Y are $[0, \infty]$-r.v., and $a, b \in [0, \infty]$, then

$$\mathbb{E}(aX + bY) = \lim_{n \to \infty} \uparrow \mathbb{E}(a[X]_n + b[Y]_n)$$

$$= a \lim_{n \to \infty} \uparrow \mathbb{E}([X]_n) + b \lim_{n \to \infty} \uparrow \mathbb{E}([Y]_n)$$

$$= a\mathbb{E}(X) + b\mathbb{E}(Y).$$

Fatou's Lemma   An important corollary of the monotone convergence theorem is the following.

Lemma   A.3.3 (Fatou   Lemma) *For   any   sequence $(X_n)_{n \geq 0}$ of $[0, \infty]$-valued r.v.,*

$$\mathbb{E}\left(\liminf_{n \to \infty} X_n \right) \leq \liminf_{n \to \infty} \mathbb{E}(X_n).$$

Proof: Recall that $\liminf_{n \to \infty} u_n := \sup_{n \geq 0} \inf_{k \geq n} u_n$ in $[0, \infty]$. Then,

$$\mathbb{E}\left(\sup_{n \geq 0} \inf_{k \geq n} X_k \right) = \sup_{n \geq 0} \mathbb{E}\left(\inf_{k \geq n} X_k \right) \leq \sup_{n \geq 0} \inf_{k \geq n} \mathbb{E}(X_k)$$

using the monotone convergence theorem and monotonicity.   ∎

Let us finish with a quite useful result.

Lemma A.3.4 *If X is a $[0, \infty]$-valued r.v., then*

$$\mathbb{E}(X) < \infty \Rightarrow \mathbb{P}(X = \infty) = 0, \qquad \mathbb{E}(X) = 0 \Longleftrightarrow \mathbb{P}(X = 0) = 0.$$

Proof: We give the proof of the only nontrivial implication, by contradiction. By monotone limit (Lemma A.3.1),

$$\mathbb{P}(X > 0) = \lim_{n \to \infty} \uparrow \mathbb{P}(X > 1/n).$$

Thus, if $\mathbb{P}(X > 0) > 0$, then there exists $\varepsilon > 0$ and $\eta > 0$ s.t. $\mathbb{P}(X > \varepsilon) > \eta$ and then $\mathbb{E}(X) > \varepsilon\eta > 0$.   ∎

A.3.3.3 Real-valued random variables and integrability

Let X be an r.v. with values in $[-\infty, \infty] := \mathbb{R} \cup \{-\infty, \infty\}$. Let

$$X^+ = \min(X, 0), \qquad X^- = \min(-X, 0),$$

so that

$$X = X^+ - X^-, \qquad |X| = X^+ + X^-.$$

The natural extension to $\mathbb{R} \cup \{-\infty, \infty\}$ of the operations on $\mathbb{R}$ lead to setting, except if the indeterminacy $\infty - \infty$ occurs,

$$\mathbb{E}(X) = \int_\Omega X(\omega)\mathbb{P}(d\omega) = \mathbb{E}(X^+) - \mathbb{E}(X^-) \in [-\infty, \infty].$$

This definition is *monotone* and *linear*: if all is well defined in $[-\infty, \infty]$, then

$$X \geq Y \Rightarrow \mathbb{E}(X) \geq \mathbb{E}(Y),$$

$$a, b \in [-\infty, \infty] \Rightarrow \mathbb{E}(aX + bY) = a\mathbb{E}(X) + b\mathbb{E}(Y) \in [0, \infty].$$

A.3.3.4 Integrable random variables

In particular,

$$\mathbb{E}(|X|) < \infty \Rightarrow \mathbb{E}(X) \in \mathbb{R}$$

and the latter is well defined. This is the most useful case and is extended by linearity to define $\mathbb{E}(X)$ for X with values in $\mathbb{R}^d$ satisfying $\mathbb{E}(|X|) < \infty$ for some (and then every) norm $|\cdot|$. Then, X is said to be *integrable*. The integrable random variables form a vector space

$$L^1 = L^1(\Omega, \mathbb{P}) = L^1(\Omega, \mathbb{R}^d, \mathbb{P}).$$

It is a simple matter to check that if X is an r.v. with values in $\mathcal{V}$ and $f : \mathcal{V} \to \mathbb{R}^d$ is measurable then

$$\mathbb{E}(f(X)) = \int_\Omega f(X(\omega))\mathbb{P}(d\omega) = \int_\mathcal{V} f(x)\mathbb{P}_X(dx)$$

in all cases in which one of these expressions can be defined, and then all can.

The expectation has good properties w.r.t. the a.s. convergence of random variables. The monotone convergence theorem has already been seen. Its corollary the Fatou lemma will be used to prove an important result.

A sequence of $\mathbb{R}^d$-valued random variables $(X_n)_{n\geq 0}$ is said to be *dominated* by an r.v. Y if

$$|X_n| \leq Y, \qquad \forall n \geq 0,$$

and to be *dominated* in L^1 by Y if moreover $Y \in L^1$. The sequence is thus dominated in L^1 if and only if

$$\sup_{n\geq 0} |X_n| \in L^1.$$

Theorem A.3.5 (Dominated convergence theorem) *If* $(X_n)_{n\geq 0}$ *is a sequence of random variables dominated in* L^1, *and if* $X = \lim_{n\to\infty} X_n$ *a.s., then*

$$\lim_{n\to\infty} \mathbb{E}(|X_n - X|) = 0, \qquad \lim_{n\to\infty} \mathbb{E}(X_n) = \mathbb{E}(X).$$

Proof: As

$$\sup_{n\geq 0} |X_n - X| \leq 2 \sup_{n\geq 0} |X_n|,$$

we may assume that

$$0 \leq X_n \leq Y, \qquad \mathbb{E}(Y) < \infty \qquad \lim_{n\to\infty} X_n = 0.$$

The Fatou lemma yields that $0 \leq \lim\inf_{n\to\infty} \mathbb{E}(X_n)$ and that

$$\mathbb{E}(Y) \leq \lim_{n\to\infty}\inf \mathbb{E}(Y - X_n) = \mathbb{E}(Y) - \lim_{n\to\infty}\sup \mathbb{E}(X_n),$$

so that $\lim\sup_{n\to\infty} \mathbb{E}(X_n) \leq 0$. Hence, $\lim_{n\to\infty} \mathbb{E}(X_n) = 0$. ∎

This theorem can be extended to the case when

$$X = \lim_{n\to\infty} X_n \quad \text{in probability.}$$

Indeed, the Borel–Cantelli lemma implies that then, from each subsequence, a further subsubsequence converging a.s. can be extracted. Applying Theorem A.3.5 to this a.s. converging sequence yields that the only accumulation point in $[-\infty, \infty]$ for $(\mathbb{E}(|X_n - X|))_{n\geq 0}$ is 0, and hence that $\lim_{n\to\infty} \mathbb{E}(|X_n - X|) = 0$.

A.3.3.5   Convexity inequalities and L^p spaces

Lemma A.3.6 (Jensen inequality) *Let* X *be a* $\mathbb{R}^d$*-valued integrable r.v.. If* $\phi : \mathbb{R}^d \to \mathbb{R}$ *is convex, then*

$$\phi(\mathbb{E}(X)) \leq \mathbb{E}(\phi(X)) \in \mathbb{R} \cup \{\infty\}.$$

If equality holds, then ϕ *is linear a.s. for the law of* X*, and in particular if* ϕ *is strictly convex, then* $\mathbb{P}(X = \mathbb{E}(X)) = 1$.

Proof: Let $m = \mathbb{E}(X)$. Convexity of ϕ yields that there exists $a \in \mathbb{R}^d$ s.t.

$$\phi(x) - \phi(m) \geq a \cdot (x - m), \qquad \forall x \in \mathbb{R}^d.$$

Taking expectations yields that

$$\mathbb{E}(\phi(X)) - \phi(m) \geq a \cdot \mathbb{E}(X - m) = 0.$$

Moreover, if $\mathbb{E}(\phi(X)) = \phi(m)$, then $\mathbb{P}(\phi(X) = \phi(m) + a \cdot (x - m)) = 1$. ∎

For $p \in [1, \infty[$, we will check that the set of all $\mathbb{R}^d$-valued random variables X s.t. $\mathbb{E}(|X|^p) < \infty$ forms a Banach space, denoted by

$$L^p = L^p(\Omega, \mathbb{P}), \qquad \|X\|_p = \mathbb{E}(|X|^p)^{1/p},$$

if two a.s. equal random variables are identified (i.e., on the quotient space). In particular, L^2 is a Hilbert space with scalar product

$$(X, Y) \mapsto \mathbb{E}(X \cdot Y).$$

Lemma A.3.7 (Hölder inequality) *Let X and Y be $\mathbb{R}^d$-valued random variables. If $p, q > 1$ satisfy $\frac{1}{p} + \frac{1}{q} = 1$, then*

$$\|XY\|_1 = \mathbb{E}(|XY|) \leq \mathbb{E}(|X|^p)^{1/p}\mathbb{E}(|Y|^q)^{1/q} = \|X\|_p\|Y\|_q \in [0, \infty].$$

If $\mathbb{E}(|XY|) < \infty$, then equality implies that X^p and Y^q are proportional, a.s.

Proof: We may assume that $X \geq 0$ and $Y \geq 0$. Possibly interchanging X and Y, we may assume that $0 < \mathbb{E}(X^p) < \infty$, else the result is obvious. Then,

$$\mathbb{E}(XY) = \int XY dP = \mathbb{E}(X^p) \int X^{1-p}Y\frac{X^p dP}{\int X^p dP},$$

where $X^{1-p}Y$ is integrated by a probability measure. The convexity of $x \mapsto x^q$ and the Jensen inequality yield that

$$\left(\int X^{1-p}Y\frac{X^p dP}{\int X^p dP} \right)^q \leq \int (X^{1-p}Y)^q \frac{X^p dP}{\int X^p dP}.$$

As $(1 - p)q = -p$ and $1 - 1/q = 1/p$, all this yields that

$$\mathbb{E}(XY) \leq \mathbb{E}(X^p)^{1/p}\mathbb{E}(Y^q)^{1/q}.$$

As $x \mapsto x^q$ is strictly convex, equality yields that $X^{1-p}Y$ is constant $X^p dP$-a.s., using the equality result in Lemma (A.3.6). Symmetrically, XY^{1-q} is constant $Y^q dP$-a.s. Hence, $\mathbb{P}(XY = 0) = 1$ and then

$$\mathbb{E}(X^p)^{1/p}\mathbb{E}(Y^q)^{1/q} = \mathbb{E}(XY) = 0$$

and thus $X = 0$ a.s. or $Y = 0$ a.s., or else X and Y are proportional on $\{(X, Y) \neq (0, 0)\}$, and hence everywhere, P-a.s. ∎

This proof remains valid if P is replaced by an arbitrary positive measure. The case $p = q = 1/2$ is a special case of the Cauchy–Schwarz inequality.

Lemma A.3.8 (Minkowski inequality) *Let X and Y be $\mathbb{R}^d$-valued random variables, and $p \geq 1$. Then, they satisfy the triangular inequality*

$$\|X + Y\|_p \leq \|X\|_p + \|Y\|_p.$$

Proof: We assume that $\mathbb{E}(|X + Y|^p) < \infty$, as else it is a simple matter to prove that $\mathbb{E}(|X|^p) = \infty$ or $\mathbb{E}(|Y|^p) = \infty$. Then,

$$\mathbb{E}(|X + Y|^p) \leq \mathbb{E}(|X||X + Y|^{p-1}) + \mathbb{E}(|Y||X + Y|^{p-1}),$$

which is enough if $p = 1$. Else, as $(p - 1)q = p$, the Hölder inequality yields that

$$\mathbb{E}(|X||X + Y|^{p-1}) \leq \mathbb{E}(|X|^p)^{1/p}\mathbb{E}(|X + Y|^p)^{1/q},$$

and similarly $\mathbb{E}(|Y||X + Y|^{p-1}) \leq \mathbb{E}(|Y|^p)^{1/p}\mathbb{E}(|X + Y|^p)^{1/q}$. Thus,

$$\mathbb{E}(|X + Y|^p) \leq (\mathbb{E}(|X|^p)^{1/p} + \mathbb{E}(|Y|^p)^{1/p})\mathbb{E}(|X + Y|^p)^{1/q}$$

and the conclusion follows using $\mathbb{E}(|X + Y|^p) < \infty$ and $1 - (1/q) = 1/p$. ∎

The Jensen inequality yields that if $1 \leq a \leq b < \infty$, then $\|X\|_a \leq \|X\|_b$. The linear form $\mathbb{E} : L^p \to \mathbb{R}$ hence has operator norm 1, as

$$|\mathbb{E}(X)| \leq \mathbb{E}(|X|) = \|X\|_1 \leq \|X\|_p$$

with equality for constant X.

Remark A.3.9 *The Lebesgue integration theory is streamlined and powerful. Its difficulty resides in constructing measures with desired properties, such as the Lebesgue measure on $\mathbb{R}$ or $[0, 1]$ which associates to any interval $[a, b]$ s.t. $a \leq b$ its length $b - a$. The fundamental difficult result allowing such constructions is the Caratheodory extension theorem. The following subsection is devoted to this problem in the context of Markov chains.*

A.3.4   Random sequences and Kolmogorov extension theorem

Let us go back to Section 1.1. Let be given a family of laws $\pi_{n_1,\ldots,n_k}$ on $\mathcal{V}^k$, for $k \geq 1$ and $0 \leq n_1 < \cdots < n_k$ in $\mathbb{N}$. Two natural questions arise:

- Does there exist a probability space $(\Omega, \mathcal{F}, \mathbb{P})$, a σ-field on $\mathcal{V}^{\mathbb{N}}$, and an r.v. $(X_n)_{n \geq 0} : \omega \in \Omega \to (X_n(\omega))_{n \geq 0} \in \mathcal{V}^{\mathbb{N}}$, satisfying that

$$\pi_{n_1,\ldots,n_k} = \mathcal{L}(X_{n_1}, \ldots, X_{n_k})$$

  that is, this family of laws are the finite-dimensional marginals of $(X_n)_{n \geq 0}$?

- If it is so, is the law of $(X_n)_{n \geq 0}$ unique, that is, is it characterized by its finite-dimensional marginals?

Clearly, the $\pi_{n_1,\ldots,n_k}$ must be consistent, or compatible: if $0 \leq m_1 < \cdots < m_j$ is a j-tuple included in the k-tuple $0 \leq n_1 < \cdots < n_k$, then $\pi_{m_1,\ldots,m_j}$ must be equal to the corresponding marginal of $\pi_{n_1,\ldots,n_k}$.

It is natural and "economical" to take $\Omega = \mathcal{V}^{\mathbb{N}}$, called the canonical space, the process $(X_n)_{n \geq 0}$ given by the canonical projections

$$X_n : \omega = (\omega_0, \omega_1, \dots) \in \mathcal{V}^{\mathbb{N}} \mapsto X_n(\omega) = \omega_n \in \mathcal{V},$$

called the canonical process, and to furnish $\mathcal{V}^{\mathbb{N}}$ with the smallest σ-field s.t. each X_n and hence each $(X_{n_1}, \dots, X_{n_k})$ is an r.v.: the product σ-field

$$\mathcal{F} := \sigma(X_n : n \geq 0).$$

Note that if $E_0, E_1, \dots$ is a sequence of subsets of the discrete space $\mathcal{V}$, then

$$E_0 \times E_1 \times \cdots = \{X_0 \in E_0\} \cap \{X_1 \in E_1\} \cap \cdots \in \mathcal{F} := \sigma(X_n : n \geq 0),$$

and that events of this form are sufficient to characterize convergence in results such as the pointwise ergodic theorem (Theorem 4.1.1). See also Section 2.1.1.

By construction, $(X_n)_{n \geq 0}$ is measurable and hence an r.v. on $\mathcal{V}^{\mathbb{N}}$ furnished with the product σ-field, and if this space is furnished with a probability measure $\mathbb{P}$, then $(X_n)_{n \geq 0}$ has law $\mathbb{P}$.

The following result is fundamental. It is relatively easy to show the uniqueness part: any two laws on the product σ-field with the same finite-dimensional marginals are equal. The difficult part is the existence result, which relies on the Caratheodory extension theorem.

Theorem A.3.10 (Kolmogorov extension theorem) *Let be given a consistent family of probability measures*

$$\pi_{n_1,\dots,n_k} \in \mathcal{M}_+^1(\mathcal{V}^k), \qquad k \geq 1, \ 0 \leq n_1 < \cdots < n_k \in \mathbb{N}.$$

There exists a unique probability measure $\mathbb{P}$, on the canonical space $\mathcal{V}^{\mathbb{N}}$ with the canonical process $(X_n)_{n \geq 0}$ and product σ-field $\sigma((X_n)_{n \geq 0})$, s.t. the $\pi_{n_1,\dots,n_k}$ are the finite-dimensional marginals of $(X_n)_{n \geq 0}$, that is, s.t.

$$\mathcal{L}(X_{n_1}, \dots, X_{n_k}) = \pi_{n_1,\dots,n_k}, \qquad k \geq 1, \ 0 \leq n_1 < \cdots < n_k \in \mathbb{N}.$$

The explicit form given in Definition 1.2.1, in terms of the initial law and the transition matrix P, allows to check easily that these probability measures are consistent. The Kolmogorov extension theorem then yields the existence and uniqueness of the law of the Markov chain on the product space. This yields the mathematical foundation for all the theory of Markov chains.

Corollary A.3.11 *Let $\mathcal{V}$ be a discrete space and P a transition matrix on $\mathcal{V}$. Then, for every probability measure π_0 on $\mathcal{V}$, there exists a unique law denoted by $\mathbb{P}_{\pi_0}$ on the canonical space $\Omega = \mathcal{V}^{\mathbb{N}}$ s.t. the canonical process $(X_n)_{n \geq 0}$ is a Markov chain on $\mathcal{V}$ with initial law π_0 and transition matrix P.*

References

Chung, K.L. (1967) Markov Chains with Stationary Transition Probabilities, Die Grundlehren der mathematischen Wissenschaften, Band 104, 2nd edn, Springer-Verlag New York, Inc., New York.

Duflo, M. (1996) *Algorithmes stochastiques, Mathématiques & Applications (Berlin) [Mathematics & Applications]*, Vol. 23, Springer-Verlag, Berlin.

Feller, W. (1968) *An Introduction to Probability Theory and Its Applications*, 3rd edn, Vol. I, John Wiley & Sons, Inc., New York.

Kelly, F.P. (2011) *Reversibility and Stochastic Networks*, Cambridge Mathematical Library, Cambridge University Press, Cambridge, Revised edition of the 1979 original with a new preface.

Propp, J.G. and Wilson, D.B. (1996) Exact sampling with coupled Markov chains and applications to statistical mechanics. Proceedings of the 7th International Conference on Random Structures and Algorithms (Atlanta, GA, 1995), Vol. 9, pp. 223–252.

Robert, P. (2003) *Stochastic Networks and Queues*, Applications of Mathematics (New York), Vol. 52, Stochastic Modelling and Applied Probability, Springer-Verlag, Berlin.

Rudin, W. (1991) *Functional Analysis, International Series in Pure and Applied Mathematics*, 2nd edn, McGraw-Hill Inc., New York.

Saloff-Coste, L. (1997) Lectures on finite Markov chains, *Lectures On Probability Theory and Statistics (Saint-Flour, 1996), Lecture Notes in Mathematics*, Vol. 1665, Springer, Berlin, pp. 301–413.

Williams, D. (1991) *Probability with Martingales*, Cambridge Mathematical Textbooks, Cambridge University Press, Cambridge.

Solutions for the exercises

Solutions for Chapter 1

1.1 This constitutes a Markov chain on $\{0, 1, \ldots, 6\}$ with matrix

$$P = (P(x, y))_{x,y \in \{0,1,\ldots,6\}} = \begin{pmatrix} 0 & \frac{1}{6} & \frac{1}{6} & \frac{1}{6} & \frac{1}{6} & \frac{1}{6} & \frac{1}{6} \\ \frac{1}{5} & 0 & \frac{2}{5} & 0 & 0 & 0 & \frac{2}{5} \\ \frac{1}{5} & \frac{2}{5} & 0 & \frac{2}{5} & 0 & 0 & 0 \\ \frac{1}{5} & 0 & \frac{2}{5} & 0 & \frac{2}{5} & 0 & 0 \\ \frac{1}{5} & 0 & 0 & \frac{2}{5} & 0 & \frac{2}{5} & 0 \\ \frac{1}{5} & 0 & 0 & 0 & \frac{2}{5} & 0 & \frac{2}{5} \\ \frac{1}{5} & \frac{2}{5} & 0 & 0 & 0 & \frac{2}{5} & 0 \end{pmatrix}$$

from which the graph is readily deduced. The astronaut can reach any module from any module in a finite number of steps, and hence, the chain is irreducible, and as the state space is finite, this yields that there exists a unique invariant measure π. Moreover, $\pi(0) = \frac{1}{5}(\pi(1) + \cdots + \pi(6))$ and by uniqueness and symmetry, $\pi(1) = \cdots = \pi(6)$, and hence, $\pi(1) = \cdots = \pi(6) = \frac{5}{6}\pi(0)$. By normalization, we conclude that $\pi(0) = \frac{1}{6}$ and $\pi(1) = \cdots = \pi(6) = \frac{5}{36}$.

Markov Chains: Analytic and Monte Carlo Computations, First Edition. Carl Graham.
© 2014 John Wiley & Sons, Ltd. Published 2014 by John Wiley & Sons, Ltd.

1.2 This constitutes a Markov chain on $\{1, \dots, 6\}$ with matrix

$$P = (P(x,y))_{x,y \in \{1,\dots,6\}} = \begin{pmatrix} 0 & \frac{1}{2} & 0 & 0 & 0 & \frac{1}{2} \\ \frac{1}{2} & 0 & 0 & 0 & \frac{1}{4} & \frac{1}{4} \\ 0 & 0 & 0 & 0 & 1 & 0 \\ 0 & 0 & 0 & 0 & 1 & 0 \\ 0 & \frac{1}{3} & \frac{1}{3} & \frac{1}{3} & 0 & 0 \\ \frac{2}{3} & \frac{1}{3} & 0 & 0 & 0 & 0 \end{pmatrix}$$

from which the graph is readily deduced. The mouse can reach one room from any other room in a finite number of steps, and hence, the chain is irreducible, and as the state space is finite, this yields that there exists a unique invariant measure π. Solving a simple linear system and normalizing the solution yield $\pi = (\frac{1}{4}, \frac{1}{4}, \frac{1}{16}, \frac{1}{16}, \frac{3}{16}, \frac{3}{16})$.

1.3.a The uniform measure is invariant if and only if the matrix is doubly stochastic.

1.3.b The uniform measure is again invariant for P^n for all $n \geq 1$.

1.3.c Then, $\sum_{x \in V} P(x,y) = \sum_{x \in V} p(y-x) = \sum_{z \in V} p(z) = 1$, where p is the law of the jumps.

1.4.a The non zero terms are $Q(\emptyset, \emptyset) = 1$ and for $k \geq 2$ and $n_1, n_2, \dots, n_k \in \mathbb{N}$,

$$Q(n_1, \emptyset) = Q(n_1 \cdots n_k, n_2 \cdots n_{k-1}) = p \, ,$$
$$Q(n_1, n_1 n_1) = Q(n_1 \cdots n_k, n_1 \cdots n_k(n_1 + n_k)) = 1 - p \, .$$

Any wager lost during the game is inscribed in the list and will be wagered again in the future. When the list is empty, the gambler would have won the initial sum of all the terms on the list S, with the other gains cancelling precisely the losses occurred during the game.

1.4.b Then, $\mathbb{P}(X_0 = k_0, \dots, X_m = k_m)$ can be written in terms of $(L_n)_{n \geq 0}$ as

$$\sum_{n_{0,1}, \dots, n_{0,k_0}, \dots, n_{m,1}, \dots, n_{m,k_m}} \mathbb{P}(L_0 = n_{0,1} \cdots n_{0,k_0}, \cdots, L_m = n_{m,1} \cdots n_{m,k_m})$$

and the Markov property for $(L_n)_{n \geq 0}$ yields that the terms of this sum write

$$\mathbb{P}(L_0 = n_{0,1} \cdots n_{0,k_0}, \cdots, L_{m-1} = n_{m-1,1} \cdots n_{m-1,k_{m-1}})$$
$$\times \sum_{n_{m,1} \cdots n_{m,k_m}} Q(n_{m-1,1} \cdots n_{m-1,k_{m-1}}, n_{m,1} \cdots n_{m,k_m})$$

and hence,

$$\mathbb{P}(X_0 = k_0, \ldots, X_m = k_m) = \mathbb{P}(X_0 = k_0, \ldots, X_{m-1} = k_{m-1})P(k_{m-1}, k_m),$$

where the non zero terms of P are $P(0,0) = 1$ and $P(1,0) = P(k, k-2) = p$ for $k \geq 2$ and $P(k, k+1) = 1 - p$ for $k \geq 1$.

1.5.a A natural state space is the set of permutations of $\{1, 2, 3\}$, of the form $\sigma = (\sigma_1, \sigma_2, \sigma_3)$, which has cardinal 6. By definition

$$Q((\sigma_1, \sigma_2, \sigma_3), (\sigma_1, \sigma_3, \sigma_2)) = p , \qquad Q((\sigma_1, \sigma_2, \sigma_3), (\sigma_2, \sigma_1, \sigma_3)) = 1 - p .$$

Clearly, Q is irreducible. As the state space is finite, this implies existence and uniqueness for the invariant law ρ. Intuition (the matrix is doubly stochastic) or solving a simple linear system shows that ρ is the uniform law, with density $1/6$.

1.5.b A natural state space is the set $\{1, 2, 3\}$ of cardinal 3, and

$$P = \begin{pmatrix} p & 1-p & 0 \\ 1-p & 0 & p \\ 0 & p & 1-p \end{pmatrix}$$

is clearly irreducible; hence, there is a unique invariant law. The invariant law is the uniform law, with density $1/3$.

1.5.c The characteristic polynomial of P is

$$\begin{aligned}
\det(XI - P) &= X(X - p)(X + p - 1) - p^2(X - p) - (p - 1)^2(X + p - 1) \\
&= X^3 - X^2 + (-3p^2 + 3p - 1)X + 3p^2 - 3p + 1 \\
&= (X - 1)(X^2 - 3p^2 + 3p - 1)
\end{aligned}$$

in which $\frac{1}{4} \leq 3p^2 - 3p + 1 < 1$ with equality on the left for $p = 1/2$. Hence, $\det(XI - P)$ has three distinct roots

$$1 , \qquad r(p) = \sqrt{3p^2 - 3p + 1} , \qquad -r(p) = -\sqrt{3p^2 - 3p + 1} .$$

Hence, $P^n = a_n P^2 + b_n P + c_n I$, where

$$a_n + b_n + c_n = 1 ,$$

$$r(p)^2 a_n + r(p)b_n + c_n = r(p)^n ,$$

$$r(p)^2 a_n - r(p)b_n + c_n = (-r(p))^n .$$

If n is even, then $b_n = 0$, and $c_n = 1 - a_n$ so that $r(p)^2 a_n + 1 - a_n = r(p)^n$ yields $a_n = \frac{r(p)^n - 1}{r(p)^2 - 1}$.

If n is odd, then $b_n = r(p)^{n-1}$, and $c_n = 1 - a_n - b_n$ so that $r(p)^2 a_n + r(p)^n + 1 - a_n - r(p)^{n-1} = r(p)^n$ yields $a_n = \frac{r(p)^{n-1}-1}{r(p)^2-1}$.

As computing P^2 is quite simple, this yields an explicit expression for P^n.

The law of X_n converges to the uniform law at rate $r(p)^n$, which is maximal for $p = 1/2$ and then takes the value $r(1/2)^n = 1/2^n$.

1.6   The transition matrix is given by

$$
P = \begin{pmatrix} 0 & 0 & 1 \\ \frac{1}{2}\frac{1}{10} & \frac{1}{2}\frac{9}{10} & \frac{1}{2} \\ \frac{1}{4}\frac{1}{10} & \frac{1}{4}\frac{9}{10} & \frac{3}{4} \end{pmatrix} = \begin{pmatrix} 0 & 0 & 1 \\ \frac{1}{20} & \frac{9}{20} & \frac{1}{2} \\ \frac{1}{40} & \frac{9}{40} & \frac{3}{4} \end{pmatrix}
$$

and the graph can easily be deduced from it. Clearly, P is irreducible. As the state space is finite, it implies that there is a unique invariant law π. This law solves

$$
\pi(1) = \frac{1}{20}\pi(2) + \frac{1}{40}\pi(3) ,
$$

$$
\pi(2) = \frac{9}{20}\pi(2) + \frac{9}{40}\pi(3) ,
$$

$$
\pi(3) = \pi(1) + \frac{1}{2}\pi(2) + \frac{3}{4}\pi(3) ,
$$

hence $\pi(2) = \frac{9}{22}\pi(3)$, then $\pi(1) = (\frac{9}{440} + \frac{1}{40})\pi(3) = \frac{1}{22}\pi(3)$. As $\frac{1}{22} + \frac{9}{22} + 1 = \frac{16}{11}$, normalization yields $\pi(1) = \frac{1}{32}$, $\pi(2) = \frac{9}{32}$ and $\pi(3) = \frac{11}{16}$.
The characteristic polynomial of P is

$$
\begin{vmatrix} X & 0 & -1 \\ -\frac{1}{20} & X - \frac{9}{20} & -\frac{1}{2} \\ -\frac{1}{40} & -\frac{9}{40} & X - \frac{3}{4} \end{vmatrix}
$$

$$
= X\left(X - \frac{9}{20}\right)\left(X - \frac{3}{4}\right) - \frac{9}{800} - \frac{1}{40}\left(X - \frac{9}{20}\right) - \frac{9}{80}X
$$

$$
= X^3 - \frac{6}{5}X^2 + \frac{1}{5}X = X(X-1)\left(X - \frac{1}{5}\right) .
$$

Hence, $P^n = a_n P^2 + b_n P + c_n I$ with $a_n + b_n + c_n = 1$, $c_n = 0$, and $\frac{1}{5^2}a_n + \frac{1}{5}b_n + c_n = \frac{1}{5^n}$. Thus, $a_n = \frac{5}{4}(1 - \frac{1}{5^{n-1}})$ and $b_n = \frac{1}{4}(\frac{1}{5^{n-2}} - 1)$. The law of X_n converges to π at rate $\frac{1}{5^n}$.

1.7.a   States $(9, j, 1)$ for $j < 8$, $(10, 8, 1)$, and $(10, 9, 1)$ are wins for Player A and states $(i, 9, 2)$ for $i < 8$, $(8, 10, 2)$, and $(9, 10, 2)$ are wins for Player B, and they are the absorbing states.

Let $i \leq 8$ and $j \leq 8$, or $8 \leq i \leq 9$ and $8 \leq j \leq 9$.

Considering all rallies, transitions from (i,j,A) to $(i+1,j,A)$ have probability a, from (i,j,A) to (i,j,B) probability $1-a$, and symmetrically from (i,j,B) to $(i,j+1,B)$ probability b and from (i,j,B) to (i,j,A) probability $1-b$.

Considering only the points scored, transitions from (i,j,A) to $(i+1,j,A)$ have probability $a \sum_{k=0}^{\infty}(1-a)(1-b) = \frac{a}{a+b-ab}$, from (i,j,A) to $(i,j+1,B)$ probability $(1-a)b \sum_{k=0}^{\infty}(1-b)(1-a) = \frac{b-ab}{a+b-ab}$, and symmetrically from (i,j,B) to $(i,j+1,B)$ probability $\frac{b}{a+b-ab}$ and from (i,j,B) to $(i+1,j,A)$ probability $\frac{a-ab}{a+b-ab}$.

1.7.b Straightforward.

1.7.c We use the transition for scored points. Player B wins in 9 points if he or she scores first, and we have seen that this happens with probability $\frac{b-ab}{a+b-ab} := B_9$.

Player B wins in 10 points if Player A scores 1 point and then Player B scores 2, if Player B scores 1 and then Player A scores 1 and then Player B scores 1, or if Player B scores 2 points in a row, which happens with probability

$$\frac{a(b-ab)b}{(a+b-ab)^3} + \frac{(b-ab)(a-ab)(b-ab)}{(a+b-ab)^3} + \frac{(b-ab)b}{(a+b-ab)^2}$$

$$= \frac{b-ab}{(a+b-ab)^3}(3ab + b^2 - 2ab^2 - a^2b + a^2b^2) .$$

Then,

$$B_{10} - B_9 = \frac{a(b-ab)}{(a+b-ab)^3}(ab - a + b) .$$

The hyperbole $b = \frac{a}{a+1}$ divides the square $0 < a, b < 1$ into two subsets. In the first subset, in which $b > \frac{a}{a+1}$, Player B should go to 10 points (this is the largest subset and contains the diagonal, which is tangent at 0 to the hyperbole). In the other, in which $b < \frac{a}{a+1}$, Player B should go to 9 points.

1.8.a The microscopic representation $(a_i)_{1 \leq i \leq N}$ yields the macroscopic representation

$$\left(\sum_{i=1}^{N} \mathbb{1}_{\{a_i=1\}}, \ldots, \sum_{i=1}^{N} \mathbb{1}_{\{a_i=K\}} \right) .$$

1.8.b Synchronous: the transition from $(a_i)_{1 \leq i \leq N}$ to $(b_j)_{1 \leq j \leq N}$ and the transition from $(m_1, \ldots, m_K)$ to $(n_1, \ldots, n_K)$ have probabilities

$$\frac{\prod_{j=1}^{N} c(b_j) \sum_{i=1}^{N} \mathbb{1}_{\{a_i=b_j\}}}{\left(\sum_{i=1}^{N} c(a_i) \right)^N} , \qquad \frac{K!}{n_1! \cdots n_K!} \frac{(c(1)m_1)^{n_1} \cdots (c(K)m_K)^{n_K}}{(c(1)m_1 + \cdots + c(K)m_K)^N} .$$

Asynchronous: for $1 \leq k \leq N$, the transition from $(a_i)_{1 \leq i \leq N}$ to the vector in which a_k is replaced by b and, for $1 \leq i, j \leq K$, the transition from $(m_1, \cdots, m_K)$ if $i = j$ to $(m_1, \cdots, m_K)$ and if $i \neq j$ to the vector in which the i th coordinate is replaced by $m_i - 1$ and the j th by $m_j + 1$, have probabilities

$$\frac{c(b) \sum_{i=1}^{N} \mathbb{1}_{\{a_i = b\}}}{N \sum_{i=1}^{N} c(a_i)} , \qquad \frac{m_i}{N} \frac{c(j) m_j}{c(1) m_1 + \cdots + c(K) m_K} .$$

The absorbing states are the pure states, constituting of populations carrying a single allele.

1.9.a For instance,

$$\mathbb{P}(R_3 = 2 \mid R_2 = 1, R_1 = 0) = p \neq \mathbb{P}(R_3 = 2 \mid R_2 = 1, R_1 = 1) = 0 .$$

1.9.b As then $D_{n+1} = (D_n + 1)\mathbb{1}_{\{X_{n+1}=1\}}$, Theorem 1.2.3 yields that $(D_n)_{n \geq 0}$ is a Markov chain on $\mathbb{N}$ with matrix given by $P(x, x + 1) = p$ and $P(x, 0) = 1 - p$ for $x \geq 0$. This matrix is clearly irreducible, but the state space is infinite and we cannot conclude now on existence and uniqueness for invariant law.

As an invariant measure, π, satisfies the equation $\pi = \pi P$, which develops into

$$\pi(0) = \sum_{x \geq 0} \pi(x)(1 - p) , \qquad \pi(x) = p \pi(x - 1) , \quad x \geq 1 ,$$

so that necessarily $\pi(x) = p^x \pi(0)$, and we check that $\pi(0) = \sum_{x \geq 0} p^x \pi(0)$ $(1 - p)$. Moreover, $\sum_{x \geq 0} p^x = \frac{1}{1-p}$ and hence, $\pi(x) = (1 - p)p^x$ for $x \geq 0$, which is a geometric law on $\mathbb{N}$.

1.9.c As $Z_{n+1} = (Z_n + 1)\mathbb{1}_{\{X_{n+1}=1, Z_n < k\}} + k\mathbb{1}_{\{Z_n = k\}}$, Theorem 1.2.3 yields that $(Z_n)_{n \geq 0}$ is a Markov chain on $\{0, \ldots, k\}$, with matrix given by $P_k(k, k) = 1$, $P_k(x, x + 1) = p$, and $P_k(x, 0) = 1 - p$ for $x = 0, 1, \ldots, k - 1$.

1.9.d Then, $\mathbb{P}(R_n \geq k) = \mathbb{P}(Z_n = k)$, and as $Z_0 = 0$, $\mathbb{P}(R_n \geq k) = P_k^n(0, k)$. Hence, for $k \in \mathbb{N}$,

$$\mathbb{P}(R_n = k) = \mathbb{P}(Z_n \geq k) - \mathbb{P}(Z_n \geq k + 1) = P_k^n(0, k) - P_{k+1}^n(0, k + 1) .$$

1.9.e This probability is $P_5^{100}(0, 5) = 0.81011$.

1.10.a The non zero terms of P are $P(E, E_{i,j}) = 1/N^2$ for $E \subset \mathcal{V}$, $i \in E$, and $j \in \mathcal{V} - E$, where $E_{i,j} \in \mathcal{V}$ is obtained from $E \in \mathcal{V}$ by interchanging i and j. The matrix is clearly irreducible. As the state space is finite, this implies that there exists a unique invariant law π. Intuition (the matrix is doubly stochastic) or a simple computation shows that the uniform law with density $1/\mathrm{Card}(\mathcal{V}) = 1/\binom{2N}{N}$ is invariant.

1.10.b The non zero terms of Q are, for $x \in \{0, 1, \ldots, N\}$,

$$Q(x, x+1) = \frac{(N-x)^2}{N^2}, \quad Q(x, x-1) = \frac{x^2}{N^2}, \quad Q(x, x) = 2\frac{x(N-x)}{N^2}.$$

The matrix is clearly irreducible. As the state space is finite, this implies that there exists a unique invariant law σ. A combinatorial computation starting from π yields that $\sigma(x) = \binom{N}{x}\binom{N}{N-x} / \binom{2N}{N} = \binom{N}{x}^2 / \binom{2N}{N}$ for $x \in \{0, 1, \ldots, N\}$. This is a hypergeometric law.

1.10.c The Markov property yields that

$$\mathbb{P}(i \in A_n) = \sum_{E \in \mathcal{V}} \mathbb{P}(A_{n-1} = E, i \in A_n) = \sum_{E \in \mathcal{V}} \mathbb{P}(A_{n-1} = E)\mathbb{P}_E(i \in A_1)$$

$$= \sum_{E \in \mathcal{V} \,:\, i \in E} \mathbb{P}(A_{n-1} = E)\left(1 - \frac{1}{N}\right) + \sum_{E \in \mathcal{V} \,:\, i \notin E} \mathbb{P}(A_{n-1} = E)\frac{1}{N}$$

$$= \left(1 - \frac{2}{N}\right)\mathbb{P}(i \in A_{n-1}) + \frac{1}{N},$$

and this affine recursion is solved by

$$\mathbb{P}(i \in A_n) = \left(1 - \frac{2}{N}\right)^n\left(\mathbb{P}(i \in A_0) - \frac{1}{2}\right) + \frac{1}{2}.$$

Moreover, $\mathbb{E}(S_n) = \sum_{i=1}^{N} \mathbb{P}(i \in A_n)$ and hence,

$$\mathbb{E}(S_n) = \left(1 - \frac{2}{N}\right)^n\left(\mathbb{E}(S_0) - \frac{N}{2}\right) + \frac{N}{2}.$$

Then, $\lim_{n\to\infty}\mathbb{P}(i \in X_n) = \lim_{n\to\infty}\frac{1}{N}\mathbb{E}(S_n) = \frac{1}{2}$ at rate $\left(1 - \frac{2}{N}\right)^n$.

1.11.a Theorem 1.2.3 yields that this is a Markov chain.

1.11.b Computations, quite similar to those for branching, show that

$$G_n(s) = h(s)G_{n-1}(g(s)) = \cdots = h(s)h(g(s))\cdots h(g^{\circ n-1}(s))G_0(g^{\circ n}(s)).$$

1.11.c Similarly, or by a Taylor expansion,

$$\mathbb{E}(X_n) = z + x\mathbb{E}(X_{n-1}) = \cdots = z(1 + x + \cdots + x^{n-1}) + x^n\mathbb{E}(X_0),$$

which takes the value $z\frac{1-x^n}{1-x} + x^n\mathbb{E}(X_0)$ if $x \neq 1$ or else $zn + \mathbb{E}(X_0)$ if $x = 1$.

1.12.a Theorem 1.2.3 yields that this is a Markov chain. The irreducibility condition is obvious.

1.12.b Then, $g_n(s) = \mathbb{E}(s^{X_n})$ can be written, using independence, as

$$g_n(s) = a(s)\mathbb{E}(s^{(X_{n-1}-1)^+}) = a(s)\mathbb{E}(s^{X_{n-1}-1} + (1 - s^{-1})\mathbb{1}_{\{X_{n-1}=0\}})$$

$$= a(s)s^{-1}g_{n-1}(s) + a(s)(1 - s^{-1})g_{n-1}(0).$$

1.12.c Then, necessarily

$$g(s) = a(s)s^{-1}g(s) + a(s)(1 - s^{-1})\pi(0)$$

and thus, $g(s)(s - a(s)) = \pi(0)(s - 1)a(s)$. For $s = 1 + \varepsilon$,

$$g(1 + \varepsilon)(1 + \varepsilon - a(1 + \varepsilon)) = \pi(0)\varepsilon a(1 + \varepsilon), \quad a(1 + \varepsilon) = 1 + m\varepsilon + o(\varepsilon),$$

and identification of the ε terms yields that $1 - m = \pi(0)$.

1.12.d As $\pi \geq 0$, necessarily $m \geq 1$, and if $m = 1$, then $\pi(0) = 0$, and the chain cannot be irreducible (as then $\pi > 0$) and thus $\mathbb{P}(A_n = 0)\mathbb{P}(A_n \geq 2) = 0$ and $\mathbb{P}(A_n = 1) = 1$ as $\mathbb{E}(A_n) = 1$.

1.12.e If $\sigma^2 = \mathrm{Var}(A_1) < \infty$, then

$$\mathbb{E}(A_1(A_1 - 1)) = \mathbb{E}(A_1^2) - \mathbb{E}(A_1) = \sigma^2 + m^2 - m$$

and hence, $a(1 + \varepsilon) = 1 + m\varepsilon + \frac{\sigma^2 + m^2 - m}{2}\varepsilon^2 + o(\varepsilon^2)$, and using $g(1 + \varepsilon) = 1 + \mu\varepsilon + o(\varepsilon)$ and identifying the terms in ε^2 in the above-mentioned Taylor expansion yields that $\mu = \frac{1}{2}(m + \frac{\sigma^2}{1-m})$.

1.13.a By definition and the basic properties of the total variation norm, $\rho_n \leq 1$. For all x and y, if f is such that $||f||_\infty = 1$, then $P^n(x, \cdot)f - P^n(y, \cdot)f \leq 2\rho_n$ and hence,

$$(\mu P^n - \mu P^n)f = \sum_{x,y \in V} \mu(x)\mu(y)(P^n(x, \cdot)f - P^n(y, \cdot)f) \leq 2\rho_n,$$

so that $||\mu P^n - \mu P^n|| \leq 2\rho_n$.

1.13.b Then, $(\mu - \mu)P^{n+m}f = (\mu P^n - \mu P^n)(P^m f - c)$ for all laws μ and μ, all f such that $||f||_\infty = 1$, and all $c \in \mathbb{R}$, and

$$\inf_{c \in \mathbb{R}} \sup_{x \in V} |g(x) - c| \leq \frac{1}{2} \sup_{x,y \in V} |g(x) - g(y)|$$

implies that $\inf_{c \in \mathbb{R}} ||P^m f - c||_\infty \leq \rho_m$. This yields that $(\mu - \mu)P^{n+m}f \leq \rho_n \rho_m$. Then, it is a simple matter to obtain that $\rho_{n+m} \leq \rho_n \rho_m$ and then that $\rho_n \leq \rho_k^{\lfloor n/k \rfloor}$.

1.13.c Taking $\mu = \mu P$, it holds that $||\mu P^n - \mu P^{n+1}|| \leq 2\rho_n \leq 2\rho_k^{\lfloor n/k \rfloor}$, which forms a geometrically convergent series, hence classically $(\mu P^n)_{n \in \mathbb{N}}$ is Cauchy, and as the metric space is complete, there is a limit π, which is π invariant. Then, $||\mu P^n - \pi P^n|| = ||\mu P^n - \pi||_{\mathrm{Var}} \leq 2\rho_n$.

1.13.d For all x, y, and f such that $||f||_\infty = 1$, it holds that $P^k(x, \cdot)f - P^k(y, \cdot)f = (P^k(x, \cdot) - \varepsilon\hat{\pi})f - (P^k(y, \cdot) - \varepsilon\hat{\pi})f \leq 2(1 - \varepsilon)$. It is a simple matter to conclude.

1.13.e For all x, y, and f such that $||f||_\infty = 1$, it holds that

$$P^k(x, \cdot)f - P^k(y, \cdot)f = \mathbb{E}(f(X_k^x) - f(X_k^y)) \leq 2\mathbb{P}(X_k^x \neq X_k^y) = 2\mathbb{P}(T_{x,y} > k) .$$

It is a simple matter to conclude.

Solutions for Chapter 2

2.1 Then, $\{S \leq n\} = \bigcup_{i=1}^n \{X_i \geq \max\{X_0, \dots, X_{i-1}\} + k\}$, $\{T \leq n\} = \bigcup_{j=k}^n \{X_j > \max_{0 \leq i \leq k} X_i\}$, and $\{U \leq n\} = \bigcup_{i=k}^n \{X_i > X_{i-1} > \cdots > X_{i-k}\}$.

2.2.a By definition of stopping times, $\{S \leq n\}$ and $\{T \leq n\}$ belong to $\mathcal{F}_n$. Thus, by definition of the σ-field $\mathcal{F}_n$,

$$\{S \wedge T \leq n\} = \{S \leq n\} \cup \{T \leq n\} , \quad \{S \vee T \leq n\} = \{S \leq n\} \cap \{T \leq n\} ,$$

also belong to $\mathcal{F}_n$, and $S \wedge T$ and $S \vee T$ are also stopping times. Moreover,

$$\{S + \theta_S T = n\} = \bigcup_{k=0}^n \{S = k\} \cap \{\theta_k T = n - k\} ,$$

where $\{S = k\} \in \mathcal{F}_k \subset \mathcal{F}_n$, and $\{T = n - k\} \in \mathcal{F}_{n-k}$ can be written as $\{(X_0, \dots, X_{n-k}) \in E\}$ so that $\{\theta_k T = n - k\} = \{(X_k, \dots, X_n) \in E\} \in \mathcal{F}_n$. Hence, $\{S + \theta_S T = n\} \in \mathcal{F}_n$, and thus, $S + \theta_S T$ is a stopping time.

2.2.b If $S \leq T$ and $A \in \mathcal{F}_S$, then

$$A \cap \{T \leq n\} = A \cap \{S \leq n\} \cap \{T \leq n\} \in \mathcal{F}_n$$

and hence $A \in \mathcal{F}_T$. Applying this to $S \wedge T \leq S$ and $S \wedge T \leq T$ yields that $\mathcal{F}_{S \wedge T} \subset \mathcal{F}_S \cap \mathcal{F}_T$. If $A \in \mathcal{F}_S \cap \mathcal{F}_T$, then

$$A \cap \{S \wedge T \leq n\} = (A \cap \{S \leq n\}) \cup (A \cap \{T \leq n\}) \in \mathcal{F}_n$$

and hence, $A \in \mathcal{F}_{S \wedge T}$. Thus, $\mathcal{F}_{S \wedge T} = \mathcal{F}_S \cap \mathcal{F}_T$.

2.2.c Then, $S \leq S \vee T$ and $T \leq S \vee T$ and thus, $\mathcal{F}_S \cup \mathcal{F}_T \subset \mathcal{F}_{S \vee T}$, which is a σ-field, and thus, $\sigma(\mathcal{F}_S \cup \mathcal{F}_T) \subset \mathcal{F}_{S \vee T}$. Conversely, let $B \in \mathcal{F}_{S \vee T}$. Then,

$$B \cap \{S \leq T\} \cap \{T = n\} = B \cap \{S \leq n\} \cap \{T = n\} \in \mathcal{F}_n$$

and thus, $B \cap \{S \leq T\} \in \mathcal{F}_T$, and similarly $B \cap \{T \leq S\} \in \mathcal{F}_S$, hence

$$B = (B \cap \{S \leq T\}) \cup (B \cap \{T \leq S\}) \in \sigma(\mathcal{F}_S \cup \mathcal{F}_T) .$$

We conclude that $\mathcal{F}_{S \vee T} \subset \sigma(\mathcal{F}_S \cup \mathcal{F}_T)$.

2.3.a   The matrix Q is Markovian as

$$\frac{P(x,y)\mathbb{1}_{\{x\neq y\}}}{1-P(x,x)} \geq 0 , \quad \sum_{y\in\mathcal{V}} \frac{P(x,y)\mathbb{1}_{\{x\neq y\}}}{1-P(x,x)} = \frac{1-P(x,x)}{1-P(x,x)} = 1 .$$

Moreover,

$$P(x,x_1)P(x_1,x_2)\cdots P(x_n,y) > 0 \iff Q(x,x_1)Q(x_1,x_2)\cdots Q(x_n,y) > 0 .$$

2.3.b   As $\{S_k \leq n\} = \{\sum_{i=1}^{n} \mathbb{1}_{\{X_i\neq X_{i-1}\}} \geq k\} \in \mathcal{F}_n$, the S_k are stopping times. Moreover,

$$\mathbb{P}(\exists k \geq 1 : S_k = \infty) \leq \sum_{n\geq 0, x\in\mathcal{V}} \mathbb{P}(X_n = X_{n+1} = \cdots = x) = 0$$

as

$$\mathbb{P}(X_n = X_{n+1} = \cdots = x) \leq \mathbb{P}(X_n = \cdots = X_{n+m} = x) \leq P(x,x)^m , \quad \forall m \geq 0 .$$

2.3.c   Let $k \geq 0$, $y_0, \ldots, y_k \in \mathcal{V}$, and $n_1 \geq 1, \ldots, n_k \geq 1$. If $y_0 \neq y_1, \ldots, y_{k-1} \neq y_k$, then

$$\mathbb{P}(D_0 = 0, Y_0 = y_0, D_1 = n_1, Y_1 = y_1, \ldots, Y_{k-1} = y_{k-1}, D_k = n_k, Y_k = y_k)$$
$$= \mathbb{P}(X_0 = \cdots = X_{n_1-1} = y_0, X_{n_1} = y_1, \ldots,$$
$$X_{n_1+\cdots+n_{k-1}} = \cdots = X_{n_1+\cdots+n_k-1} = y_{k-1}, X_{n_1+\cdots+n_k} = y_k)$$
$$= \mathbb{P}(X_0 = y_0)P(y_0,y_0)^{n_1-1}P(y_0,y_1)\cdots P(y_{k-1},y_{k-1})^{n_k-1}P(y_{k-1},y_k)$$
$$= \mathbb{P}(D_0 = 0, Y_0 = y_0)g_{y_0}(n_1)Q(y_0,y_1)\cdots g_{y_{k-1}}(n_k)Q(y_{k-1},y_k)$$

or else the first and last terms in the previous equation are both zero and hence equal. Thus, $(D_k, Y_k)_{k\in\mathbb{N}}$ is a Markov chain with the said transition matrix.

Summation over $n_1 \geq 1, \cdots, n_k \geq 1$ yields that

$$\mathbb{P}(Y_0 = y_0, Y_1 = y_1, \cdots, Y_k = y_k) = \mathbb{P}(Y_0 = y_0)Q(y_0,y_1)\cdots Q(y_{k-1},y_k) ,$$

and thus, $(Y_k)_{k\in\mathbb{N}}$ is a Markov chain with matrix Q. The Markov property yields that

$$\mathbb{P}(Y_k = y, D_{k+1} = n, Y_{k+1} = z)$$
$$= \sum_{n\geq 1} \mathbb{P}(D_k = n, Y_k = y, D_{k+1} = n, Y_{k+1} = z)$$
$$= \sum_{n\geq 1} \mathbb{P}(D_k = n, Y_k = y)g_y(n)Q(y,z) = \mathbb{P}(Y_k = y)g_y(n)Q(y,z)$$

and hence,

$$\mathbb{P}(D_{k+1} = n, Y_{k+1} = z \mid Y_k = y) = \frac{\mathbb{P}(Y_k = y, D_{k+1} = n, Y_{k+1} = z)}{\mathbb{P}(Y_k = y)}$$

$$= g_y(n) Q(y, z) \, .$$

Thus,

$$\mathbb{E}(D_{k+1} \mid Y_k = y) = \sum_{n \geq 1} n g_y(n) = \mathbb{E}(D_{k+1} \mid D_k = m, Y_k = y) = \frac{1}{1 - P(y,y)} \, .$$

2.3.d Then, $\{S_U = n\} = \bigcup_{k \geq 0} \{U = k, S_k = n\}$. By definition of filtrations, $\{S_k = n\} \in \mathcal{F}_n$ can be written as $\{(X_0, \dots, X_n) \in E_k\}$ and $\{U = k\} \in \mathcal{G}_k$ as $\{(D_0, Y_0, \dots, D_k, Y_k) \in F_k\}$. If $S_k = n$, then $(D_0, Y_0, \dots, D_k, Y_k)$ can be written in terms of $(X_0, \dots, X_n)$, *i.e.* that is, there exists a deterministic function h_k such that $(D_0, Y_0, \dots, D_k, Y_k) = h_k(X_0, \dots, X_n)$. Hence,

$$\{S_U = n\} = \bigcup_{k \geq 0} \{h_k(X_0, \dots, X_n) \in F_k, (X_0, \dots, X_n) \in E_k\} \in \mathcal{F}_n \, ,$$

so that S_U is a stopping time for $(X_n)_{n \in \mathbb{N}}$.

2.4.a Then, $\{T > n\} = \{X_0^1 \neq X_0^2, \dots, X_n^1 \neq X_n^2\} \in \mathcal{F}_n$. Actually, T is the first hitting time of the diagonal $\mathcal{V} \times \mathcal{V}$ by $(X_n^1, X_n^2)_{n \geq 0}$.

2.4.b Then,

$$\mathbb{P}(Z_0^1 = x_0^1, Z_0^2 = x_0^2, \dots, Z_n^1 = x_n^1, Z_n^2 = x_n^2)$$

$$= \mathbb{P}(T < n, Z_0^1 = x_0^1, Z_0^2 = x_0^2, \dots, Z_n^1 = x_n^1, Z_n^2 = x_n^2)$$

$$+ \mathbb{P}(T \geq n, X_0^1 = x_0^1, X_0^2 = x_0^2, \dots, X_n^1 = x_n^1, X_n^2 = x_n^2)$$

and the first r.h.s. term can be expressed as the sum over $k = 0, \dots, n - 1$ of

$$\mathbb{P}(T = k, X_0^1 = x_0^1, X_0^2 = x_0^2, \dots, X_k^1 = X_k^2 = x_k^1 = x_k^2, \dots, X_n^2 = x_n^1, X_n^1 = x_n^2)$$

and the fact that $(X_n^2, X_n^1)_{n \geq 0}$ also has matrix Q, $\{T = k\} \in \mathcal{F}_k$, and the Markov property (Theorem 2.1.1) yields that this expression can be written as

$$\mathbb{P}(T = k, X_0^1 = x_0^1, X_0^2 = x_0^2, \dots, X_k^1 = X_k^2 = x_k^1 = x_k^2)$$
$$\times Q((x_k^2, x_k^1), (x_{k+1}^2, x_{k+1}^1)) \cdots Q((x_{n-1}^2, x_{n-1}^1), (x_n^2, x_n^1))$$

$$= \mathbb{P}(T = k, X_0^1 = x_0^1, X_0^2 = x_0^2, \dots, X_k^1 = X_k^2 = x_k^1 = x_k^2)$$
$$\times Q((x_k^1, x_k^2), (x_{k+1}^1, x_{k+1}^2)) \cdots Q((x_{n-1}^1, x_{n-1}^2), (x_n^1, x_n^2))$$

$$= \mathbb{P}(T = k, X_0^1 = x_0^1, X_0^2 = x_0^2, \dots, X_n^1 = x_n^1, X_n^2 = x_n^2) \, .$$

By summing all these terms, we find that

$$\mathbb{P}(Z_0^1 = x_0^1, Z_0^2 = x_0^2, \dots, Z_n^1 = x_n^1, Z_n^2 = x_n^2)$$
$$= \mathbb{P}(X_0^1 = x_0^1, X_0^2 = x_0^2, \dots, X_n^1 = x_n^1, X_n^2 = x_n^2)$$

and hence, $(Z_n^1, Z_n^2)_{n \geq 0}$ has same law as $(X_n^1, X_n^2)_{n \geq 0}$.

Thus,

$$||\mathcal{L}(X_n^1) - \mathcal{L}(X_n^2)||_{\text{Var}} = ||\mathcal{L}(Z_n^1) - \mathcal{L}(X_n^2)||_{\text{Var}} = \sup_{||f||_\infty \leq 1} \mathbb{E}(f(Z_n^1) - f(X_n^2))$$
$$\leq 2\mathbb{P}(Z_n^1 \neq X_n^2) = 2\mathbb{P}(T > n)$$

2.4.c All this is straightforward to check.

2.4.d For $n \geq 1$, it holds that $\{T > n - 1\} \in \mathcal{F}_{n-1}$, and the Markov property (Theorem 2.1.1) yields that

$$\mathbb{P}(T > n) = \sum_{x^1 \neq x^2} \mathbb{P}(T > n - 1, X_{n-1}^1 = x^1, X_{n-1}^2 = x^2, X_n^1 \neq X_n^2)$$
$$= \sum_{x^1 \neq x^2} \mathbb{P}(T > n - 1, X_{n-1}^1 = x^1, X_{n-1}^2 = x^2) \mathbb{P}_{(x^1, x^2)}(X_1^1 \neq X_1^2)$$
$$\leq (1 - \varepsilon) \sum_{x^1 \neq x^2} \mathbb{P}(T > n - 1, X_{n-1}^1 = x^1, X_{n-1}^2 = x^2)$$
$$\leq (1 - \varepsilon)\mathbb{P}(T > n - 1)$$

and we conclude by iteration, considering that $\mathbb{P}(T > 0) = \mathbb{P}(X_0^1 \neq X_0^2)$.

2.4.e By assumption,

$$\mathbb{P}((X_0^1, X_0^2) = (x_0^1, x_0^2), \dots, (X_n^1, X_n^2) = (x_n^1, x_n^2))$$
$$= \mathbb{P}((X_0^1, X_0^2) = (x_0^1, x_0^2)) \prod_{i=1}^{n} Q((x_{i-1}^1, x_{i-1}^2), (x_i^1, x_i^2))$$

and we conclude by summing over $x_0^2, \dots, x_n^2$ and then over $x_0^1, \dots, x_n^1$.

2.4.f As $\mu P^n = \mathcal{L}(X_n^1)$ and $\mu P^n = \mathcal{L}(X_n^2)$, we conclude by the previous results.

2.5.a Let $u(x) = \mathbb{P}_x(S_4 < S_0)$. We are interested in $u(1)$.

By symmetry, $u(2) = u(6)$, and the "one step forward" method (Theorem 2.2.2) yields that $u(0) = 0$ and $u(4) = 1$ and $u(1) = \frac{4}{5}u(2)$ and $u(2) = \frac{2}{5}(u(1) + u(3))$ and $u(3) = \frac{2}{5}(u(2) + 1)$. Hence,

$$u(1) = \frac{4}{5}\frac{2}{5}u(1) + \frac{4}{5}\frac{2}{5}\frac{2}{5}\frac{5}{4}u(1) + \frac{4}{5}\frac{2}{5}\frac{2}{5} = \frac{12}{25}u(1) + \frac{16}{125}$$

and thus $u(1) = \frac{16}{65}$.

2.5.b The visit consists in reaching module 4 from module 1 by one side, go $k \geq 0$ times back and forth from module 4 to module 1 on that side, then either go to module 1 by the other side or go to module 1 by the same side, and then reach module 4 from module 1 by the other side, all this without visiting module 0.

The Markov property (Theorem 2.1.3) and symmetry arguments yield that the probability of this event is

$$u(1)\sum_{k\geq 0}\left(\frac{u(1)^2}{4}\right)^k\left(\frac{u(1)}{2}+\frac{u(1)^2}{4}\right)=\frac{2u(1)^2+u(1)^3}{4-u(1)^2}=\frac{9344}{270465}.$$

2.5.c Let $g_x(s)=\mathbb{E}_x(s^{S_4}\mathbb{1}_{\{S_4<S_0\}})$ for $0\leq s\leq 1$. We are interested in g_1.

By symmetry, $g_2=g_6$, and the "one step forward" method (Theorem 2.2.5) yields that $g_0(s)=0$ and $g_4(s)=1$, $g_1(s)=\frac{4}{5}sg_2(s)$, and $g_2(s)=\frac{2}{5}s(g_1(s)+g_3(s))$, and $g_3(s)=\frac{2}{5}s(g_2(s)+1)$. Hence,

$$g_1(s)=\frac{4}{5}\frac{2}{5}s^2g_1(s)+\frac{4}{5}\frac{2}{5}\frac{2}{5}\frac{5}{4}s^2g_1(s)+\frac{4}{5}\frac{2}{5}\frac{2}{5}s^3=\frac{12}{25}s^2g_1(s)+\frac{16}{125}s^3$$

so that $g_1(s)=\frac{16s^3}{125-60s^2}$.

We again find that $u(1)=g_1(1)=\frac{16}{65}$, and by identification

$$EE_1(s^{S_4}\mid S_4<S_0)=\frac{g_1(s)}{u(1)}=\frac{13s^3}{25-12s^2}=\frac{13}{25}\sum_{k\geq 0}\left(\frac{12}{25}\right)^k s^{2k+3},$$

$$\mathbb{P}_1(S_4=2k+3\mid S_4<S_0)=\frac{13}{25}\left(\frac{12}{25}\right)^k.$$

Moreover,

$$\frac{13(1+h)^3}{25-12(1+h)^2}=\frac{13(1+3h)}{25-12(1+2h)}+o(h)$$

$$=(1+3h)\left(1+\frac{24}{13}h\right)+o(h)=1+\frac{63}{13}h+o(h)$$

and hence, $\mathbb{E}_1(S_4=2k+3\mid S_4<S_0)=\frac{63}{13}$ by identification.

2.6 We are interested in $g(s)=\mathbb{E}_1(s^{S_4}\mathbb{1}_{\{S_4<R_1\}})$ for $s\in[0,1]$ and $\mathbb{P}_1(S_4<R_1)=g(1)$. Let $g_x(s)=G^{\{1,4\}}(x,4,s)$.

Then, $g(s)=\frac{s}{2}(g_2(s)+g_6(s))$ by the "one step forward" method.

This method also yields that (Theorem 2.2.5) $g_1=0$ and $g_4=1$, $g_2(s)=\frac{s}{4}(g_5(s)+g_6(s))$, $g_3(s)=sg_5(s)$, $g_5(s)=\frac{s}{3}(g_2(s)+g_3(s)+1)$, $g_6(s)=\frac{s}{3}g_2(s)$. Thus, $g(s)=\frac{3s+s^2}{6}g_2(s)$ and $g_2(s)=\frac{3s}{12-s^2}g_5(s)$ and then,

$$\frac{12-s^2}{3s}g_2(s)=\frac{s}{3}g_2(s)+\frac{s^2}{3}\frac{12-s^2}{3s}g_2(s)+\frac{s}{3}$$

and hence, $g_2(s) = \frac{3s^2}{36 - 18s^2 + s^4}$, and finally,

$$g(s) = \frac{1}{2} \frac{3s^3 + s^4}{36 - 18s^2 + s^4} = \frac{1}{2}\left(1 - \frac{36 - 18s^2 - 3s^3}{36 - 18s^2 + s^4}\right).$$

Then, $\mathbb{P}_1(S_4 < R_1) = g(1) = \frac{2}{19}$, the conditional generating function is given by $\frac{g(s)}{g(1)} = \frac{19}{4}\left(1 - \frac{36 - 18s^2 - 3s^3}{36 - 18s^2 + s^4}\right)$, and a Taylor expansion at 1 yields

$$\frac{g(1 + h)}{g(1)} = \frac{19}{4}\left(1 - \frac{15 - 45h}{19 - 32h}\right) + o(h)$$

$$= \frac{19}{4} - \frac{1}{4}(15 - 45h)\left(1 + \frac{32}{19}h\right) + o(h) = 1 + \frac{375}{76}h + o(h)$$

and thus, by identification, the conditional expectation is $\frac{375}{76}$.

2.7 If $g_x(s) = \mathbb{E}(s^{S_1})$ for $0 \leq s \leq 1$, then $g(s) = g_3(s)$. The "one step forward" method (Theorem 2.2.5) yields that $g_1 = 1$ and

$$\left(1 - \frac{9}{20}s\right)g_2(s) - \frac{s}{2}g_3(s) = \frac{s}{20}, \quad -\frac{9}{40}sg_2(s) + \left(1 - \frac{3}{4}s\right)g_3(s) = \frac{s}{40},$$

so that

$$g(s) = \frac{s}{9s^2 - 48s + 40}$$

$$= \frac{1}{18}\left(\frac{1 + \frac{2\sqrt{6}}{3}}{s - \frac{8 + 2\sqrt{6}}{3}} + \frac{1 - \frac{2\sqrt{6}}{3}}{s - \frac{8 - 2\sqrt{6}}{3}}\right)$$

$$= -\frac{1}{18}\left(\frac{3 + 2\sqrt{6}}{8 + 2\sqrt{6}} \frac{1}{1 - \frac{3s}{8 + 2\sqrt{6}}} + \frac{3 - 2\sqrt{6}}{8 - 2\sqrt{6}} \frac{1}{1 - \frac{3s}{8 - 2\sqrt{6}}}\right).$$

By identification,

$$\mathbb{P}(L = n) = -\frac{1}{36}\frac{3^n}{2^n}\left(\frac{3 + 2\sqrt{6}}{4 + \sqrt{6}}\left(\frac{1}{4 + \sqrt{6}}\right)^n + \frac{3 - 2\sqrt{6}}{4 - \sqrt{6}}\left(\frac{1}{4 - \sqrt{6}}\right)^n\right).$$

Moreover, the Taylor expansion

$$g(1 + h) = \frac{1 + h}{1 - 30h} + o(h) = (1 + h)(1 + 30h) = 1 + 31h + o(h)$$

yields that $\mathbb{E}(L) = 31$.

2.8.a This equation writes, for $0 < x < N$,

$$\left(1 - \frac{x^2 c(1) + (N-x)^2 c(2)}{N(xc(1) + (N-x)c(2))}\right) u(x)$$
$$= \frac{x(N-x)(c(2)u(x-1) + c(1)u(x+1))}{N(xc(1) + (N-x)c(2))},$$

is of the form $a(x)u(x) = c(2)u(x-1) + c(1)u(x+1)$ and has 1 as a solution, hence the result. (The direct computation is simple.)

2.8.b This is the same equation found in the Dirichlet problem in gambler's ruin when the gain probability at each toss for Gambler A is $p = \frac{c(1)}{c(1)+c(2)}$ (see Section 2.3.1). We refer to that section for the computation of $\mathbb{P}_x(Z_0 < Z_N)$ and $\mathbb{P}_x(Z_N < Z_0)$.

2.8.c Transitions are according to a binomial law. The equation is, for $0 < x < N$,

$$u(x) = \sum_{y=0}^{N} \binom{N}{y} \left(\frac{xc(1)}{xc(1) + (N-x)c(2)}\right)^y \left(\frac{(N-x)c(2)}{xc(1) + (N-x)c(2)}\right)^{N-y} u(y).$$

If $c(1) = c(2)$, then $\frac{xc(1)}{xc(1)+(N-x)c(2)} = \frac{x}{N}$ and classically (total mass and expectation of a binomial law) for $u(x) = \alpha + \beta x$, the r.h.s. of the equation takes the value $\alpha 1 + \beta N \frac{x}{N} = u(x)$, and hence, as for gambler's ruin, $\mathbb{P}_x(Z_0 < Z_N) = \frac{N-x}{N}$ and $\mathbb{P}_x(Z_N < Z_0) = \frac{x}{N}$.

2.9.a The graph can be found in Section 3.1.3. Clearly, P is irreducible.

2.9.b If $X_0 = x \neq y$, then $R_y = S_y$. For $x < y$, a simple spatial translation allows to use the results in Section 2.3.2. For $x > y$, it suffices to interchange x and y as well as p and q and use the previous result.

2.9.c By the "one step forward" method, $\mathbb{P}_x(R_x < \infty) = p\mathbb{P}_{x+1}(R_x < \infty) + q\mathbb{P}_{x-1}(R_x < \infty)$ and $\mathbb{E}_x(R_x) = 1 + p\mathbb{E}_{x+1}(R_x) + q\mathbb{E}_{x-1}(R_x)$. Hence, the results follow from the previous ones.

2.9.d The strong Markov property (Theorem 2.1.3) and the preceding results allow to compute $\mathbb{P}_0(N_x \geq k)$. Moreover, $\mathbb{P}_0(N_x = \infty) = \lim_{k \to \infty} \downarrow \mathbb{P}_0(N_x \geq k)$.

2.9.e If $p \neq 1/2$, then $\mathbb{P}_0(N_x = \infty) = 0$ for all x and the chain goes to infinity, a.s. More precisely, if $p > 1/2$, then as previously shown $\mathbb{P}_x(R_0 < \infty) = 1$ for $x < 0$, and thus $\lim_{n \to \infty} X_n = \infty$. Similarly, if $p < 1/2$, then $\lim_{n \to \infty} X_n = -\infty$.

2.9.f Then, $\mathbb{P}_0(D = \infty) = \mathbb{P}_0(N_0 = \infty)$ and we conclude by a previous result.

If $D < \infty$ a.s., then $(X_{D+n})_{n\geq 0}$ cannot hit 0, and hence does not have same law as $(X_n)_{n\geq 0}$. By contradiction, D cannot be a stopping time, as then the strong Markov property would have applied.

2.9.g Then,

$$\mathbb{P}_0(M = \infty) = \lim_{x\to\infty} \downarrow \mathbb{P}_0(M \geq x), \qquad \mathbb{P}_0(M \geq x) = \mathbb{P}_0(R_x < \infty),$$

and previous results allow to conclude.

If $\mathbb{P}_0(M = \infty) = 0$, then $(X_{M+n})_{n\geq 0}$ cannot reach a state greater than its initial value, and hence does not have same law as $(X_n)_{n\geq 0}$. By contradiction, M cannot be a stopping time, as then the strong Markov property would have applied.

2.9.h If $x > 0$, then $\mathbb{E}_x(s^{R_0}) = \mathbb{E}_x(s^{S_0})$, hence the result. The result for $x < 0$ is obtained by symmetry.

2.9.i The "one step forward" method yields that

$$\mathbb{E}_0(s^{R_0}) = p\mathbb{E}_1(s^{R_0}) + q\mathbb{E}_{-1}(s^{R_0}) = 1 - \sqrt{1 - 4pqs^2}$$

$$= \sum_{k\geq 1} \frac{1}{2k - 1}\binom{2k}{k} p^k q^k s^{2k}$$

in which we use the classic Taylor expansion provided at the end of Section 2.3.2. By identification, $\mathbb{P}_0(R_0 = 2k) = \frac{1}{2k-1}\binom{2k}{k}p^k q^k$ for $k \geq 1$.

2.10.a We have $L_n = X_{T\wedge n}$ for $n \geq 0$.

2.10.b Straightforward, notably P is clearly irreducible.

2.10.c The "one step forward" method (Theorem 2.2.2) yields the equation. Its characteristic polynomial is $(1 - p)X^3 - X^2 + p$, and its roots are 1 and λ_- and λ_+. This yields the general solution, considering the case of multiple roots.

2.10.d We have only two boundary conditions, whereas the space of general solutions is of dimension three, so we must use the minimality result in Theorem 2.2.2 to find the solution of interest.

2.10.e We use Theorem 2.2.6 and seek the least solution with values in $[0, \infty]$. We use the above-mentioned general solution for the associated linear equation, and a particular solution of the form ax when 1 is a simple root and ax^2 if it is a double.

2.10.f We may use Theorem 2.2.5, but we do not have a trivial solution for the characteristic polynomial of degree three $(1 - p)sX^3 - X^2 + ps$ for the linear recursion.

Solutions for Chapter 3

3.1.a The convergences follow from $\sum_n \mathbb{P}_x(R_y = n) < \infty$ and $P^n(x, y) \leq 1$. The monotone convergence theorem or the Abel theorem allows to conclude.

3.1.b The first result follows from the strong Markov property (Theorem 2.1.3). The second from classic result on products of power series and convolutions.

3.1.c For the first, use

$$\mathbb{P}_x(R_x < \infty) = \lim_{s \uparrow 1} \frac{H_{x,x}(s) - 1}{H_{x,x}(s)}, \qquad \mathbb{E}_x(N_x) = \lim_{s \uparrow 1} H_{x,x}(s) .$$

For the second, use $\mathbb{E}_x(N_y) = \lim_{s \uparrow 1} H_{x,y}(s)$ is equal to

$$I(x, y) + \mathbb{P}_x(R_y < \infty) \lim_{s \uparrow 1} H_{y,y}(s) = I(x, y) + \mathbb{P}_x(R_y < \infty) \lim_{s \uparrow 1} \mathbb{E}_y(N_y) .$$

3.1.d If n is odd, then $P^n(x, x) = 0$, if n is even, then $P^n(x, x) = \binom{n}{n/2} p^{n/2} (1 - p)^{n/2}$, and

$$H_{x,x}(s) = \sum_{k \in \mathbb{N}} \binom{2k}{k} p^k (1 - p)^k s^{2k} = \frac{1}{\sqrt{1 - 4p(1 - p)s^2}} .$$

As $p(1 - p) \leq 1/4$ with equality if and only if $p = 1/2$, then $\mathbb{E}_x(N_x) = \frac{1}{\sqrt{1 - 4p(1-p)}} < \infty$ for $p \neq 1/2$ and $\mathbb{E}_x(N_x) = \infty$ for $p = 1/2$. Thus, the random walk is recurrent if and only if $p = 1/2$.

3.2 As in Section 3.1.3, $P^{2n+1}(x, x) = 0$ and, using the Stirling formula,

$$P^{2n}(x, x) = \binom{2n}{n} \frac{1}{2^n} \frac{1}{2^n} \simeq \frac{1}{(\pi n)^{1/2}} .$$

Hence, $\sum_{k \geq 0} P^k(x, x) = \infty$ and $\sum_{k \geq 0} (P \otimes P)^k(x, x) = \sum_{k \geq 0} P^k(x, x)^2 = \infty$, whereas $\sum_{k \geq 0} (P \otimes P \otimes P)^k(x, x) = \sum_{k \geq 0} P^k(x, x)^3 < \infty$. We conclude by the potential matrix criterion.

3.3.a Then, $\mathbb{P}_x(N_x < \infty) \geq \mathbb{P}_x(S_y < \infty, \theta_{S_y} N_x < \infty)$, and the strong Markov property yields that

$$\mathbb{P}_x(S_y < \infty, \theta_{S_y} N_x < \infty) = \mathbb{P}_x(S_y < \infty, X_{S_y} = y, \theta_{S_y} N_x < \infty)$$

$$= \mathbb{P}_x(S_y < \infty) \mathbb{P}_y(N_x < \infty) .$$

Moreover, $\mathbb{P}_x(N_x < \infty) = 0$ and $\mathbb{P}_x(S_y < \infty) > 0$ and hence, $\mathbb{P}_y(N_x < \infty) = 0$. If $i, j \geq 1$ are such that $P^i(x, y) > 0$ and $P^j(y, x) > 0$, and if $\sum_{k \geq 0} P^k(x, x) = \infty$, then $\sum_{n \geq 0} P^n(y, y) \geq \sum_{k \geq 0} P^j(y, x) P^k(x, x) P^i(x, y) = \infty$. Thus, y is recurrent and $y \to x$, and hence, $\mathbb{P}_x(N_y = \infty) = 1$ by interchanging the roles of x and y.

3.3.b Then, $\mathbb{P}_x(S_y = \infty) = \mathbb{P}_x(R_x < \infty, S_y \geq R_x, X_{R_x} = x, S_y = \infty)$, and as $\{S_y \geq R_x\} \in \mathcal{F}_{R_x}$, the strong Markov property yields the first result.

By contradiction, if $\mathbb{P}_x(S_y \geq R_x) = 1$, then the strong Markov property and $R_x \geq 1$ yield that

$$\mathbb{P}_x(S_y \leq k) = \mathbb{P}_x(R_x < \infty, S_y \geq R_x, X_{R_x} = x, S_y \leq k) \leq \mathbb{P}_x(S_y \leq k-1) ,$$

iteratively $\mathbb{P}_x(S_y \leq k) = 0$ for all k, and thus

$$\mathbb{P}_x(S_y < \infty) = \lim_{k\to\infty} \uparrow \mathbb{P}_x(S_y \leq k) = 0 ,$$

which is a contradiction as $x \to y$.

The two first results imply that $\mathbb{P}_x(S_y = \infty) = 0$.

The strong Markov property and $\mathbb{P}_x(S_y < \infty) = 1$ yield that

$$0 = \mathbb{P}_x(R_x = \infty) \geq \mathbb{P}_x(S_y < \infty, \theta_{S_y} S_x = \infty) = \mathbb{P}_y(S_x = \infty) .$$

Similarly, $\mathbb{P}_y(R_y < \infty) \geq \mathbb{P}_y(S_x < \infty)\mathbb{P}_x(S_y < \infty) = 1$ and

$$\infty = \mathbb{P}_y(N_y = \infty) = \mathbb{P}_y(S_x < \infty, N_y = \infty) = \mathbb{P}_x(N_y = \infty) .$$

3.3.c We reason as in the previous question. Simplifying by $\mathbb{P}_x(S_y < R_x) > 0$ yields that $\mathbb{P}_x(N_y \geq k) = \mathbb{P}_x(N_y \geq k-1)$, and by iteration $\mathbb{P}_x(N_y \geq k) = \mathbb{P}_x(N_y \geq 0) = 1$. We conclude with $\mathbb{P}_x(N_y = \infty) = \lim_{k\to\infty} \downarrow \mathbb{P}_x(N_y \geq k)$.

3.4 Graph: the transient class is $\{1, 5, 6\}$, and the recurrent classes are $\{2, 3, 4\}$, $\{7, 8\}$, and $\{9, 10, 11, 12\}$. First matrix: the transient class is $\{2, 3, 5\}$, and the recurrent class is $\{1, 4\}$. Second matrix: the transient class is $\{2, 4\}$, and the recurrent class is $\{1, 3, 5\}$.

3.5 The transient class is constituted of all populations in which there are individuals with at least two different alleles, and the recurrent classes are constituted each of an absorbing state corresponding to populations with a single allele. As the transient class is finite, and as each state in it can be visited only a finite number of times by the chain, the chain will eventually end in an absorbing state.

3.6 Then, $\mu = \mu P$ and hence, $\mu(A) = \mu P(A)$, that is,

$$\sum_{x\in A} \mu(x) = \sum_{x\in A}\sum_{y\in\mathcal{V}} \mu(y)P(y, x) = \sum_{y\in A} \mu(y)P(y, A) + \sum_{y\in\mathcal{V}-A} \mu(y)P(y, A)$$

and changing y into x and using $1 - P(x, A) = P(x, \mathcal{V} - A)$ yields the result.

3.7 The "one step forward" method and the invariance of π yield that

$$\mathbb{P}_\pi(R_x = \infty) = \sum_{z\in\mathcal{V}} \pi(z)\mathbb{P}_z(R_x = \infty)$$

$$= \sum_{z \in \mathcal{V}} \pi(z) \sum_{y \in \mathcal{V}} \mathbb{P}_z(X_1 = y, R_x = \infty)$$

$$= \sum_{y \in \mathcal{V}} \sum_{z \in \mathcal{V}} \pi(z) P(z, y) \mathbb{P}_y(S_x = \infty)$$

$$= \sum_{y \in \mathcal{V}} \pi(y) \mathbb{P}_y(S_x = \infty) = \mathbb{P}_\pi(S_x = \infty) .$$

As $S_x = 0$ if $X_0 = x$ and else $S_x = R_x$, necessarily $\pi(x)\mathbb{P}(R_x = \infty) = 0$ and thus $\pi(x) = 0$ or $\mathbb{P}(R_x = \infty) = 0$.

3.8.a This is obvious.

3.8.b By global balance, for x in $\mathcal{V}$,

$$\mu(x)(1 - P(x, x)) = \sum_{y \neq x} \mu(y)P(y, x) = \sum_{y \neq x} \mu(y)(1 - P(y, y))Q(y, x) ,$$

and $v(x) = \mu(x)(1 - P(x, x))$ is an invariant measure for Q. Conversely, $v(x) = \sum_{y \neq x} v(y)Q(y, x)$ can be written as

$$\frac{v(x)}{1 - P(x, x)}(1 - P(x, x)) = \sum_{y \neq x} \frac{v(y)}{1 - P(y, y)} P(y, x)$$

and $\mu(x) = \frac{v(x)}{1 - P(x,x)}$ is an invariant measure for P.

3.8.c Use the invariant law criterion. If P is positive recurrent, then it has an invariant law μ, the invariant measure $\mu := (\mu(x)(1 - P(x, x)))_{x \in \mathcal{V}}$ of Q satisfies $\sum_{x \in \mathcal{V}} \mu(x)(1 - P(x, x)) \leq \sum_{x \in \mathcal{V}} \mu(x) = 1 < \infty$, thus Q is positive recurrent.

3.8.d Clearly, Q is given by $Q(x, x - 1) = q$ and $Q(x, x + 1) = p$ for $x \geq 1$ and $Q(0, 1) = 1$ (reflected random walk).

Its invariant measure μ solves the local balance equations $\mu(x - 1)p = \mu(x)q$ for $x \geq 2$ and $\mu(0) = \mu(1)q$, and taking $\mu(0) = p$ yields that $\mu(x) = (p/q)^x$ for $x \geq 1$, which is summable, and thus Q is positive recurrent.

Hence, P is recurrent. Its invariant measure μ is given by $\mu(0) = p$ and $\mu(x) = 1$ for $x \geq 1$, which is not summable, and hence, P is not positive recurrent.

3.9.a Clearly, P is irreducible, and 0 is the only solution of the reversibility equations.

3.9.b The global balance equations are given by $\mu(x) = a\mu(x - 1)$ for $x \notin k\mathbb{N}$ and

$$a\mu(0) = (1 - a) \sum_{x=1}^{k} \mu(x), \quad \mu(ik) = a\mu(ik-1) + (1-a) \sum_{x=ik+1}^{(i+1)k} \mu(x), \quad i \geq 1.$$

For $j \geq 0$, it holds that $\mu(x) = a^{x-jk}\mu(jk)$ for $jk \leq x < (j+1)k$ and

$$(1-a)\sum_{x=jk+1}^{(j+1)k}\mu(x) = (a-a^k)\mu(jk) + (1-a)\mu((j+1)k) .$$

Then, $a\mu(0) = (a-a^k)\mu(0) + (1-a)\mu(k)$ and thus $\mu(k) = \frac{a^k}{1-a}\mu(0)$, so that $\mu(k) = a\mu(k-1) + (a-a^k)\mu(k) + (1-a)\mu(2k)$, that is,

$$\frac{a^k}{1-a}\mu(0) = a^k\mu(0) + (a-a^k)\frac{a^k}{1-a}\mu(0) + (1-a)\mu(2k) ,$$

and hence, $\mu(2k) = (\frac{a^k}{1-a})^2\mu(0)$, and similarly we check that $\mu(jk) = (\frac{a^k}{1-a})^j\mu(0)$ for $j \geq 0$. Hence, the invariant measure is unique, and given by $\mu(jk+m) = \mu(0)(\frac{a^k}{1-a})^j a^m$ for $j \geq 0$ and $0 \leq m < k$.

3.9.c The invariant measure has finite total mass if and only if $a^k < 1 - a$, and we conclude by the invariant law criterion. Specifically, then

$$\sum_{j\geq 0}\sum_{m=0}^{k-1}\left(\frac{a^k}{1-a}\right)^j a^m = \frac{1}{1-\frac{a^k}{1-a}}\frac{1-a^k}{1-a} = \frac{1-a^k}{1-a-a^k} ,$$

$$\pi(jk+m) = \frac{1-a-a^k}{1-a^k}\left(\frac{a^k}{1-a}\right)^j a^m , \qquad j \geq 0, \ 0 \leq m < k .$$

Moreover, $\mathbb{E}_{jk}(R_{jk}) = 1/\pi(jk) = \frac{1-a^k}{1-a-a^k}(\frac{1-a}{a^k})^j$.

3.9.d As $\{S_i \leq n\} = \{\sum_{j=0}^n \mathbb{1}_{\{X_n\in k\mathbb{N}\}} > i\}$, the S_i are stopping times, and it is quite simple to prove that they are finite. The strong Markov property yields that $(Y_i)_{i\geq 0}$ is a Markov chain, and its matrix is clearly given by $Q(x,x-1) = 1-a$ for $x \geq 1$ and $Q(x,x+1) = a^k$ for $x \geq 0$, and hence, $Q(x,x) = a - a^k$ for $x \geq 1$ and $Q(0,0) = 1 - a^k$.

3.9.e Simple conditional probability computations, as in Exercise 2.3), yield that $Q'(x,x-1) = \frac{1-a}{1-a+a^k}$ and $Q'(x,x+1) = \frac{a^k}{1-a+a^k}$ for $x \geq 1$ and $Q'(0,1) = 1$.

3.9.f Let $R_0 = \inf\{n \geq 1 : Z_n = 0\}$. Then, $\mathbb{P}_0(R_0 < \infty) = \mathbb{P}_1(S_0 < \infty)$ and the results on unilateral hitting times prove that $\mathbb{P}_1(S_0 < \infty) = 1$ if $a^k = 1 - a$ and $\mathbb{P}_1(S_0 < \infty) < 1$ if $a^k > 1 - a$. If state 0 is transient for $(Y_n)_{n\geq 0}$, then it is transient for $(X_n)_{n\geq 0}$ and for $(Z_n)_{n\geq 0}$, if state 0 is recurrent for $(Z_n)_{n\geq 0}$, then it is recurrent for $(Y_n)_{n\geq 0}$ and for $(X_n)_{n\geq 0}$, and the three chains have same nature.

3.10.a Clearly yes.

3.10.b Theorem 3.3.11 yields that there is a unique invariant measure and gives a formula for it, but we elect to directly solve the local balance equations instead:

$$\mu(x-1)\frac{\lambda}{\lambda+\mu(x-1)} = \mu(x)\frac{\mu x}{\lambda+\mu x}\ , \qquad\qquad 1 \le x \le K\ ,$$

$$\mu(x-1)\frac{\lambda}{\lambda+\mu K} = \mu(x)\frac{\mu K}{\lambda+\mu K}\ , \qquad\qquad x > K\ .$$

This yields by iteration that

$$\mu(x) = \mu(0)\left(1 + \frac{x}{\rho}\right)\frac{\rho^x}{x!}\ , \ x \le K\ ,$$

$$\mu(x) = \mu(0)\left(1 + \frac{K}{\rho}\right)\frac{\rho^K}{K!}\left(\frac{\rho}{K}\right)^{x-K}\ , \ x > K\ .$$

3.10.c The invariant measure has finite total mass if and only if $\rho < K$, and we conclude using the invariant law criterion.

3.10.d The necessary and sufficient condition for transience in Theorem 3.3.11 allows to conclude. This also follows from the Lamperti and Tweedie criteria (Theorems 3.3.3 and 3.3.4) with Lyapunov function $\phi : x \mapsto x$ and $E = \{0, \dots, K-1\}$.

3.10.e Then, $\mu(x) = \mu(0)(1 + \frac{x}{\rho})\frac{\rho^x}{x!}$ for $x \ge 0$. As $\sum_{x \ge 0}(1 + \frac{x}{\rho})\frac{\rho^x}{x!} = 2e^\rho$, the invariant law π is given by $\pi(x) = 2^{-1}e^{-\rho}(1 + \frac{x}{\rho})\frac{\rho^x}{x!}$. The invariant law criterion (Theorem 3.2.4) yields that the chain is positive recurrent and that $\mathbb{E}_0(R_0) = 1/\pi(0) = 2e^\rho$.

3.11.a It is clear that this random recursion corresponds to the description, and Theorem 1.2.3 yields that $(X_n)_{n\ge 0}$ is a Markov chain. This chain has positive probability of going in one step from any state $x \ge 1$ to $x-1$ as $\mathbb{P}(A_1 = 0) > 0$ and $\mathbb{P}(R_{1,1} = 1, R_{1,2} = 0, \dots, R_{1,x} = 0) = p(1-p)^{x-1} > 0$, and from any state $x \ge 0$ to $y \ge x + 2$ as $\mathbb{P}(A_1 \ge 2) > 0$, and thus the chain is irreducible.

3.11.b The Lyapunov function will be $\phi : x \mapsto x$. For $x \ge 1$,

$$\mathbb{E}_x(X_1) = x + \mathbb{E}(A_1) - \mathbb{P}\left(A_1 + \sum_{i=1}^{x} R_{1,i} = 1\right)$$

$$\ge x + \mathbb{E}(A_1) - \mathbb{P}\left(\sum_{i=1}^{x} R_{1,i} \le 1\right)$$

in which the Markov inequality yields that

$$-\mathbb{P}\left(\sum_{i=1}^{x} R_{1,i} \le 1\right) = \mathbb{P}\left(\sum_{i=1}^{x} R_{1,i} \ge 2\right) - 1$$

$$\le \frac{1}{2}\mathbb{E}\left(\sum_{i=1}^{x} R_{1,i}\right) - 1 = \frac{p}{2}x - 1 ,$$

and thus

$$\mathbb{E}_x(X_1) \ge x + \mathbb{E}(A_1) + \frac{p}{2}x - 1 .$$

It is then enough to choose $E = \{0, \dots, x_0 - 1\}$ and $\varepsilon = \mathbb{E}(A_1) + \frac{p}{2}x_0 - 1$ with $x_0 \ge 1$ large enough that $\varepsilon > 0$ and then hypothesis 1 is satisfied. Hypothesis (2) follows easily from $\mathbb{E}(A_1^2) < \infty$.

3.11.c   Let $(Y_n)_{n \ge 0}$ be constructed similarly to $(X_n)_{n \ge 0}$ with A_n replaced by $B_n = \min(A_n, 2)$. Then, $(Y_n)_{n \ge 0}$ is transient as $\mathbb{E}(B_1^2) \le 4 < \infty$, clearly $Y_n \le X_n$, and hence, $(X_n)_{n \ge 0}$ is transient.

3.12.a   Let T^x denote an r.v. with same law as T_n conditional on $X_{n-1} = x$ and independent of the rest and $S_n = T_1 + \cdots + T_n$. For $B \in F_{n-1}$, the i.i.d. property of the A_i yields that

$$\mathbb{P}(B, X_{n-1} = x, X_n = y) = \sum_{k \in \mathbb{N}} \mathbb{P}(S_{n-1} = k, B, X_{n-1} = x)\mathbb{P}\left(\sum_{i=k+1}^{k+T_n} A_i = y\right)$$

$$= \mathbb{P}(B, X_{n-1} = x)\mathbb{P}\left(\sum_{i=1}^{T^x} A_i = y\right)$$

and thus $(X_n)_{n \ge 0}$ is a Markov chain with matrix $P(x, y) = \mathbb{P}(\sum_{i=1}^{T^x} A_i = y)$.

3.12.b   As $\mathbb{P}(A_1 = 0) > 0$ implies that $P(x, 0) > 0$, the set of reachable states from 0 is the unique closed irreducible class, and the other states are transient.

3.12.c   We use the Foster criterion (Theorem 3.3.6) with Lyapunov function ϕ : $x \mapsto x$. There exists $\varepsilon > 0$ and $x_0 \ge 0$ such that $\alpha\mathbb{E}_x(T_1) \le x - \varepsilon$ for $x > x_0$. Classically, $\mathbb{E}_x(X_1) = \sum_{k=1}^{\infty} \mathbb{E}(\sum_{i=1}^{k} A_i)\mathbb{P}(T_1 = k) = \alpha\mathbb{E}_x(T_1)$, which proves (1) for $F = \{0, \dots, x_0\}$. Moreover, (2) follows easily from the assumption that $\mathbb{E}(T_n \mid X_{n-1} = x) < \infty$ for all x.

3.12.d   Let $R_0 = \inf\{n \ge 1 : X_n = 0\}$ and $\eta = \mathbb{P}(T_1 = 1, A_1 = 0)$. Then,

$$\mathbb{P}(R_0 > k) = \mathbb{P}(X_1 \ge 1, \dots, X_k \ge 1) \le (1 - \eta)^k , \quad \mathbb{E}(R_0) = \sum_{k \ge 0} \mathbb{P}(R_0 > k) < \infty.$$

Moreover, $\alpha\mathbb{E}(T_1 \mid X_0 = x) = x + \alpha/2$ and hypothesis (1) of the Lamperti criterion is true, hence also hypothesis (1) of the Tweedie criterion is true.

As the chain is positive recurrent, hypothesis (2) of these criteria cannot be true, which can easily be checked directly.

3.13.a From every state x, it is possible to reach 0 in a finite number of steps, and it is possible to reach $x + 1$ in one step, hence the irreducibility.

3.13.b Use the Foster criterion (Theorem 3.3.6) with Lyapunov function $\phi = \ln$:

$$\mathbb{E}_x(\ln X_1) - \ln x = \mathbb{E}_x \left(\ln \frac{X_1}{x} \right) = \alpha \ln \frac{x+1}{x} + \alpha \ln 2 + (1 - 2\alpha) \ln \frac{\lfloor x/2 \rfloor}{x}$$

$$\xrightarrow[x \to \infty]{} (3\alpha - 1) \ln 2 ,$$

and if $\alpha < 1/3$, there is $x_0 < \infty$ such that if $x > x_0$, then $\mathbb{E}_x(\ln X_1) - \ln x \geq \varepsilon > 0$.

3.13.c Similarly, hypothesis (1) of the Lamperti criterion (Theorem 3.3.3) is true for the Lyapunov function $\phi = \ln$, and $\mathbb{E}_x((\ln X_1 - \ln x)^2) = \alpha(\ln \frac{x+1}{x})^2 + \alpha(\ln 2)^2 + (1 - 2\alpha)(\ln \frac{\lfloor x/2 \rfloor}{x})^2$ is uniformly bounded in x and thus hypothesis (2) is true.

3.14.a The "one step forward" method yields the first result. The second follows by iteration.

3.14.b Classically, $\mathbb{E}_x(T) = 1 + \sum_{k \geq 1} \mathbb{P}_x(T > k) \leq 1 + \phi(x) \sum_{k \geq 1} \rho^k = 1 + \frac{\rho}{1-\rho} \phi(x)$, and Lemma A.1.3 yields that $\mathbb{E}_x(s^T) = s + (s - 1) \sum_{k \geq 1} \mathbb{P}(T > k) s^k \leq s + (s - 1) \phi(x) \sum_{k \geq 1} (\rho s)^k = s + \frac{\rho s(s-1)}{1-\rho s} \phi(x)$ for $1 < s < \frac{1}{\rho}$.

3.14.c Let $h(s) = g(s) - s^K$ for $0 \leq s < R$. Then, $h(1) = 0$ and $h'(1) = g'(1) - K < 0$, hence there exists $\beta > 1$ satisfying $h(\beta) < 0$ and then $\rho := g(\beta)/\beta^K > 1$. If $x \notin E = \{0, \ldots, K - 1\}$, then $\mathbb{E}_x(\beta^{X_1}) = \mathbb{E}(\beta^{x-K+A_1}) \leq \rho \beta^x$, and $\phi(x) := \beta^x$.

3.14.d Let $\phi : x \mapsto \mathbb{1}_{\{x \notin E\}}$. If $x \notin E$, then $P\phi(x) = \sum_{y \notin E} P(x, y) = 1 - P(x, E) \leq \rho \phi(x)$ for $\rho = 1 - \inf_{x \notin E} P(x, E) < 1$. These are the renewal processes such that $\sup_{x \geq 1} p_x \leq \rho < 1$.

3.15 The adjoint w.r.t. the reversible measure μ is P itself, and the adjoint w.r.t. the uniform measure is obtained by interchanging p and q.

3.16.a No, as $\mu(x)P(x, x + 1) = \mu(x + 1)P(x + 1, x)$ implies that $\mu(x) = 0$.

3.16.b The global balance equations are $\mu(x) = p\mu(x + 2) + (1 - p)\mu(x - 1)$ for $x \in \mathbb{Z}$. The characteristic polynomial $pX^3 - X + 1 - p$ has 1 as a root, and factors into $p(X - 1)(X^2 + X - \frac{1-p}{p})$, and its other roots are $\rho_\pm$. As $\rho_- < 0$ and $\rho_+ \geq 0$ and $\rho_+ = 1 \Longleftrightarrow p = 1/3$, the general form of solutions is the one given.

3.16.c   Then, $\rho_+ < |\rho_-|$ and $\rho_- < -1$, hence if $\alpha_- \neq 0$, then for sufficiently large x of appropriate sign we would have $\mu(x) < 0$. Similarly, if $p = 1/3$ necessarily $\alpha_+ \geq 0$ by letting $x \to \infty$ and $\alpha_+ \leq 0$ by letting $x \to -\infty$, and hence, $\alpha_+ = 0$. For $p \neq 1/3$, there is no uniqueness of the invariant measure for this irreducible chain, which hence cannot be recurrent (Theorem 3.2.3).

3.16.d   The global balance equations for $x \geq 1$ are the same as the previous ones, and as they involve $\mu(0)$ for $x = 1$, they have same general solution for $x \in \mathbb{N}$. The same reasoning as above shows that $\alpha_- = 0$. Moreover, the equation for $x = 0$ is $(1 - p)\mu(0) = p(\mu(1) + \mu(2))$, and we use this on the general solutions. If $p \neq 1/3$, then $(1 - p)(\alpha_+ + \beta) = p(\alpha_+\rho_+ + \alpha_+\rho_+^2 + 2\beta)$ and thus $\beta(1 - 3p) = \alpha_+ p(\rho_+^2 + \rho_+ - \frac{1-p}{p}) = 0$ and then $\beta = 0$ as $1 - 3p \neq 0$. If $p = 1/3$, then $(1 - p)\beta = p(3\alpha_+ + 2\beta)$ and hence, $(1 - 3p)\beta = 3p\alpha_+$ and thus $\alpha_+ = 0$.

3.16.e   Use the invariant law criterion (Theorem 3.2.4). If $p > 1/3$, then $\rho_+ < 1$ and $\sum_{x \geq 0}\rho_+^x = \frac{1}{1-\rho_+}$ and $\pi(x) = (1 - \rho_+)\rho_+^x$. If $p \leq 1/3$, then the invariant law has infinite total mass. Moreover, $\mathbb{E}_0(R_0) = 1/\pi(0)$, and the "one step forward" method yields that $\mathbb{E}_0(R_0) = p + (1 - p)\mathbb{E}_1(S_0)$, hence $\mathbb{E}_1(S_0) = \frac{1}{(1-p)(1-\rho_+)} - \frac{p}{1-p}$ if $p > 1/3$ and else $\mathbb{E}_1(S_0) = \infty$.

3.16.f   Use the Foster and Lamperti criteria (Theorems 3.3.3 and 3.3.6), with Lyapunov function $\phi : x \mapsto x$ and $E = \{0, 1\}$. If $x \geq 2$ then $\mathbb{E}_x(X_1) - x = -2p + 1 - p = 1 - 3p$, which yields the hypotheses (1). The hypotheses (2) are obviously true.

3.17.a   Then, $P(x, y) = 1/d(x)$ if $\{x, y\} \in \mathcal{L}$ and else $P(x, y) = 0$. The chain is irreducible if and only if the graph is connected.

3.17.b   Take for $V(x)$ the set of nearest neighbors of x (other than x itself) in $\mathbb{Z}^d$ or $\{0, 1\}^N$, for the ℓ^1 distance.

3.17.c   The local balance equations write $\mu(x)/d(x) = \mu(y)/d(y)$ for $\{x, y\} \in \mathcal{L}$, and their solution is given by $\mu(x) = d(x)$. The invariant law criterion (Theorem 3.2.4) yields that the chain is positive recurrent if and only if $\sum_{x \in V}d(x) < \infty$, which happens if and only if V is finite, and then $\pi(x) = d(x)/\sum_{x \in V}d(x)$.

3.17.d   Then, $d(x) = 3$ and the uniform measure is reversible. Moreover, the first coordinate follows a random walk on $\mathbb{Z}$ with probability $2/3$ of going from k to $k + 1$ and $1/3$ from k to $k - 1$, and the reversible measure for this random walk is given by $(2^k)_{k \in \mathbb{Z}}$, and it is a simple matter to check that $(2^k, 1)_{k \in \mathbb{Z}, n \in \mathbb{N}}$ is an invariant measure for P. Theorem 3.2.3 yields that P cannot be recurrent.

3.18.a   It is possible to go from x to $x + 1$ in one step and from x to 0 in a finite number of steps, hence the chain is irreducible. Moreover, as $\mathbb{E}_x(X_1) - x \leq \theta^x - (1 - \theta^x)x/2$, the Foster criterion (Theorem 3.3.6) with Lyapunov function

$\phi : x \mapsto x$ yields positive recurrence. We conclude by the invariant law criterion.

3.18.b These are $\pi(0) = (1 - \theta)\pi(1)$ and

$$\pi(x) = \theta^{x-1}\pi(x-1) + (1 - \theta^{2x})\pi(2x) + (1 - \theta^{2x+1})\pi(2x+1), \quad x \geq 1.$$

3.18.c Use Exercise 3.6 with $A = \{0, \ldots, x\}$, or sum up the equations, to obtain $\pi(x)\theta^x = \sum_{y=x+1}^{2x+1} \pi(y)(1 - \theta^y)$. The lower bound only retains the last two terms.

3.18.d Use the global balance equations and the previous upper bound and an immediate recursion.

3.18.e Then, $L = \exp\left(-\sum_{y=1}^{\infty} \ln(1 - \theta^y)\right)$ and the series converges.

3.18.f The global balance equations yield that

$$\beta(x)\theta^{\frac{x(x-1)}{2}} \geq \theta^{x-1}\beta(x-1)\theta^{\frac{(x-1)(x-2)}{2}} = \beta(x-1)\theta^{\frac{x(x-1)}{2}}$$

and hence that $\beta(x) \geq \beta(x-1)$. Then, $\pi(0) = \beta(0) \leq \beta(x)$ and the previous results yield that $\beta(x) \leq \pi(0)L$, and hence, $\pi(0)\theta^{\frac{x(x-1)}{2}} \leq \pi(x) \leq \pi(0)L\theta^{\frac{x(x-1)}{2}}$. Summing over $x \geq 0$ yields $\pi(0)Z \leq 1 \leq \pi(0)LZ$, thus $Z^{-1}L^{-1} \leq \pi(0) \leq Z^{-1}$.

Solutions for Chapter 4

4.1.a The pointwise ergodic theorem (Theorem 4.1.1) yields that these are given by the probabilities of the events for the invariant laws, and specifically by $\frac{1}{6}, \frac{1}{4}$, and $\frac{1}{6}$.

4.1.b The spatial station and the mouse: the period is a divisor of 2 and 3, and thus is 1. Three-card Monte: it is 2 for the chain on permutations (consider the signature) and 1 for the ace of spades (which can remain in the same position).

4.1.c The Kolmogorov ergodic theorem (Theorem 4.2.9) yields that this is well approximated by the probability of the event for the invariant law, namely $\frac{5}{36}$.

4.1.d The Kolmogorov ergodic theorem (Theorem 4.2.9) yields that this is given by the probability of the event for the invariant law, specifically $\frac{3}{16}$.

4.1.e Use Corollary 4.2.10 of the Kolmogorov ergodic theorem. The period is 2, and there are two aperiodic classes constituted of the three permutations with even signature and of the three permutations with odd signature. The approximation of the probability after 1000 steps is given by $\frac{1}{3}$. The probability after 1001 steps is 0.

4.1.f Use the Kolmogorov ergodic theorem. The probability that he wins is approximated by the probability $\frac{1}{3}$ under the invariant law, and the expectation is approximated by $\frac{100}{3}$. For $p = 1/2$, it was shown that the error after n steps has a bound of order $1/2^n$, and for $n = 10$, this yields $1/1024$.

4.2 In the first case, the period is a divisor of 2 and 3 and thus is 1, and the Kolmogorov ergodic theorem yields that the limit is $\sum_{j\geq 0}\pi(jk) = \frac{1-a}{1-a^k}$. In the second case, the pointwise ergodic theorem applied to this null recurrent chain yields that this limit is given by the ratio of the invariant measures, and specifically by $1/\sum_{m=1}^{k-1} a^m = \frac{1-a}{a-a^k} = \frac{1-a}{2a-1}$.

4.3 The pointwise ergodic theorem and $\sum_{x\geq K}(\frac{\rho}{K})^{K-x} = \frac{K}{K-\rho}$ yield that

$$\pi([K,\infty[) = \frac{\left(1 + \frac{K}{\rho}\right)\frac{K}{K-\rho}\frac{\rho^K}{K!}}{2\sum_{x=0}^{K-2}\frac{\rho^x}{x!} + \frac{\rho^{K-1}}{(K-1)!} + \left(1 + \frac{K}{\rho}\right)\frac{K}{K-\rho}\frac{\rho^K}{K!}} .$$

4.4.a As $\mathbb{E}(\xi_1) = 0$ and $\mathbb{P}(\xi_1 = 0) \neq 1$ and $\mathbb{P}(\xi_1 < -1) = 0$, necessarily $\mathbb{P}(\xi_1 = -1) > 0$ and $\mathbb{P}(\xi_1 > 0) > 0$, and it is a simple matter to conclude that the chain is irreducible. It is immediate to check that the uniform measure is invariant for a random walk, and the recurrence and irreducibility yield that the invariant measure is unique. As $\{U_k \leq n\} = \{\sum_{i=1}^{n} \mathbb{1}_{X_i \geq X_{i-1}} \geq k\}$, the U_k are stopping times.

4.4.b As the chain cannot decrease of more than 1 at each step, if $x \leq 0$, then

$$\sum_{n=0}^{\infty} \mathbb{P}_0(U_1 > n, X_n = x) = \sum_{n=0}^{\infty} \mathbb{P}_0(R_0 > n, X_n = x) = \mu_0(y) = 1 ,$$

in which the last equality follows from the uniqueness of the invariant measure and $\mu_0(0) = 1$ (see Section 3.2.2).

4.4.c As $y \geq 0$, the strong Markov property and the previous result yield

$$\mathbb{P}_0(X_{U_1} = y) = \sum_{n=0}^{\infty}\sum_{x\leq 0} \mathbb{P}_0(U_1 = n+1, X_n = x, X_{n+1} = y)$$

$$= \sum_{n=0}^{\infty}\sum_{x\leq 0} \mathbb{P}_0(U_1 > n, X_n = x, X_{n+1} = y)$$

$$= \sum_{n=0}^{\infty}\sum_{x\leq 0} \mathbb{P}_0(U_1 > n, X_n = x)P(x,y)$$

$$= \sum_{x\leq 0} P(x,y) := \sum_{x\leq 0} \mathbb{P}(\xi_1 = y - x) = \mathbb{P}(\xi_1 \geq y) .$$

4.4.d Follows from the strong Markov property.

4.4.e   Use $\frac{1}{n}X_{U_n} = \frac{1}{n}\sum_{k=1}^{n}(X_{U_k} - X_{U_{k-1}})$ and the previous result and the strong law of large numbers (Theorem A.1.5) and

$$\mathbb{E}_0(X_{U_1}) = \sum_{y\geq 0}\mathbb{P}_0(X_{U_1} = y)y = \sum_{y\geq 0}\mathbb{P}(\xi_1 \geq y)y = \sum_{y\geq 0}\sum_{x\geq y}\mathbb{P}(\xi_1 = x)y$$

$$= \sum_{x\geq 0}\mathbb{P}(\xi_1 = x)\sum_{y=0}^{x}y = \sum_{x\geq 0}\mathbb{P}(\xi_1 = x)\frac{x(x+1)}{2} = \frac{\mathbb{E}_0(\xi_1^2) + \mathbb{E}_0(\xi_1)}{2} = \frac{\sigma^2}{2}.$$

4.5   It is a simple matter to check that the aperiodic class decomposition for $\tilde{P}$ is given by the aperiodic class decomposition of P in opposite order. These aperiodic classes are closed and irreducible for $\tilde{P}P$ and $P\tilde{P}$, hence these matrices cannot be irreducible.

4.6.a   The irreducibility is clear. The period is a divisor of $m + 1$ and $m + 2$ and hence is 1. There is an invariant law, given by

$$\pi(x) = \frac{p_0 \cdots p_{x-1}}{\sum_{y\geq 0}p_0 \cdots p_{y-1}} = \frac{\min(1, 2^{m-x})}{m + 2},$$

and the invariant law criterion yields that the chain is positive recurrent (a direct computation can be made).

4.6.b   The pointwise ergodic theorem yields that the limit is given by the mean value for the invariant law, specifically

$$\sum_{x\in\mathbb{N}}\pi(x)x = \sum_{x\in\mathbb{N}}\pi(]x, \infty[)$$

$$= \frac{1}{m + 2}\sum_{x=0}^{m}(m + 1 - x) + \frac{1}{m + 2}\sum_{k=1}^{\infty}2^{-k} = m + 1 + \frac{1}{m + 2}.$$

4.6.c   The Kolmogorov ergodic theorem yields that the limit is $\frac{1}{m+2}\sum_{x\geq m}2^{x-m} = \frac{2}{m+2}$. Lemma 1.4.1 yields that the distance is bounded by $2/2^{11} = 1/1024$.

4.6.d   If $1 \leq k \leq m$, then the recurrent classes of $P^k\tilde{P}^k$ are $\{0\}, \ldots, \{m - k\}, \{m - k + 1, m - k + 2, \ldots\}$, and those of $\tilde{P}^kP^k$ are $\{k\}, \ldots, \{m\}, \{0, \ldots, k - 1, m + 1, m + 2, \ldots\}$. If $k \geq m + 1$, then there is a single recurrent class $\mathbb{N}$. Thus, these matrices are irreducible if and only if $k \geq m + 1$.

4.7.a   The invariance of π yields that

$$\mu P^n f - \mu(\mathcal{V})\pi f = \sum_{x\in\mathcal{V}}\frac{\mu P^k(x) - \mu(\mathcal{V})\pi(x)}{\sqrt{\pi(x)}} \times \sqrt{\pi(x)}P^{n-k}(f - \pi f)(x)$$

and the Cauchy–Schwarz inequality allows to conclude.

4.7.b As π is an invariant law for $\tilde{P}$,

$$||\delta_x P^n - \pi||_{L^2(\pi)^*}^2 = \langle \delta_x P^n, \delta_x P^n \rangle_{L^2(\pi)^*} - 2\langle \delta_x P^n, \pi \rangle_{L^2(\pi)^*} + \langle \pi, \pi \rangle_{L^2(\pi)^*}$$

$$= \langle \delta_x, \delta_x P^n \tilde{P}^n \rangle_{L^2(\pi)^*} - 2\langle \delta_x, \pi \rangle_{L^2(\pi)^*} + 1 ,$$

and the conclusion follows using the explicit expression for the scalar products.

4.8.a 1. Clearly, $\mathcal{A}(f,f) \geq 0$ with equality if f is constant, if $\mathcal{A}(f,f) = 0$, then $f(y) = f(x)$ whenever $P(x,y) > 0$ and thus f is constant as P is irreducible,

$$\mathcal{A}(f,f) = \frac{1}{2} \sum_{x,y \in \mathcal{V}} \mu(x)P(x,y)f(y)^2 + \frac{1}{2} \sum_{x,y \in \mathcal{V}} \mu(x)P(x,y)f(x)^2$$

$$- \sum_{x,y \in \mathcal{V}} \mu(x)P(x,y)f(y)f(x)$$

$$= \sum_{x \in \mathcal{V}} \mu(x)f(x)^2 - \sum_{x \in \mathcal{V}} \mu(x)Pf(x)f(x)$$

$$= \sum_{x \in \mathcal{V}} \mu(x)(f(x) - Pf(x))f(x) ,$$

and in particular $\mathcal{A}(f,f) \leq \sum_{x \in \mathcal{V}} \mu(x)f(x)^2 = ||f||_{L^2(\pi)}^2 < \infty$.

4.8.b Such a function f is such that $\mathcal{A}(f,f) = \sum_{x \in \mathcal{V}} \mu(x)(f(x) - Pf(x))f(x) \leq 0$ and hence that $\mathcal{A}(f,f) = 0$, and thus is constant. If f is lower bounded, then there exists a constant c such that $f + c \geq 0$, and if $f \leq Pf$, then $f + c \leq P(f + c)$, and thus $f + c$ is constant, and so is f.

4.8.c As $f = Pf$, by linearity $f = f^+ - f^-$ yields that $f = Pf^+ - Pf^-$. As $f = f^+ - f^-$ is the minimal decomposition of f as a difference of non negative functions and Pf^+ and Pf^- are both non negative, $f^+ \leq Pf^+$ and $f^- \leq Pf^-$. Moreover, $f^+ = (|f| + f)/2$ and $f^- = (|f| - f)/2$ are in $L^2(\mu)$ and non negative. Hence, f^+ and f^- are constant, and so is f.

4.9.a Using $\mathcal{A}_P(f,f) = \langle f - Pf, f \rangle_{L^2(\pi)}$ and the Cauchy–Schwarz inequality,

$$(\langle f,f \rangle_{L^2(\pi)} - \mathcal{A}_P(f,f))^2 = \langle Pf,f \rangle_{L^2(\pi)}^2 \leq \langle Pf, Pf \rangle_{L^2(\pi)} \langle f,f \rangle_{L^2(\pi)} ,$$

and thus

$$\left(1 - \frac{\mathcal{A}_P(f,f)}{\langle f,f \rangle_{L^2(\pi)}}\right)^2 \leq \frac{\langle Pf, Pf \rangle_{L^2(\pi)}}{\langle f,f \rangle_{L^2(\pi)}} = 1 - \frac{\langle f - \tilde{P}Pf, f \rangle_{L^2(\pi)}}{\langle f,f \rangle_{L^2(\pi)}}$$

and it is a simple matter to conclude.

4.9.b It is obvious that R is an irreducible transition matrix. Moreover,

$$\check{P}P = (1-c)^2 \tilde{Q}Q + c(\check{P} + P) - c^2 I$$

and hence, $\mathcal{A}_{\check{P}P} = (1-c)^2 \mathcal{A}_{\tilde{Q}Q} + 2c\mathcal{A}_{\frac{P+\check{P}}{2}}$. This implies that $\mathcal{A}_{\check{P}P} \geq 2c\mathcal{A}_{\frac{P+\check{P}}{2}}$ and thus that $\lambda_{\check{P}P} \geq 2c\lambda_P$.

4.9.c Then, $(1-\lambda_P)^2 \leq 1 - \lambda_{P\check{P}} \leq 1 - \lambda_P$ and thus $\lambda_P \leq 1$ and Lemma 4.3.7 allows to conclude.

4.10.a Apply Theorem 4.3.5 to P^k, as well as the fact that the operator norm of P on $L^2(\pi)$ is 1.

4.10.b Reason as in the proof of Theorem 4.2.4, and use that if a transition matrix Q is irreducible on a finite space, then $\lambda_Q < \infty$ (Theorem 4.3.4).

4.11.a Developing the square and using that $\pi(\mathcal{V}) = 1$ yields that

$$\frac{1}{2} \sum_{x,y \in \mathcal{V}} \pi(x)\pi(y)(f(y) - f(x))^2$$

$$= \sum_{x \in \mathcal{V}} \pi(x)f(x)^2 - \left(\sum_{x \in \mathcal{V}} \pi(x)f(x) \right)^2 = \mathrm{Var}_\pi(f) .$$

The i.i.d. chain with common law π.

4.11.b Clearly,

$$f(y) - f(x) = \sum_{(x',y') \in \gamma(x,y)} (f(y') - f(x')) = \sum_{(x',y') \in \gamma(x,y)} \sqrt{L(x',y')} \frac{f(y') - f(x')}{\sqrt{L(x',y')}}$$

and the Cauchy–Schwarz inequality yields that

$$(f(y) - f(x))^2 \leq |\gamma(x,y)|_L \sum_{(x',y') \in \gamma(x,y)} \frac{(f(y') - f(x'))^2}{L(x',y')} .$$

Hence, using the previous result,

$$\mathrm{Var}_\pi(f) \leq \frac{1}{2} \sum_{x \neq y} \pi(x)\pi(y)|\gamma(x,y)|_L \sum_{(x',y') \in \gamma(x,y)} \frac{(f(y') - f(x'))^2}{L(x',y')}$$

and the conclusion follows by interchanging the summation order and x and y with x' and y'.

4.11.c Use the Poincaré inequality (Theorem 4.3.4) and

$$\mathrm{Var}_\pi(f) \leq \frac{1}{2} A_{L,\Gamma} \sum_{(x,y) \in G} \pi(x)P(x,y)(f(y) - f(x))^2 = A_{L,\Gamma} \mathcal{A}_P(f,f) .$$

4.12.a   This is obvious using $p = \frac{r}{1+r}$.

4.12.b   Then,

$$A_{L,\Gamma} = \frac{1-r^2}{r}\sup_{x\in\mathbb{N}}\left(r^{-x/2}\sum_{y\le x<z}r^{y+z}\sum_{y\le a<z}r^{-a/2}\right)$$

$$= \frac{1-r^2}{r^{1/2}-r}\sup_{x\in\mathbb{N}}\left(r^{-x/2}\sum_{y\le x<z}(r^y r^{z/2} - r^{y/2}r^z)\right)$$

$$= \frac{1-r^2}{r^{1/2}-r}\sup_{x\in\mathbb{N}}\left(\frac{r^{1/2}(1-r^{\frac{x+1}{2}})}{(1-r)(1-r^{1/2})}\right) = \frac{1+r}{(1-r^{1/2})^2}$$

and we conclude with the result of Exercise 4.11.

4.13.a   The continuity of h is obvious. As $h'(x) = \log\, x + 1$ is increasing, h is strictly convex. By dominated or monotone convergence, $H(\cdot\mid\pi)$ is continuous. For laws μ and $\underline\mu$, and $a, b \ge 0$ such that $a + b = 1$,

$$H(a\mu + b\underline\mu \mid \pi) = \pi h\left(\frac{a\mu + b\underline\mu}{\pi}\right) \le \pi\left(ah\left(\frac{\mu}{\pi}\right) + bh\left(\frac{\underline\mu}{\pi}\right)\right)$$

$$= aH(\mu\mid\pi) + bH(\underline\mu\mid\pi)$$

and as h is strictly convex and $\pi > 0$, equality happens if and only if $ab = 0$, and hence, $H(\cdot\mid\pi)$ is strictly convex. The Jensen inequality yields

$$H(\mu\mid\pi) = \pi h\left(\frac{\mu}{\pi}\right) \ge h\left(\pi\frac{\mu}{\pi}\right) = h(1) = 0 ,$$

which together with the strict convexity of H implies that $H(\mu\mid\pi) = 0$ if and only if $\mu = \pi$. (Else use the equality condition for the Jensen inequality.)

4.13.b   Then, $h(\frac{\mu P}{\pi}) = h(\tilde P\frac{\mu}{\pi}) \le \tilde P h(\frac{\mu}{\pi})$ using the duality (4.3.8) and the Jensen inequality for the laws $\tilde P(x, \cdot)$, and integrating by $\tilde P$ yields $H(\mu P\mid\pi) \le H(\mu\mid\pi)$. Further,

$$H(\mu P\mid\pi) = \sum_{y\in\mathcal{V}}\pi(y)h\left(\tilde P(y,\cdot)\frac{\mu}{\pi}\right)$$

so that $\pi > 0$ and the equality condition for the Jensen inequality yield that for all y it holds that $\frac{\mu}{\pi}$ is constant, $\tilde P(y, \cdot)$-a.s., hence that $\frac{\mu}{\pi}$ is constant on the set of all $x \in \mathcal{V}$ satisfying $P(x, y) > 0$. Hence, if there exists $y \in \mathcal{V}$ such that $P(x, y) > 0$ for all x, then μ and π are proportional, and hence equal since as they are probability measures.

4.13.c   Use Theorem 4.2.4. The sequence $(\mu P^n)_{n\ge 0}$ is bounded by 1 in the finite-dimensional normed vector space $\mathcal{M}$.

4.13.d   As $n \mapsto H(\mu P^n \mid \pi)$ is a non negative non increasing sequence, it converges, and hence necessarily $\lim_{n \to \infty} H(\mu P^n \mid \pi) = \lim_{n \to \infty} H(\mu P^n P^k \mid \pi)$. By passing to the limit along a converging subsequence (the sums defining H are finite), $H(\mu^* \mid \pi) = H(\mu^* P^k \mid \pi)$ by continuity, and as $P^k > 0$, the previous result yields that $\mu^* = \pi$. Classically, a relatively compact sequence with a unique accumulation point must converge to it.

Index

σ-field, 47, 183–6, 192
  engendered, 47
  generated, 184
  product, 5, 47, 192

a.s., 185
absorbing, 14, 24–5, 43, 83, 86
algorithm
  Metropolis, 163–6
  Propp-Wilson, *see* exact simulation
almost sure, almost surely, *see* a.s.
aperiodic, *see* period, 133

birth and death, *see* chain
branching, 25–6, 67–71, 86

canonical space, process, 5, 48, 192
cemetery state, 52–3, 136
Chapman-Kolmogorov formula, 9
class
  aperiodic, 130–131, 134
  closed irreducible, 80–83
  recurrent, 81–3
coalescence, 167–9
communication, 79–80
  Condition
  Doeblin condition, 15–17, 35, 46,
    74, 129, 163, 168
  Kolmogorov
    Condition, 106
conditioning on the first step, *see* one
    step forward
convergence
  weak, in law, 182

counting automaton, *see* word
    search
coupling, 122, 132–6
  Doeblin, 74
  maximal, 179
criterion
  Foster, 102
  invariant law, 91, 108
  Lamperti, 99
  potential matrix, 81, 84–5,
    87, 108, 112
  transience-recurrence, 97–105
  Tweedie, 100
curse of dimensionality, 158

Dirichlet form, 143–6, 148
Dirichlet problem, 54–7, 97, 99,
    155–161
distribution
  hitting, 53–60
  stationary, *see* invariant law
dog-flea model, *see* Ehrenfest
Doob, 98
duality
  Hilbert space, 140–150
  measure-function, 6–7, 11–15,
    141, 180–183

Ehrenfest, 27–33, 86, 95–6, 106,
    109, 127, 131–2,
    139, 147
eigenspace, *see* spectrum
eigenvalues, *see* spectrum

Markov Chains: Analytic and Monte Carlo Computations, First Edition. Carl Graham.
© 2014 John Wiley & Sons, Ltd. Published 2014 by John Wiley & Sons, Ltd.

WILEY SERIES IN PROBABILITY AND STATISTICS
ESTABLISHED BY WALTER A. SHEWHART AND SAMUEL S. WILKS

Editors: *David J. Balding, Noel A. C. Cressie, Garrett M. Fitzmaurice, Geof H. Givens, Harvey Goldstein, Geert Molenberghs, David W. Scott, Adrian F. M. Smith, Ruey S. Tsay, Sanford Weisberg*
Editors Emeriti: *J. Stuart Hunter, Iain M. Johnstone, Joseph B. Kadane, Jozef L. Teugels*

The *Wiley Series in Probability and Statistics* is well established and authoritative. It covers many topics of current research interest in both pure and applied statistics and probability theory. Written by leading statisticians and institutions, the titles span both state-of-the-art developments in the field and classical methods.

Reflecting the wide range of current research in statistics, the series encompasses applied, methodological and theoretical statistics, ranging from applications and new techniques made possible by advances in computerized practice to rigorous treatment of theoretical approaches.

This series provides essential and invaluable reading for all statisticians, whether in academia, industry, government, or research.

*Now available in a lower priced paperback edition in the Wiley Classics Library.
†Now available in a lower priced paperback edition in the Wiley–Interscience Paperback Series.

BARTOSZYNSKI and NIEWIADOMSKA-BUGAJ · Probability and Statistical Inference, *Second Edition*

BASILEVSKY · Statistical Factor Analysis and Related Methods: Theory and Applications

BATES and WATTS · Nonlinear Regression Analysis and Its Applications

BECHHOFER, SANTNER, and GOLDSMAN · Design and Analysis of Experiments for Statistical Selection, Screening, and Multiple Comparisons

BEH and LOMBARDO · Correspondence Analysis: Theory, Practice and New Strategies

BEIRLANT, GOEGEBEUR, SEGERS, TEUGELS, and DE WAAL · Statistics of Extremes: Theory and Applications

BELSLEY · Conditioning Diagnostics: Collinearity and Weak Data in Regression

† BELSLEY, KUH, and WELSCH · Regression Diagnostics: Identifying Influential Data and Sources of Collinearity

BENDAT and PIERSOL · Random Data: Analysis and Measurement Procedures, *Fourth Edition*

BERNARDO and SMITH · Bayesian Theory

BHAT and MILLER · Elements of Applied Stochastic Processes, *Third Edition*

BHATTACHARYA and WAYMIRE · Stochastic Processes with Applications

BIEMER, GROVES, LYBERG, MATHIOWETZ, and SUDMAN · Measurement Errors in Surveys

BILLINGSLEY · Convergence of Probability Measures, *Second Edition*

BILLINGSLEY · Probability and Measure, *Anniversary Edition*

BIRKES and DODGE · Alternative Methods of Regression

BISGAARD and KULAHCI · Time Series Analysis and Forecasting by Example

BISWAS, DATTA, FINE, and SEGAL · Statistical Advances in the Biomedical Sciences: Clinical Trials, Epidemiology, Survival Analysis, and Bioinformatics

BLISCHKE and MURTHY (editors) · Case Studies in Reliability and Maintenance

BLISCHKE and MURTHY · Reliability: Modeling, Prediction, and Optimization

BLOOMFIELD · Fourier Analysis of Time Series: An Introduction, *Second Edition*

BOLLEN · Structural Equations with Latent Variables

BOLLEN and CURRAN · Latent Curve Models: A Structural Equation Perspective

BONNINI, CORAIN, MAROZZI and SALMASO · Nonparametric Hypothesis Testing: Rank and Permutation Methods with Applications in R

BOROVKOV · Ergodicity and Stability of Stochastic Processes

BOSQ and BLANKE · Inference and Prediction in Large Dimensions

BOULEAU · Numerical Methods for Stochastic Processes

* BOX and TIAO · Bayesian Inference in Statistical Analysis

BOX · Improving Almost Anything, *Revised Edition*

* BOX and DRAPER · Evolutionary Operation: A Statistical Method for Process Improvement

BOX and DRAPER · Response Surfaces, Mixtures, and Ridge Analyses, *Second Edition*

BOX, HUNTER, and HUNTER · Statistics for Experimenters: Design, Innovation, and Discovery, *Second Editon*

BOX, JENKINS, and REINSEL · Time Series Analysis: Forcasting and Control, *Fourth Edition*

BOX, LUCEÑO, and PANIAGUA-QUIÑONES · Statistical Control by Monitoring and Adjustment, *Second Edition*

* BROWN and HOLLANDER · Statistics: A Biomedical Introduction

CAIROLI and DALANG · Sequential Stochastic Optimization

CASTILLO, HADI, BALAKRISHNAN, and SARABIA · Extreme Value and Related Models with Applications in Engineering and Science

CHAN · Time Series: Applications to Finance with R and S-Plus®, *Second Edition*

CHARALAMBIDES · Combinatorial Methods in Discrete Distributions

CHATTERJEE and HADI · Regression Analysis by Example, *Fourth Edition*

*Now available in a lower priced paperback edition in the Wiley Classics Library.
†Now available in a lower priced paperback edition in the Wiley–Interscience Paperback Series.

*Now available in a lower priced paperback edition in the Wiley Classics Library.

†Now available in a lower priced paperback edition in the Wiley–Interscience Paperback Series.

HEDAYAT and SINHA · Design and Inference in Finite Population Sampling
HEDEKER and GIBBONS · Longitudinal Data Analysis
HELLER · MACSYMA for Statisticians
HERITIER, CANTONI, COPT, and VICTORIA-FESER · Robust Methods in
  Biostatistics
HINKELMANN and KEMPTHORNE · Design and Analysis of Experiments, Volume 1:
  Introduction to Experimental Design, *Second Edition*
HINKELMANN and KEMPTHORNE · Design and Analysis of Experiments, Volume 2:
  Advanced Experimental Design
HINKELMANN (editor) · Design and Analysis of Experiments, Volume 3: Special
  Designs and Applications
HOAGLIN, MOSTELLER, and TUKEY · Fundamentals of Exploratory Analysis of
  Variance
* HOAGLIN, MOSTELLER, and TUKEY · Exploring Data Tables, Trends and Shapes
* HOAGLIN, MOSTELLER, and TUKEY · Understanding Robust and Exploratory Data
  Analysis
HOCHBERG and TAMHANE · Multiple Comparison Procedures
HOCKING · Methods and Applications of Linear Models: Regression and the Analysis of
  Variance, *Third Edition*
HOEL · Introduction to Mathematical Statistics, *Fifth Edition*
HOGG and KLUGMAN · Loss Distributions
HOLLANDER, WOLFE, and CHICKEN · Nonparametric Statistical Methods, *Third
  Edition*
HOSMER and LEMESHOW · Applied Logistic Regression, *Second Edition*
HOSMER, LEMESHOW, and MAY · Applied Survival Analysis: Regression Modeling
  of Time-to-Event Data, *Second Edition*
HUBER · Data Analysis: What Can Be Learned From the Past 50 Years
HUBER · Robust Statistics
† HUBER and RONCHETTI · Robust Statistics, *Second Edition*
HUBERTY · Applied Discriminant Analysis, *Second Edition*
HUBERTY and OLEJNIK · Applied MANOVA and Discriminant Analysis, *Second
  Edition*
HUITEMA · The Analysis of Covariance and Alternatives: Statistical Methods for
  Experiments, Quasi-Experiments, and Single-Case Studies, *Second Edition*
HUNT and KENNEDY · Financial Derivatives in Theory and Practice, *Revised Edition*
HURD and MIAMEE · Periodically Correlated Random Sequences: Spectral Theory and
  Practice
HUSKOVA, BERAN, and DUPAC · Collected Works of Jaroslav Hajek— with
  Commentary
HUZURBAZAR · Flowgraph Models for Multistate Time-to-Event Data
JACKMAN · Bayesian Analysis for the Social Sciences
† JACKSON · A User's Guide to Principle Components
JOHN · Statistical Methods in Engineering and Quality Assurance
JOHNSON · Multivariate Statistical Simulation
JOHNSON and BALAKRISHNAN · Advances in the Theory and Practice of Statistics: A
  Volume in Honor of Samuel Kotz
JOHNSON, KEMP, and KOTZ · Univariate Discrete Distributions, *Third Edition*
JOHNSON and KOTZ (editors) · Leading Personalities in Statistical Sciences: From the
  Seventeenth Century to the Present
JOHNSON, KOTZ, and BALAKRISHNAN · Continuous Univariate Distributions,
  Volume 1, *Second Edition*
JOHNSON, KOTZ, and BALAKRISHNAN · Continuous Univariate Distributions,
  Volume 2, *Second Edition*
JOHNSON, KOTZ, and BALAKRISHNAN · Discrete Multivariate Distributions

*Now available in a lower priced paperback edition in the Wiley Classics Library.
†Now available in a lower priced paperback edition in the Wiley–Interscience Paperback Series.

JUDGE, GRIFFITHS, HILL, LÜTKEPOHL, and LEE · The Theory and Practice of Econometrics, *Second Edition*

JUREK and MASON · Operator-Limit Distributions in Probability Theory

KADANE · Bayesian Methods and Ethics in a Clinical Trial Design

KADANE AND SCHUM · A Probabilistic Analysis of the Sacco and Vanzetti Evidence

KALBFLEISCH and PRENTICE · The Statistical Analysis of Failure Time Data, *Second Edition*

KARIYA and KURATA · Generalized Least Squares

KASS and VOS · Geometrical Foundations of Asymptotic Inference

† KAUFMAN and ROUSSEEUW · Finding Groups in Data: An Introduction to Cluster Analysis

KEDEM and FOKIANOS · Regression Models for Time Series Analysis

KENDALL, BARDEN, CARNE, and LE · Shape and Shape Theory

KHURI · Advanced Calculus with Applications in Statistics, *Second Edition*

KHURI, MATHEW, and SINHA · Statistical Tests for Mixed Linear Models

* KISH · Statistical Design for Research

KLEIBER and KOTZ · Statistical Size Distributions in Economics and Actuarial Sciences

KLEMELÄ · Smoothing of Multivariate Data: Density Estimation and Visualization

KLUGMAN, PANJER, and WILLMOT · Loss Models: From Data to Decisions, *Third Edition*

KLUGMAN, PANJER, and WILLMOT · Loss Models: Further Topics

KLUGMAN, PANJER, and WILLMOT · Solutions Manual to Accompany Loss Models: From Data to Decisions, *Third Edition*

KOSKI and NOBLE · Bayesian Networks: An Introduction

KOTZ, BALAKRISHNAN, and JOHNSON · Continuous Multivariate Distributions, Volume 1, *Second Edition*

KOTZ and JOHNSON (editors) · Encyclopedia of Statistical Sciences: Volumes 1 to 9 with Index

KOTZ and JOHNSON (editors) · Encyclopedia of Statistical Sciences: Supplement Volume

KOTZ, READ, and BANKS (editors) · Encyclopedia of Statistical Sciences: Update Volume 1

KOTZ, READ, and BANKS (editors) · Encyclopedia of Statistical Sciences: Update Volume 2

KOWALSKI and TU · Modern Applied U-Statistics

KRISHNAMOORTHY and MATHEW · Statistical Tolerance Regions: Theory, Applications, and Computation

KROESE, TAIMRE, and BOTEV · Handbook of Monte Carlo Methods

KROONENBERG · Applied Multiway Data Analysis

KULINSKAYA, MORGENTHALER, and STAUDTE · Meta Analysis: A Guide to Calibrating and Combining Statistical Evidence

KULKARNI and HARMAN · An Elementary Introduction to Statistical Learning Theory

KUROWICKA and COOKE · Uncertainty Analysis with High Dimensional Dependence Modelling

KVAM and VIDAKOVIC · Nonparametric Statistics with Applications to Science and Engineering

LACHIN · Biostatistical Methods: The Assessment of Relative Risks, *Second Edition*

LAD · Operational Subjective Statistical Methods: A Mathematical, Philosophical, and Historical Introduction

LAMPERTI · Probability: A Survey of the Mathematical Theory, *Second Edition*

LAWLESS · Statistical Models and Methods for Lifetime Data, *Second Edition*

LAWSON · Statistical Methods in Spatial Epidemiology, *Second Edition*

LE · Applied Categorical Data Analysis, *Second Edition*

LE · Applied Survival Analysis

*Now available in a lower priced paperback edition in the Wiley Classics Library.
†Now available in a lower priced paperback edition in the Wiley–Interscience Paperback Series.

*Now available in a lower priced paperback edition in the Wiley Classics Library.

†Now available in a lower priced paperback edition in the Wiley–Interscience Paperback Series.

---

*Now available in a lower priced paperback edition in the Wiley Classics Library.

†Now available in a lower priced paperback edition in the Wiley–Interscience Paperback Series.

RYAN · Modern Regression Methods, *Second Edition*
RYAN · Sample Size Determination and Power
RYAN · Statistical Methods for Quality Improvement, *Third Edition*
SALEH · Theory of Preliminary Test and Stein-Type Estimation with Applications
SALTELLI, CHAN, and SCOTT (editors) · Sensitivity Analysis
SCHERER · Batch Effects and Noise in Microarray Experiments: Sources and Solutions
* SCHEFFE · The Analysis of Variance
SCHIMEK · Smoothing and Regression: Approaches, Computation, and Application
SCHOTT · Matrix Analysis for Statistics, *Second Edition*
SCHOUTENS · Levy Processes in Finance: Pricing Financial Derivatives
SCOTT · Multivariate Density Estimation: Theory, Practice, and Visualization
* SEARLE · Linear Models
† SEARLE · Linear Models for Unbalanced Data
† SEARLE · Matrix Algebra Useful for Statistics
† SEARLE, CASELLA, and McCULLOCH · Variance Components
SEARLE and WILLETT · Matrix Algebra for Applied Economics
SEBER · A Matrix Handbook For Statisticians
† SEBER · Multivariate Observations
SEBER and LEE · Linear Regression Analysis, *Second Edition*
† SEBER and WILD · Nonlinear Regression
SENNOTT · Stochastic Dynamic Programming and the Control of Queueing Systems
* SERFLING · Approximation Theorems of Mathematical Statistics
SHAFER and VOVK · Probability and Finance: It's Only a Game!
SHERMAN · Spatial Statistics and Spatio-Temporal Data: Covariance Functions and
    Directional Properties
SILVAPULLE and SEN · Constrained Statistical Inference: Inequality, Order, and Shape
    Restrictions
SINGPURWALLA · Reliability and Risk: A Bayesian Perspective
SMALL and McLEISH · Hilbert Space Methods in Probability and Statistical Inference
SRIVASTAVA · Methods of Multivariate Statistics
STAPLETON · Linear Statistical Models, *Second Edition*
STAPLETON · Models for Probability and Statistical Inference: Theory and Applications
STAUDTE and SHEATHER · Robust Estimation and Testing
STOYAN · Counterexamples in Probability, *Second Edition*
STOYAN and STOYAN · Fractals, Random Shapes and Point Fields: Methods of
    Geometrical Statistics
STREET and BURGESS · The Construction of Optimal Stated Choice Experiments:
    Theory and Methods
STYAN · The Collected Papers of T. W. Anderson: 1943–1985
SUTTON, ABRAMS, JONES, SHELDON, and SONG · Methods for Meta-Analysis in
    Medical Research
TAKEZAWA · Introduction to Nonparametric Regression
TAMHANE · Statistical Analysis of Designed Experiments: Theory and Applications
TANAKA · Time Series Analysis: Nonstationary and Noninvertible Distribution Theory
THOMPSON · Empirical Model Building: Data, Models, and Reality, *Second Edition*
THOMPSON · Sampling, *Third Edition*
THOMPSON · Simulation: A Modeler's Approach
THOMPSON and SEBER · Adaptive Sampling
THOMPSON, WILLIAMS, and FINDLAY · Models for Investors in Real World Markets
TIERNEY · LISP-STAT: An Object-Oriented Environment for Statistical Computing
    and Dynamic Graphics
TROFFAES and DE COOMAN · Lower Previsions
TSAY · Analysis of Financial Time Series, *Third Edition*
TSAY · An Introduction to Analysis of Financial Data with R

---

*Now available in a lower priced paperback edition in the Wiley Classics Library.
†Now available in a lower priced paperback edition in the Wiley–Interscience Paperback Series.

*Now available in a lower priced paperback edition in the Wiley Classics Library.
†Now available in a lower priced paperback edition in the Wiley–Interscience Paperback Series.

Printed and bound by CPI Group (UK) Ltd, Croydon, CR0 4YY

27/10/2024

14580349-0003